Metal Matrix Composites

THERMOMECHANICAL BEHAVIOR

Pergamon Titles of Related Interest

ASHBY & JONES
Engineering Materials 1 & 2

BEE & GARRETT
Materials Engineering

BEVER
Encyclopedia of Materials Science & Engineering

CAHN
Encyclopedia of Materials Science & Engineering, 1st supplement

GIBSON & ASHBY
Cellular Solids

HEARN
Mechanics of Materials

HULL & BACON
Introduction to Dislocations, 3rd edition

KETTUNEN et al
Strength of Metals & Alloys (ICSMA 8)

SALAMA
Fracture 89 (ICF 7)

YAN et al
Mechanical Behaviour of Materials V (ICM 5)

Pergamon Related Journals

(free specimen copy gladly sent on request)

Acta Mechanica Solida Sinica

Acta Metallurgica

Canadian Metallurgical Quarterly

Engineering Fracture Mechanics

Fatigue & Fracture of Engineering Materials & Structures

International Journal of Solids & Structures

Journal of the Mechanics & Physics of Solids

Materials & Society

Materials Research Bulletin

Progress in Materials Science

Scripta Metallurgica

Metal Matrix Composites

THERMOMECHANICAL BEHAVIOR

MINORU TAYA

University of Washington, Seattle, WA, USA

RICHARD J. ARSENAULT

University of Maryland, College Park, MD, USA

PERGAMON PRESS

OXFORD · NEW YORK · BEIJING · FRANKFURT

SÃO PAULO · SYDNEY · TOKYO · TORONTO

U.K.	Pergamon Press plc, Headington Hill Hall, Oxford OX3 OBW, England
U.S.A.	Pergamon Press, Inc., Maxwell House, Fairview Park, Elmsford, New York 10523, U.S.A.
PEOPLE'S REPUBLIC OF CHINA	Pergamon Press, Room 4037, Qianmen Hotel, Beijing, People's Republic of China
FEDERAL REPUBLIC OF GERMANY	Pergamon Press GmbH, Hammerweg 6, D-6242 Kronberg, Federal Republic of Germany
BRAZIL	Pergamon Editora Ltda, Rua Eca de Queiros, 346, CEP 04011, Paraiso, São Paulo, Brazil
AUSTRALIA	Pergamon Press Australia Pty Ltd., P.O. Box 544, Potts Point, N.S.W. 2011, Australia
JAPAN	Pergamon Press, 5th Floor, Matsuoka Central Building, 1-7-1 Nishishinjuku, Shinjuku-ku, Tokyo 160, Japan
CANADA	Pergamon Press Canada Ltd., Suite No. 271, 253 College Street, Toronto, Ontario, Canada M5T 1R5

First edition 1989

Library of Congress Cataloging-in-Publication Data

Taya, Minoru.
Metal matrix composites.
Includes bibliographies and index.
1. Metallic composites—Thermal properties.
2. Metallic composites—Mechanical properties.
I. Arsenault, R. J. II. Title.
TA481.T38 1989 620.1'1896 88-34488

British Library Cataloguing in Publication Data

Taya, Minoru
Metal matrix composites
1. Structures. Metallic composite materials
I. Title II. Arsenault, Richard J.
624.1'82

ISBN 0-08-036984-7 Hardcover
ISBN 0-08-036983-9 (Flexicover)

Printed in Great Britain by BPCC Wheaton Ltd., Exeter

Contents

provide such a subject area, i.e., the thermomechanical behavior of metal matrix composites, which we believe is suited not only to a textbook in a classroom, but also as a reference book at a work desk.

If the book is used as a textbook, the users of the book are encouraged to follow Chapters 1 through 5, in that order, with the remaining chapters subject to choice. To assist readers, we have added problem sections at the end of each chapter. Although the book is designed primarily for use as a textbook, readers may also use it as a reference data book, because it contains complete information on the thermomechanical data of various metal matrix composite systems (Chapter 3 through 5 and Appendices). We also have included a chapter (Chapter 6) which includes several engineering topics which are considered to be emerging subject areas within the framework of the thermomechanical behavior of metal matrix composites.

M. Taya would like to express his appreciation to the Royal Society for supporting his semi-sabbatical leave at the University of Oxford. The initial layout of this work was done in early summer of 1986 at Oxford and was then immediately passed to R. J. Arsenault, who agreed to join as the co-author. Taya also wishes to acknowledge his past and present graduate students, research associates, and his colleagues who have worked on several different aspects of the characterization and modeling of metal matrix composites, the results of which are used extensively in this book. Among these people, Dr. H. S. Yoon deserves our special thanks for his help in making some of the appendices complete. He also wishes to express many thanks to Ms Linda Kager and Ms Bianca Plank for their skillful typing, and patience with the slow process of writing. Finally, M. Taya would like to acknowledge government agencies (NSF, ARO, ONR, NASA, AFOSR) and industries (Honda R&D, Toray Inc., and Alcan International) for their support in this subject area.

R. J. Arsenault would like to acknowledge his past and present students who have contributed their time and tireless effort in conducting research on discontinuous metal matrix composites. Also, he would like to acknowledge the assistance of Ms C. Arsenault and Mrs J. Anderson for their help in preparing portions of the manuscript. Finally, R. J. A. would like to thank Dr S. Fishman of the Office of Naval Research for his continued support and encouragement.

Fall 1988

MINORU TAYA at Seattle
RICHARD J. ARSENAULT at College Park

Credits to Publishers and Authors

We would like to acknowledge the permission to reproduce the figures and photographs used in this book, which were granted by the following publishers:

Academic Press
American Ceramic Society
American Institute of Aeronautics and Astronautics
American Institute of Physics
American Society for Testing and Materials
American Society of Mechanical Engineers
Arthur G. Metcalfe and Associates
Capman and Hall Ltd.
The Electrochemical Society
Elsevier Science Publishers
Kluwer Academic Publishers (Martinus Nijhoff Publishers)
The Metallurgical Society of AIME
Oxford University Press
Pergamon Press
Prenum Publishing Corporation
Society of Automotive Engineers
Technomic Publishing Co.

Our special thanks to the following people who sent us original photographs:

Dr. D. M. Aylor, David W. Taylor Naval Ship Research and Development Center
Mr. A. Daimaru, Honda Research and Development Company
Dr. M. Hashish, Flow Industries
Mr. M. Akiyama, Tokai Carbon Co.
Professor A. Kohyama, University of Tokyo
Mr. T. Kyono, Toray Inc.
Dr. J. J. Lee, Boeing Aerospace Company
Dr. D. J. Lloyd, Alcan International
Dr. B. K. Min, Lockheed Palo Alto Research Laboratory
Dr. D. M. Schuster, Dural Aluminum Composites Corp.
Dr. J. Stringer, Electric Power Research Institute
Dr. N. Tsangarakis, Army Materials Technology Laboratory
Mr. S. Utsunomiya, Mitsubishi Electric Corp.
Professor K. Wakashima, Tokyo Institute of Technology

CHAPTER 1

Introduction

1.1 Definition of Metal Matrix Composite

If we consider the term composite with its broadest meaning, then a common piece of (polycrystalline) metal is a composite of many grains. Therefore, it is necessary to restrict the definition of the term composite and we shall use the following criteria to designate a composite [1, 2].

- It must be man-made.
- It must be a combination of at least two chemically distinct materials with a distinct interface separating the constituents.
- The separate materials forming the composite must be combined three-dimensionally. (Laminates such as clad metals or honeycomb sandwiches are not considered basic composite materials if the same metal is used throughout.)
- It should be created to obtain properties which would not otherwise be achieved by any of the individual constituents.

There are three basic types of composite materials that meet the above-mentioned criteria:

- Dispersion-strengthened
- Particle-reinforced
- Fiber (whisker)-reinforced

Metal matrix composites, in general, consist of at least two components: one obviously is the metal matrix (in most cases, an alloy is the metal matrix), and the second component is a reinforcement (in general, an intermetallic compound, an oxide, a carbide or a nitride). The distinction of metal matrix composites from other two or more phase* alloys comes about from the processing of the composite. In the production of the composite, the matrix and the reinforcement are mixed together. This is to distinguish a composite from a two or more phase alloy, where the second phase forms as a particulate, eutectic or eutectoid reaction, etc. In other words, a composite initially begins as separate components, i.e., the metal matrix and the re-

* The term 'phase' is used interchangeably with component. A phase can be defined in the following manner. A phase is defined as a homogeneous body of matter that is physically distinct.

inforcement. In all cases the matrix is defined as a metal, but a pure metal is rarely used as the matrix; it is generally an alloy.

Each type of metal matrix composite is defined as follows:

- *Dispersion-strengthened*—This composite is characterized by a microstructure consisting of an elemental matrix within which fine particles are uniformly dispersed. The particle diameter ranges from about 0.01 to 0.1 μm, and the volume fraction of particles ranges from 1 to 15%.
- *Particle-reinforced*—This composite is characterized by dispersed particles of greater than 1.0 μm diameter with a volume fraction of 5 to 40%.
- *Fiber (whisker)-reinforced*—The reinforcing phase in fiber composite materials spans the entire size range, from 0.1 to 250 μm in length to continuous fibers, and spans the entire range of volume concentrations, from a few percent to greater than 70%. The distinguishing microstructural feature of fiber-reinforced materials is that the reinforcing fiber has one long dimension, whereas the reinforcing particles of the other two types do not.

Metal matrix composite (MMC) materials have been under development for more than 20 years. However, the initial emphasis was on continuous filament MMCs. They were first developed for applications in aerospace, followed by applications in other industries. The expansion into nonaerospace and nonmilitary fields came about slowly as the price of MMC materials was coming down. This is due mainly to the development of new low-cost fibers [3–5].

Continuous fiber-reinforced metals are a special and sophisticated class of composite materials. Fiber-reinforced metals, unlike most metals and alloys, are anisotropic. The degree of anisotropy depends primarily on the degree of fiber orientation. The prime role of the fibers is to carry the load, while the metal matrix serves to transfer and distribute the load to the fibers. The efficiency with which the loads are transferred from the matrix to the fibers depends on the bonding interface between them. Assuming high interface efficiency, the mechanical properties of the composite depend more on the properties of the fiber rather than the properties of matrix. This means that the matrix can be selected on the basis of oxidation and corrosion resistance or other required properties. Applications of continuous fiber MMC are mostly limited to some of the primary and secondary structural members of aerospace structures and military airplanes, except for an automobile connecting rod which is made of aluminum reinforced with continuous stainless-steel fibers [6] (Fig. 1.1).

In recent years, discontinuous MMCs have been investigated. Recent interest in discontinuous MMCs has been rekindled because it is more economical to produce economic production of silicon carbide (SiC) fibers (whiskers) [5], which has also led to the use of platelet or particulate SiC in a

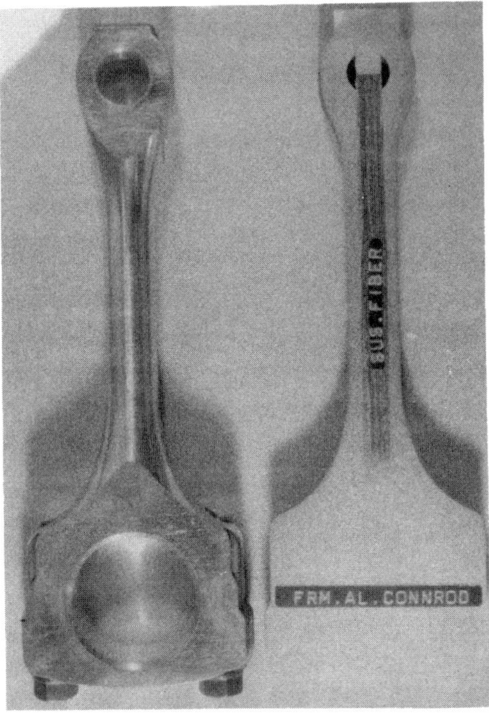

FIG. 1.1 Connecting rod made of aluminum reinforced with continuous stainless steel fibers [6] (courtesy of Honda R&D Co.).

MMC [7]. One of the advantages of discontinuous composites is that they can be shaped by standard metallurgical processes such as forging, rolling, extrusion, etc. Due to this ease of formability and relatively modest cost discontinuous MMCs have recently been used in various applications. Some of such applications are tennis rackets and heads of golf clubs, which are made of SiCp/Al composite (Fig. 1.2) [8], and automobile engine components, piston and connecting rod, which are made of SiCw/Al composite (Fig. 1.3) [9].

MMCs have several advantages that are very important for their use as structural materials. These advantages include a combination of the following properties:

- High strength
- High elastic modulus
- High toughness and impact properties
- Low sensitivity to temperature changes or thermal shock
- High surface durability and low sensitivity to surface flaws

FIG. 1.2 Tennis racket frame and golf club head made by SiCp/Al composites (courtesy of Alcan International) [8].

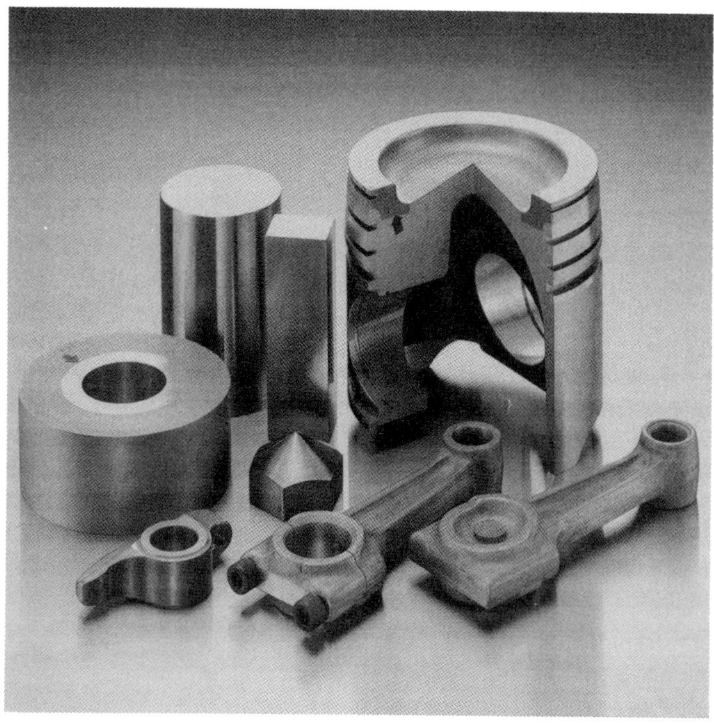

FIG. 1.3 Automobile engine components made of SiCw/Al composite (courtesy of Tokai Carbon Co.).

- High electrical and thermal conductivity
- High vacuum environment resistance

In addition to conductivity of MMCs, the most obvious advantages of MMCs are their resistance to severe environments, toughness, and retention of strength at high temperatures. For a composite structure it is possible to emphasize environmental stability of the matrix at elevated temperatures, since the required mechanical strength and stiffness can be obtained from the reinforcement. The shear strength requirements of the matrix are nominal since the matrix serves only to transfer load into the filaments in continuous filament composites. However, the matrix strength is very important in discontinuous metal matrix composites.

1.2 Matrix Metals and Reinforcements

1.2.1 Matrix metals

As stated previously, in general, pure metals are not used. The metal matrix is an alloy. The alloy can be relatively simple but, in general it is a multi-element alloy. There are a number of phases that can exist within the alloy.

Numerous metals have been used as the matrix, for example Al, Cu, Fe, Mg, Ti, and Pb. However, almost all of the structural alloy systems have been considered as matrix materials in MMCs. Tables A1, A2, and A3 (Appendix A) are lists which cover additional combinations.

1.2.2 Reinforcements

The reinforcements can be divided into two major groups, discontinuous and continuous.

The most prominent discontinuous reinforcements have been SiC, Al_2O_3, and TiB_2 in both whisker and particulate form. Table B1 (Appendix B) is a listing of some of the other candidates along with their mechanical properties. In terms of continuous reinforcements most are non-metals; some of the same compounds are listed in Table B2 (Appendix B). However, continuous metal filaments are also employed, the most prominent being W, but stainless steel is also being used as a reinforcement. Table B3 (Appendix B) is a listing of metal wire in ribbons.

1.3 Engineering Applications and their Requirements

Metal matrices and their composites have the potential to be used for many structural applications. However, due to their higher cost as compared to a monolithic metal alloy, their use will be restricted to cost-effective applications, where the strength–stiffness savings outweigh the increased cost. A major purpose of producing MMCs is to increase the strength and stiffness of the matrix alloy. However, an additional benefit that metal matrices have is that they can be tailored to produce various combinations of stiffness and strength and to produce various values of thermal coefficients of expansion. Also, MMCs can be used in wear applications, for the wear resistance of MMCs can be quite good.

Since MMCs are more expensive than the matrix alloys themselves, they will fit into places where the requirements are greater. For example, their use will be restricted to instances where a stiffness to strength ratio is advantageous. Another major potential is that they can be used at high temperatures. It is envisioned that the addition of the reinforcement does increase the high-temperature strength of the matrix alloys. Another generic advantage of metal matrix is that there can be a reduction in the density of the MMCs due to the addition of the reinforcement which has a lower density than the metal matrix. In terms of specific uses, again, that depends upon the desire to produce the material, i.e., whether the cost versus benefits ratio is advantageous. For example, if a metal matrix composite is considered for use as a turbine engine component, then its performance in the engine use environment must be carefully evaluated and compared with other possible high-temperature material systems. One of the key parameters in evaluating the performance is the strength/weight ratio (or specific strength). Figure 1.4 shows such a performance map of various high temperature materials in terms of use temperature (°F) and specific strength [10]. In this figure, metal matrix composites occupy a better region than conventional materials, but

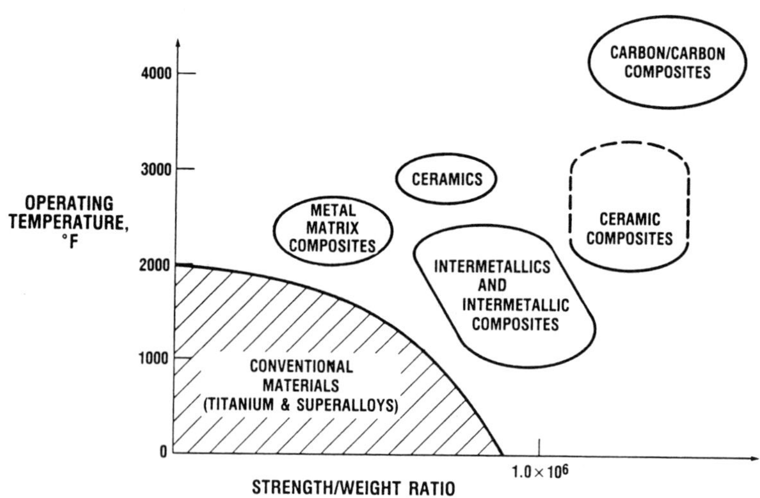

FIG. 1.4 Performance map of various high-temperature engine materials in terms of operating temperature (°F) and strength/weight ratio [10].

the specific strength appears to be lower than the other high-temperature materials. However, it should be borne in mind in Fig. 1.4 that metal matrix composites will occupy the better region if the other than mechanical behavior is focused on, for example, oxidation resistance and impact toughness. However, certain disadvantages of MMCs should be mentioned. These include the fact that MMCs, in general, have a lower fracture toughness than the matrix alloy. Secondly, MMCs are subjected to thermal fatigue which is the major thrust of this book. Metal fatigue is a phenomenon which can have deleterious effects on composites which are to be subjected to high-temperature use; especially high-temperature cyclic use as compressor blades in jet engines. An example of this is the case where the usefulness of super alloy–continuous tungsten filament composites is in doubt, due to the fatigue damage which occurs as a result of thermal cycling in this particular composite.

Despite the fact that MMCs have some disadvantages, they are still considered to be a more reliable high-temperature material system than ceramic matrix composites. This is particularly due to the fact that the science and technology of metals is much more developed than that of ceramics. Thus, application of MMCs as various components of engines in space vehicles is more realistic than that of ceramic matrix composites. Figure 1.5 [10] illustrates such a case where fiber-reinforced super alloys, a MMC system, are predicted to be applicable for use as turbine blades in the near future (mid-1990 and beyond) while ceramic matrix composites and

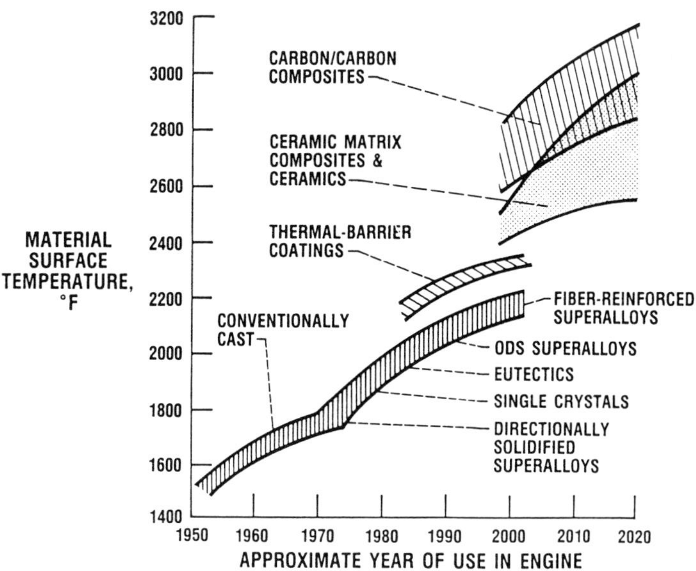

FIG. 1.5 Trends of turbine engine materials [10].

carbon/carbon composites do not appear to be ready for deployment until the early years of the twenty-first century.

Among various applications of MMCs, National Aerospace Plane (NASP) which NASA and the Air Force in the United States are developing will require the most high performing MMC systems for use not only as engine components, but also as structural components of the plane [11].

Some specific applications of metal matrix composites are listed in Tables A1, A2, and A3 (Appendix A).

1.4 Problems

(1) What would be the performance map for various high-temperature materials if fracture toughness/weight ratio is used in Fig. 1.4 instead of strength/weight ratio?
(2) Referring to Fig. 1.4, what would be the performance map if strength/weight ratio is replaced by oxidation resistance?
(3) What would be the successful non-military and non-space applications of continuous fiber metal matrix composites?
(4) What are the key engineering problems of metal matrix composites before they are fully used extensively as a structural material?

References

1. Chawla, K. K., *Composite Materials*, Springer-Verlag, New York, 1987.
2. Schoutens, J. E., *Introduction to Metal Matrix Composite Materials*, MMCIAC Tutorial Series, No. 272, 1982.
3. Fishman, S. G., A metal matrix composite requirement, more reliable mechanical property data, *MMCIAC, Current Highlights*, Vol. 1, May 1981.
4. Divecha, A. P., Fishman, S. G. and Karmarker, S. D., *J. Metals*, Vol. 9, 1981, p. 12.
5. Cutler, I. B., *Am. Ceram. Soc. Bull.*, Vol. 52, 1973, p. 425.
6. Daimaru, A., Honda R&D Co., personal communication, 1988.
7. Harrigan, W. J., Jr, DWA Composite Specialities, Chatsworth, CA, personal communication, 1982.
8. DURAL MMC, *Dural Aluminum Composites Corporation*, May 1987.
9. Tokawhisker, Tokai Carbon Co., 1988.
10. Aeropropulsion '87, Session 1—Aeropropulsion Materials Research, NASA Conf. Publ. 10003, November 1987, p. 77.
11. Johnson, W. S., *ASTM Standardization News*, October 1987, pp. 36–39.

CHAPTER 2

Foundation of Analysis

2.1 Introduction

It would be advantageous to be able to predict the thermomechanical behavior of a metal matrix composite by appropriate formulae if the thermomechanical behavior of fiber and matrix are known. This is particularly so for a composite design engineer who wants to develop a new metal matrix composite system with unique properties that can be tailored by the analytical model. Thus, the modeling of the thermomechanical behavior of a metal matrix composite is considered to be as important, as is the establishment of a reliable processing route for a new metal matrix composite system.

The thermomechanical behavior of a material can be categorized into two types: mechanical and thermal. The most important mechanical behavior can be described by a combination of elastic–plastic and creep laws, and that of the thermal behavior by thermal expansion and conduction laws. From the viewpoint of mathematical treatment the above thermomechanical behavior is regrouped into thermoelasticity, plasticity/creep and heat conduction equations.

The thermomechanical properties of a composite are often anisotropic due to the anisotropy in the fiber properties or to the morphology of fiber network (even for isotropic fiber). In this respect the previous studies have focused on the behavior along the reinforcing direction. It has been anticipated that the properties along other directions often exhibit poorer performance compared to these along the reinforcing direction. Recent studies on the thermomechanical properties of the other directions have confirmed this.

Analytical models that have been developed for other composite systems such as polymeric composites can be equally applicable to any metal matrix composite system. Thus, a number of analytical models that were originally developed for a particular composite system have been extended to other composite systems. Analytical models range from simple (law of mixtures) to rigorous; for example, the type of model based on variational principles. Simpler models usually require a smaller amount of input data (or parameters).

In the following we will briefly review the thermomechanical governing equations (both isotropic and anisotropic cases), and also various composite models. The governing equations will be expressed in index form unless otherwise noted. Examples of index forms are:

$$T_{,i} \equiv \frac{\partial T}{\partial x_i}$$

$$\dot{T}_{,j} \equiv \frac{\partial^2 T}{\partial t \partial x_j} \tag{2.1}$$

$$e_{ii} \equiv e_{11} + e_{22} + e_{33}$$

$$\sigma_{ij} e_{ij} \equiv \sigma_{11} e_{11} + \sigma_{22} e_{22} + \sigma_{33} e_{33} + 2(\sigma_{12} e_{12} + \sigma_{23} e_{23} + \sigma_{13} e_{13})$$

where repeated indices are to be summed and x_i are related to conventional coordinates (x,y,z) as $x_1 = x$, $x_2 = y$, and $x_3 = z$.

2.2 Thermomechanical Behavior of Isotropic Materials

Isotropy is defined as a state in which properties are the same in all directions, resulting in minimum number of material constants, i.e., two for elastic constants, one for coefficient of thermal expansion (CTE) and thermal conductivity. Isotropy must be distinguished from homogeneity in that the former denotes no dependence on direction and the latter no dependence on position (location) in a material. Since an isotropic material requires a minimum number of material constants—and most materials, including metals, may be approximated to exhibit such behavior—the assumption of isotropy in material constitutive equations has been used extensively in the past. In this subsection the constitutive equations to describe the thermomechanical behavior of isotropic materials, thermoelasticity, plasticity/creep and heat conduction will be reviewed. The description of plasticity and creep laws includes the concept of dislocations which are related to plasticity and creep.

2.2.1 Thermoelasticity

Consider an isotropic and linear elastic body subjected to current temperature T. Then thermoelastic constitutive equations as interpreted by Caratheodory [1, 2] are given by

$$\sigma_{ij} = \lambda \delta_{ij} e_{kk} + 2\mu e_{ij} - (3\lambda + 2\mu)\delta_{ij} \alpha (T - T_0) \tag{2.2}$$

where λ and μ are Lamé constants, σ_{ij} and e_{ij} are stress and strain tensors, δ_{ij} is the Kronecker's delta, α is the coefficient of thermal expansion and T_0 is a reference temperature at which the elastic body remains stress-free. Since $\sigma_{ij} = \sigma_{ji}$, which can be derived from the conservation of angular momentum, eq. (2.2) gives rise to six independent equations. Eq. (2.2) can be rewritten in a more general form:

$$\sigma_{ij} = C_{ijkl}(e_{kl} - e^{*}_{kl}) \tag{2.3}$$

where C_{ijkl} is the stiffness tensor and for an isotropic material it is equal to $\lambda\delta_{ij}\delta_{kl} + \mu(\delta_{ik}\delta_{jl} + \delta_{jk}\delta_{il})$, e_{kl} is the total strain which is related to displacement u_i by

$$e_{kl} = \frac{1}{2}(u_{k,l} + u_{l,k}) \tag{2.4}$$

and e^{*}_{kl} is the non-elastic strain (eigenstrain) [3, 4] which is given by

$$e^{*}_{kl} = \alpha\delta_{kl}(T - T_0) \tag{2.5}$$

where T_0 and T are reference and current temperatures, respectively. In addition to constitutive equations, eq. (2.3) (or (2.2)) and strain-displacement relation, eq. (2.4), the following equation (equation of motion) must also be satisfied

$$\sigma_{ij,j} + f_i = \rho\ddot{u}_i \tag{2.6}$$

where ρ is mass density, f_i is the ith component of body force per unit volume and u_i is the ith component of displacement vector $\mathbf{u}$. It is clear from eq. (2.2) or (2.5) that thermal strain in an isotropic matrix is dilatational. The solution procedure for thermoelasticity problems involves assuming that temperature T is given as input data, which is a function of space and time (which is termed as a decoupled thermoelasticity problem), but not of strain (coupled problem). Thus, in the case of decoupled problems, distribution of temperature must be solved first by the heat conduction equation, eq. (2.24) (see subsection 2.2.3). Using the general solutions to a heat conduction problem, one can determine the specific solution to a thermoelastic problem by approximately considering the boundary and/or initial conditions.

2.2.2 Plasticity and creep

Plasticity represents inelastic deformation of a material and it is related to the density and motion of dislocations. Plasticity (for simplicity) will be divided into two components; time-independent and time-dependent deformation. However, it is recognized that *all* plastic deformation is time-dependent, but for the present we shall define plasticity as time-independent, inelastic deformation, whereas creep denotes time-dependent deformation. In the following, basic theories of plasticity and creep will be reviewed.

Plasticity

In order to characterize the plastic deformation of a metal, one must usually conduct a simple mechanical test—for example, the uniaxial tensile

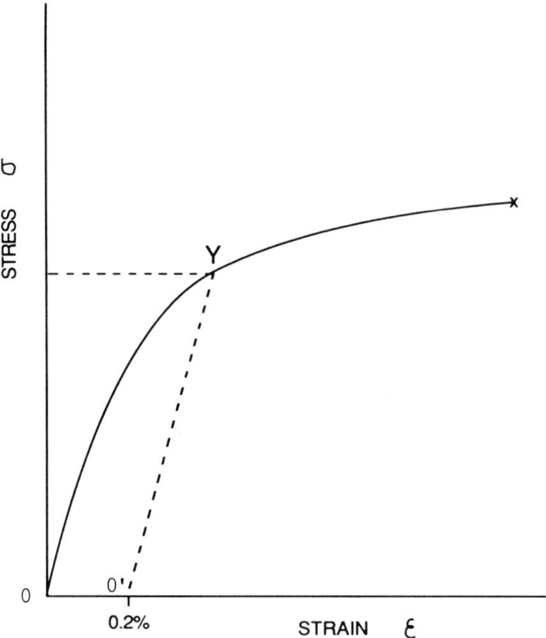

FIG. 2.1 Stress–strain curve with 0.2% offset yield stress Y.

test, the results of which are schematically shown in Fig. 2.1 and which are typical of fcc metals (aluminum, copper, etc.). σ and ε in Fig. 2.1 are true stress and strain. True stress σ is defined as load divided by the current sectional area while true strain ε can be obtained as

$$\varepsilon = \ln\left(\frac{l}{l_0}\right) \tag{2.7}$$

where l_0 and l are the initial and current lengths, respectively. In actual tests, engineering strain (e) is usually recorded, which is defined by

$$e = \frac{(l - l_0)}{l_0} \tag{2.8}$$

From eqs. (2.7) and (2.8), we have

$$\varepsilon = \ln(1 + e) \tag{2.9}$$

It is clear from eq. (2.9) that for small e ($\ll 1$)

$$\varepsilon \simeq e \tag{2.10}$$

The theory of plasticity is based on the assumptions that Bauschinger effect and hysteresis loop are neglected and hydrostatic pressure does not influence the yielding of a material (which leads to incompressibility of plastic strain). These assumptions suggest that yield criterion is given by

$$f(J_2', \kappa) = 0 \tag{2.11}$$

where

$$J_2' = \frac{1}{2}\sigma_{ij}'\sigma_{ij}' \tag{2.12}$$

and where σ_{ij}' is deviatoric stress and equal to $\sigma_{ij} - 1/3\ \sigma_{kk}\delta_{ij}$, κ is the work-hardening parameter and also equal to yield stress in shear. There exist two popular yield criteria: Tresca and Von Mises types. For the sake of simplicity we now consider principal stresses σ_1, σ_2 and σ_3 to represent the state of multi-axial stresses. Following the Tresca type yield criterion, yielding occurs when the maximum of three principal shear stresses (τ_i) reaches the yield stress in shear of a metal κ, where

$$\tau_1 = \frac{1}{2}|\sigma_2 - \sigma_3|$$

$$\tau_2 = \frac{1}{2}|\sigma_1 - \sigma_3| \tag{2.13}$$

and

$$\tau_3 = \frac{1}{2}|\sigma_1 - \sigma_2|$$

where σ_i are the principal stresses. On the other hand, yielding based on Von Mises criterion occurs when

$$J_2' = \kappa^2 \tag{2.14}$$

where J_2' can be expressed in terms of six stress components

$$J_2' = \frac{1}{6}[(\sigma_{11} - \sigma_{22})^2 + (\sigma_{22} - \sigma_{33})^2 + (\sigma_{33} - \sigma_{11})^2 + 6\{\sigma_{12}^2 + \sigma_{23}^2 + \sigma_{31}^2\}] \tag{2.15}$$

The interpretation of κ depends on the type of yield criterion used, i.e.,

$$\kappa = \begin{cases} \dfrac{1}{2}\sigma_y & \text{for Tresca} \\ \dfrac{1}{\sqrt{3}}\sigma_y & \text{for Von Mises} \end{cases} \tag{2.16}$$

where σ_y is the yield stress that can be obtained from a uniaxial tensile test (corresponding to Y in Fig. 2.1).

Following Prandtl and Reuss, the constitutive equation of an elastic–plastic material can be obtained as [5],

for a work-hardening material,

$$d\varepsilon'_{ij} = \frac{3\sigma'_{ij}\,d\bar{\sigma}}{2\bar{\sigma}H'} + \frac{d\sigma'_{ij}}{2G}$$
$$d\varepsilon_{ii} = \frac{(1-2\nu)}{E}\,d\sigma_{ii} \tag{2.17}$$

for non work-hardening material

$$d\varepsilon'_{ij} = \frac{3\sigma'_{ij}}{2\sigma_y}\,d\bar{\varepsilon}_p + \frac{d\sigma'_{ij}}{2G}$$
$$d\varepsilon_{ii} = \frac{(1-2\nu)}{E}\,d\sigma_{ii} \tag{2.18}$$
$$\sigma'_{ij}\,\sigma'_{ij} = 2\kappa^2 = \text{const.}$$

In the above equations $\bar{\sigma}$ is the equivalent stress defined by $3/2\,\sigma_{ij}\sigma_{ij}$, H' is the slope of uniaxial (or equivalent) stress vs. plastic strain ε_p curve, E and G are Young's and shear moduli, ν is the Poisson's ratio, and $d\bar{\varepsilon}_p$ is the equivalent plastic strain increment defined by $3/2\,d\varepsilon^p_{ij}\,d\varepsilon^p_{ij}$. When a material experiences unloading (such as $Y \rightarrow 0'$, in Fig. 2.1), the above constitutive equations are all reduced to an elastic equation.

$$d\varepsilon'_{ij} = \frac{d\sigma'_{ij}}{2G}$$
$$d\varepsilon_{ii} = \frac{(1-2\nu)}{E}\,d\sigma_{ii} \tag{2.19}$$

Of course, in this case one can omit the incremental symbol d in front of stress and strain.

Creep

Creep in the present case will be defined as time-dependent inelastic deformation of a material when it is subjected to a constant stress σ at a given temperature T. A typical creep strain (ε) vs. time (t) curve is shown in Fig. 2.2, which exhibits three stages, I (primary), II (secondary or steady state) and III (third or tertiary stage). Usually, the duration of the secondary creep is the longest, thus the assessment of its creep rate ($\dot{\varepsilon}_c$) is the most important task. Among various parameters, stress σ and temperature T are the major ones influencing the (secondary) creep rate $\dot{\varepsilon}_c$, and here we are concerned with

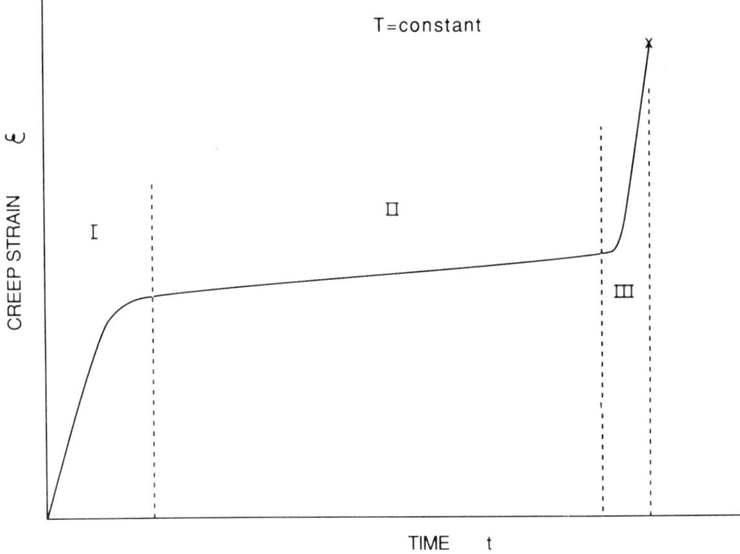

FIG. 2.2 Typical creep strain vs time curve which exhibits three stages: I (primary), II (secondary), and III (third).

creep due to climb and glide of dislocations. Frost and Ashby summarized the mechanisms of deformation of metals under various regions of temperature and stress, which is known as the deformation map [6].

Since creep is a thermally activated process, its temperature dependence is given by an Arrhenius-type equation

$$\dot{\varepsilon}_c = f(\sigma)e^{-\frac{Q_a}{RT}} \tag{2.20}$$

where $f(\sigma)$ is a known function of σ, R is the gas constant, T is the temperature in Kelvin, and Q_a is the apparent activation energy. The most popular form of $f(\sigma)$ is a power law type given by

$$f(\sigma) = A\left(\frac{\sigma}{G}\right)^n \tag{2.21}$$

where A and n are material constants, and G is the shear modulus. Although eq. (2.20) provides a basic equation for the secondary creep, a more rigorous equation is needed if the effects of parameters other than σ and T are to be examined. For example the Dorn-type equation [7] may be suited for this purpose:

$$\dot{\varepsilon}_c = A\left(\frac{\sigma}{G}\right)^n \frac{Gb}{kT} D_0 e^{-\frac{Q}{RT}} \tag{2.22}$$

where A and n are material constants, b is the Burger's vector, D_0 is the pre-exponential constant, Q is the activation energy which is constant for self-diffusion, k is Boltzmann constant and T is the absolute temperature. The values of these constants for most metals are tabulated in Ref. [7]. It is known that the stress dependence on creep rate at higher stresses cannot be described by the power law type equation (eq. (2.21)). In this case the exponential-type law is better suited; namely, eq. (2.21) is replaced by

$$f(\sigma) = A\exp(\sigma/\sigma_0) \tag{2.23}$$

where A and σ_0 are material constants.

2.2.3 Heat conduction

When an isotropic and homogeneous linear elastic solid is subjected to heat flow, but not to internal heat generation, the temperature field T satisfies the following equation,

$$kT_{,ii} = \rho c\dot{T} \tag{2.24}$$

where k, ρ and c are the thermal conductivity, specific gravity, and specific heat of the solid, respectively. In the case of a steady-state temperature field, eq. (2.24) is reduced to

$$kT_{,ii} = 0 \tag{2.25}$$

which is the well-known Laplace equation and is usually written as $k\nabla^2 T=0$ where ∇^2 is defined by in Cartesian coordinates by:

$$\nabla^2 = \frac{\partial^2}{\partial x^2} + \frac{\partial^2}{\partial y^2} + \frac{\partial^2}{\partial z^2} \tag{2.26}$$

Once a general solution of temperature is obtained with unknown coefficients, then one can determine the coefficients by using the initial and boundary conditions.

Initial conditions for temperature are usually given by the prescribed temperature at a given time, which is specified as a function of space. Boundary conditions are given on the surface of a solid or the interfaces between different phases. The most popular boundary conditions are those with [8]:

(1) a prescribed surface temperature, $T=T_0(x, y, z)$;
(2) no heat flux across the surface;

$$\partial T/\partial n = 0 \tag{2.27}$$

where n is the outward normal direction on the surface in consideration;

(3) linear heat transfer at the surface,

$$k\frac{\partial T}{\partial n} + H(T - T_0) = 0 \tag{2.28}$$

where T_0 is the temperature of the surrounding medium and H is a constant;

(4) interface between two media of different conductivities k_1 and k_2. If the continuity of heat flux and temperature is assumed (intimate contact), then

$$\begin{aligned} k_1\frac{\partial T_1}{\partial n} &= k_2\frac{\partial T_2}{\partial n} \\ T_1 &= T_2 \end{aligned} \tag{2.29}$$

In most of the interfaces, eq. (2.29) is not valid in an exact sense. Instead, the following equation must be satisfied.

$$k_1\frac{\partial T_1}{\partial n} + H(T_1 - T_2) = 0 \tag{2.30}$$

which has an identical form to eq. (2.28).

2.3 Constitutive Equations of Anisotropic Materials

The thermomechanical properties of materials including composites are strongly dependent on the morphology of microstructure, and as a result they often exhibit anisotropy, i.e., the properties are dependent on direction. In metal matrix composites, reinforcing phases are sometimes anisotropic and even if the metal matrix is isotropic (which usually is the case with most metals), the composite will become anisotropic. Or even when both reinforcing phases and matrix metal are isotropic, the arrangement of the reinforcing phases will make the macroscopic properties of the composite anisotropic. Here, we will state the basic thermomechanical constitutive equations for anisotropic materials.

2.3.1 Elastic constants

The constitutive relation of a linear elastic material can be given in general as

$$\sigma_{ij} = C_{ijkl}e_{kl} \tag{2.31}$$

where σ_{ij} is the stress tensor satisfying symmetry $\sigma_{ij} = \sigma_{ji}$ and e_{kl} is the strain tensor related to displacement u_i by eq. (2.4), and therefore it is also symmetric ($e_{kl} = e_{lk}$). C_{ijkl} is the elastic constant tensor (of rank 4) satisfying the second law of thermodynamics, which leads to

$$C_{ijkl} = C_{klij} \tag{2.32}$$

Due to the symmetry of σ_{ij} and e_{kl}, C_{ijkl} must also satisfy additional requirements,

$$\begin{aligned} C_{ijkl} &= C_{jikl} \\ C_{ijkl} &= C_{ijlk} \end{aligned} \tag{2.33}$$

The stress and strain components in the three-dimensional orthogonal coordinates (x_1, x_2 and x_3) system are sometimes expressed as variables with a single subscript, i.e.:

$$\{\sigma_{ij}\} = \begin{Bmatrix} \sigma_{11} \\ \sigma_{22} \\ \sigma_{33} \\ \sigma_{23} \\ \sigma_{31} \\ \sigma_{12} \end{Bmatrix} = \begin{Bmatrix} \sigma_1 \\ \sigma_2 \\ \sigma_3 \\ \sigma_4 \\ \sigma_5 \\ \sigma_6 \end{Bmatrix} \tag{2.34}$$

$$\{e_{ij}\} = \begin{Bmatrix} e_{11} \\ e_{22} \\ e_{33} \\ e_{23} \\ e_{31} \\ e_{12} \end{Bmatrix} = \begin{Bmatrix} e_{11} \\ e_{22} \\ e_{33} \\ \gamma_{23/2} \\ \gamma_{31/2} \\ \gamma_{12/2} \end{Bmatrix} = \begin{Bmatrix} e_1 \\ e_2 \\ e_3 \\ e_4 \\ e_5 \\ e_6 \end{Bmatrix} \tag{2.35}$$

It is noted in eq. (2.35) that the shear strain components of e_{ij} are equal to half of the corresponding engineering shear strain γ_{ij}, i.e.,

$$e_{ij} = \frac{1}{2}\gamma_{ij} \tag{2.36}$$

e_4, e_5, and e_6 that appear in eq. (2.35) are equal to $\gamma_{23/2}$, $\gamma_{13/2}$, and $\gamma_{12/2}$, respectively. With the above simplified symbols of stress (σ_i) and strain components (e_i), one can rewrite the constitutive eq. (2.31) as

$$\sigma_i = C_{ij} e_j \tag{2.37}$$

where C_{ij} are called the Voigt constant, and related to elastic tensor according to the rule given in Table 2.1 [4].

Table 2.1 *Relation between elastic tensor components C_{ijkl} and Voigt elastic constants C_{ij} [4]*

	kl					
ij	11	22	33	23	31	12
11	C_{11}	C_{12}	C_{13}	C_{14}	C_{15}	C_{16}
22	C_{21}	C_{22}	C_{23}	C_{24}	C_{25}	C_{26}
33	C_{31}	C_{32}	C_{33}	C_{34}	C_{35}	C_{36}
23	C_{41}	C_{42}	C_{43}	C_{44}	C_{45}	C_{46}
31	C_{51}	C_{52}	C_{53}	C_{54}	C_{55}	C_{56}
12	C_{61}	C_{62}	C_{63}	C_{64}	C_{65}	C_{66}

From eq. (2.32) C_{ij} must satisfy symmetry, i.e., $C_{ij}=C_{ji}$. Hence there exist 21 independent components of C_{ij}. Most material systems, however, give rise to a smaller number of independent C_{ij}'s. For example, if the material property is symmetrical with respect to two mutual orthogonal planes (say, $x_1=0$ and $x_2=0$ planes) which is called orthotropic material, the constitutive equations in principal coordinates (x_1, x_2 and x_3 axes) are given by

$$\begin{Bmatrix} \sigma_{11} \\ \sigma_{22} \\ \sigma_{33} \\ \sigma_{23} \\ \sigma_{31} \\ \sigma_{12} \end{Bmatrix} = \begin{Bmatrix} \sigma_1 \\ \sigma_2 \\ \sigma_3 \\ \sigma_4 \\ \sigma_5 \\ \sigma_6 \end{Bmatrix} = \begin{bmatrix} C_{11} & C_{12} & C_{13} & 0 & 0 & 0 \\ C_{12} & C_{22} & C_{23} & 0 & 0 & 0 \\ C_{13} & C_{23} & C_{33} & 0 & 0 & 0 \\ 0 & 0 & 0 & C_{44} & 0 & 0 \\ 0 & 0 & 0 & 0 & C_{55} & 0 \\ 0 & 0 & 0 & 0 & 0 & C_{66} \end{bmatrix} \begin{Bmatrix} e_1 \\ e_2 \\ e_3 \\ e_4 \\ e_5 \\ e_6 \end{Bmatrix} \tag{2.38}$$

where there are nine independent C_{ij}'s and normal and shear components do not couple [9]. Furthermore, if the material property is isotropic in the x_1–x_2 plane, then the material is defined as transversely isotropic and its C_{ij} matrix is given by

$$[C_{ij}] = \begin{bmatrix} C_{11} & C_{12} & C_{13} & 0 & 0 & 0 \\ C_{12} & C_{11} & C_{13} & 0 & 0 & 0 \\ C_{13} & C_{13} & C_{33} & 0 & 0 & 0 \\ 0 & 0 & 0 & C_{44} & 0 & 0 \\ 0 & 0 & 0 & 0 & C_{44} & 0 \\ 0 & 0 & 0 & 0 & 0 & (C_{11}-C_{12})/2 \end{bmatrix} \tag{2.39}$$

where only five C_{ij}'s are independent. If the material property is isotropic in the x_1–x_3 and x_2–x_3 planes in addition to the x_1–x_2 plane, i.e., infinite number of planes of material property symmetry, then the material is defined as isotropic and its C_{ij} matrix is reduced to

$$[C_{ij}] = \begin{bmatrix} C_{11} & C_{12} & C_{12} & 0 & 0 & 0 \\ C_{12} & C_{11} & C_{12} & 0 & 0 & 0 \\ C_{12} & C_{12} & C_{11} & 0 & 0 & 0 \\ 0 & 0 & 0 & (C_{11}-C_{12})/2 & 0 & 0 \\ 0 & 0 & 0 & 0 & (C_{11}-C_{12})/2 & 0 \\ 0 & 0 & 0 & 0 & 0 & (C_{11}-C_{12})/2 \end{bmatrix} \tag{2.40}$$

where C_{11} and C_{12} are related to Lamé constants defined in eq. (2.2), i.e.,

$$\begin{aligned} C_{11} &= \lambda + 2\mu \\ C_{12} &= \lambda \end{aligned} \tag{2.41}$$

and where λ and μ are related to Young's modulus (E) and Poisson's ratio (ν); $\lambda = E\nu/\{(1+\nu)(1-2\nu)\}$ and $2\mu = E/(1+\nu)$. So far, we have described the constitutive equations in terms of stress–strain relation. The constitutive equations can also be defined in terms of the strain–stress relationship. Such a relationship for a general material system corresponding to the inverse of eq. (2.37) can be expressed as

$$e_i = S_{ij}\sigma_j \tag{2.42}$$

where S_{ij} are called the elastic compliance constants and are related to the elastic compliance tensor S_{ijkl} which is defined as $e_{ij} = S_{ijkl}\sigma_{kl}$, and the relationship between them is similar to Table 2.1. The S_{ij} matrices of orthotropic, transversely isotropic materials follow:

Orthotropic with principal axes being x_1, x_2, x_3-axes

$$[S_{ij}] = \begin{bmatrix} S_{11} & S_{12} & S_{13} & 0 & 0 & 0 \\ S_{12} & S_{22} & S_{23} & 0 & 0 & 0 \\ S_{13} & S_{23} & S_{33} & 0 & 0 & 0 \\ 0 & 0 & 0 & S_{44} & 0 & 0 \\ 0 & 0 & 0 & 0 & S_{55} & 0 \\ 0 & 0 & 0 & 0 & 0 & S_{66} \end{bmatrix} \tag{2.43}$$

Transversely isotropic with the x_1–x_2 plane as the isotropic plane.

$$[S_{ij}] = \begin{bmatrix} S_{11} & S_{12} & S_{13} & 0 & 0 & 0 \\ S_{12} & S_{11} & S_{33} & 0 & 0 & 0 \\ S_{13} & S_{13} & S_{33} & 0 & 0 & 0 \\ 0 & 0 & 0 & S_{44} & 0 & 0 \\ 0 & 0 & 0 & 0 & S_{55} & 0 \\ 0 & 0 & 0 & 0 & 0 & 2(S_{11}-S_{12}) \end{bmatrix} \tag{2.44}$$

Isotropic:

$$[S_{ij}] = \begin{bmatrix} S_{11} & S_{12} & S_{12} & 0 & 0 & 0 \\ S_{12} & S_{11} & S_{12} & 0 & 0 & 0 \\ S_{12} & S_{12} & S_{11} & 0 & 0 & 0 \\ 0 & 0 & 0 & 2(S_{11}-S_{12}) & 0 & 0 \\ 0 & 0 & 0 & 0 & 2(S_{11}-S_{12}) & 0 \\ 0 & 0 & 0 & 0 & 0 & 2(S_{11}-S_{12}) \end{bmatrix} \tag{2.45}$$

The S_{ij} in the above equations are usually expressed in terms of engineering constants, Young's and shear moduli and Poisson's ratios. For example, the S_{ij} for orthotropic material (eq. (2.43)) can be rewritten in terms of these engineering constants [9].

$$\begin{bmatrix} \frac{1}{E_1}, & -\frac{\nu_{21}}{E_2}, & -\frac{\nu_{31}}{E_3}, & 0, & 0, & 0 \\ -\frac{\nu_{12}}{E_1}, & \frac{1}{E_2}, & -\frac{\nu_{32}}{E_3}, & 0, & 0, & 0 \\ -\frac{\nu_{13}}{E_2}, & -\frac{\nu_{23}}{E_2}, & \frac{1}{E_3} & 0, & 0, & 0 \\ 0, & 0, & 0, & \frac{1}{G_{23}}, & 0, & 0 \\ 0, & 0, & 0, & 0, & \frac{1}{G_{31}}, & 0 \\ 0, & 0, & 0, & 0, & 0, & \frac{1}{G_{12}} \end{bmatrix} \tag{2.46}$$

where E_j is the Young's modulus along the x_i-axis, G_{ij} is the shear modulus in the x_i–x_j plane and ν_{ij} is the Poisson's ratio defined by

$$\nu_{ij} = -\frac{e_j}{e_i} \tag{2.47}$$

and where e_i is the strain along the loading direction (x_i-axis) and e_j is the strain along its transverse direction (x_j-axis). The relation between Voigt constants C_{ij} and engineering constants can be similarly obtained and the detailed derivations are given in Ref. 9.

2.3.2 Coefficients of thermal expansion

The coefficient of thermal expansion (CTE) of a material is usually defined as the linear expansion strain per unit temperature change (with unit 1/°C or 1/K) and it is dilatational. Though CTEs can be expressed formally as a second-order tensor α_{ij}, its non-vanishing components are α_{11}, α_{22} and α_{33} (other distortional α_{ij} are zero), which correspond to the CTEs of an orthotropic material. For transversely isotropic material with x_1–x_2 plane as the isotropic plane, there exist only two components, $\alpha_{11}=\alpha_{22}$ and α_{33}. For an isotropic case $\alpha_{ij}=\delta_{ij}\,\alpha$ where δ_{ij} is Kronicker's delta, $\delta_{11}=\delta_{22}=\delta_{33}=1$ and $\delta_{ij}=0$ for $i \neq j$ and α is a scalar constant. The constitutive equation for a thermoelastic material with C_{ijkl} and α_{ij} is defined by eq. (2.3) where e^*_{kl} is given by $\alpha_{kl}\,(T-T_0)$.

2.3.3 Thermal conductivity

A theory of thermal conductivity of anisotropic materials has been developed mainly to study the anisotropic thermal conductivity behavior of crystals [10], and its essence was summarized in Ref. 8. Some of them will be reviewed here.

If we employ fundamental hypothesis that the heat flux is proportional to temperature gradient, one can obtain the governing equations in index form based on Cartesian coordinates (x_1, x_2 and x_3)

$$-q_i = K_{ij}\,T_{,j} \tag{2.48}$$

where q_i is the heat flux along the x_i-axis, $T_{,j}$ is the temperature gradient across the isothermal jth surface (i.e., the surface is perpendicular to the x_j axis), and K_{ij} is the heat conductivity tensor and its components are given by

$$[K_{ij}] = \begin{bmatrix} K_{11} & K_{12} & K_{13} \\ K_{21} & K_{22} & K_{23} \\ K_{31} & K_{32} & K_{33} \end{bmatrix} \tag{2.49}$$

It should be noted in eq. (2.49) that K_{ij} is not symmetric in general, though the majority of crystals satisfy $K_{ij}=K_{ji}$. Assuming no heat generation within an anisotropic and homogeneous solid, the requirement that the rate of gain of heat from all the surfaces of the solid is equal to the rate of gain of heat within the solid leads to

$$\rho c\dot{T} + q_{i,i} = 0 \tag{2.50}$$

where $\dot{T}$ is $\partial T/\partial t$, ρ and c are the density and specific heat of the solid.

From eqns. (2.48) and (2.50) we obtain the governing equations for heat conduction in an anisotropic homogeneous solid,

$$K_{ij}\,T_{,ij} = \rho c\dot{T} \tag{2.51}$$

Special cases of K_{ij} given by eq. (2.49) are examined here. When one plane of reflection symmetry exists, for example x_3-axis, then K_{ij} is reduced to

$$[K_{ij}] = \begin{bmatrix} K_{11} & K_{12} & 0 \\ K_{21} & K_{22} & 0 \\ 0 & 0 & K_{33} \end{bmatrix} \tag{2.52}$$

Though $K_{12} \neq K_{21}$, their values are usually much smaller than the diagonal components (K_{11}, K_{22} and K_{33}); hence one can assume that K_{ij} is often expressed by eq. (2.53). When two planes of reflection symmetry exist and the coordinates are taken along such axes (perpendicular to the planes) then the material is called orthotropic solid and its K_{ij} is given by

$$[K_{ij}] = \begin{bmatrix} K_{11} & 0 & 0 \\ 0 & K_{22} & 0 \\ 0 & 0 & K_{33} \end{bmatrix} \tag{2.53}$$

If the axes in an orthotropic solid are interchangeable, or the heat conduction is independent of direction, the material is called isotropic and K_{ij} is reduced to

$$[K_{ij}] = \begin{bmatrix} K_{11} & 0 & 0 \\ 0 & K_{11} & 0 \\ 0 & 0 & K_{11} \end{bmatrix} \tag{2.54}$$

2.4 Composite Models

A number of composite models have been developed over the last two decades with the aim of predicting the thermomechanical properties of composites for given data of constituent phases (matrix and reinforcement). These composite models can be grouped into five basic models, ranging from simplest to vigorous types, i.e., law of mixtures, shear lag, laminated plate, Eshelby's and variational principle models. The mathematical treatment of these models has recently been summarized by Mura [4] and Christensen [11] where elastic moduli of composites were focused on as a demonstration. In the following, a brief summary of these models that can predict both thermal and mechanical properties of composites will be given.

2.4.1 Law of mixtures

Consider a composite where N different reinforcing phases are distributed in a matrix. Assume that each reinforcing element has a shear modulus μ_i with a volume fraction V_i (where $i = 1, 2, \ldots N$), and the corresponding values of the shear modulus and volume fraction for the matrix are μ_0 and

V_0, respectively. According to the law of mixtures, which was originally proposed by Voigt [12], the composite shear modulus μ_c is approximated by

$$\mu_c = \sum_{i=0}^{N} V_i \mu_i \tag{2.55}$$

where

$$\sum_{i=0}^{N} V_i = 1 \tag{2.56}$$

The basic assumption of the law of mixtures is that the externally applied shear strain γ_a is equal to the shear strains in all the phases including the matrix, which can be interpreted as the average strain $\bar{\gamma}$. Since the stress in the ith phase, σ_i, is given by $\mu_i \bar{\gamma}$, the average stress in the composite $\bar{\sigma}$ can be approximated by

$$\bar{\sigma} = \sum_{i=0}^{N} V_i \sigma_i = \sum_{i=0}^{N} V_i \mu_i \bar{\gamma} \tag{2.57}$$

On the other hand, the average stress $\bar{\sigma}$ is related to the applied strain γ_a $(=\bar{\gamma})$ by

$$\bar{\sigma} = \mu_c \bar{\gamma} \tag{2.58}$$

From eqs. (2.57) and (2.58) one can obtain eq. (2.55). Hill [13] verified that the composite shear modulus predicted by law of mixtures, eq. (2.55), gives rise to the upper bound approximation of the exact shear modulus for the composite. Due to its simplicity, the law of mixtures equation, eq. (2.55), has been extended to predict other thermomechanical properties. For example, if the shear moduli μ_c and μ_i are replaced by the strength of composite (σ_c) and reinforcing phases (σ_i), one can obtain the composite strength given by

$$\sigma_c = \sum_{i=0}^{N} V_i \sigma_i \tag{2.59}$$

In the case of two-phase systems, matrix and one kind of reinforcing phase, eq. (2.59) can be simplified as

$$\sigma_c = V_0 \sigma_0 + V_1 \sigma_1$$

or

$$= V_m \sigma_m + V_f \sigma_f \tag{2.60}$$

where the matrix and reinforcing phase (usually fiber) are denoted by subscript m and f, respectively. The value predicted by the law of mixtures is an upper bound, because the strain in the reinforcement and the matrix are not the same. The law of mixtures does, however, predict reasonably well the longitudinal properties of unidirectional continuous filament composites. Therefore, the law of mixtures predictions are upper limit goals that composite design engineers and scientists attempt to obtain in the course of the development of new composite systems. In other cases involving transverse properties or the properties of a short fiber composite, the law of mixtures value is well above the exact property of the composite (see the discussion on the stiffness and strength of composites in this context in subsections, 3.2, 3.3 and 3.4 in the next chapter).

2.4.2 Shear lag model

The shear lag model was originally developed by Cox [14], and its detailed derivation is summarized by Kelly [15]. It is best suited for an aligned short fiber composite where short fibers of uniform length and diameter (hence, constant aspect ratio) are all aligned in the loading direction and distributed uniformly throughout the material (Fig. 2.3(a)). Then a 'unit cell', which is a representative short fiber surrounded by the matrix, is focused on, as shown in Fig. 2.3(b) where the other boundary of the surrounding matrix is taken as the mid-surface between two short fibers. Let this aligned short fiber composite be subjected to the applied uniaxial strain e along the z-direction (or axial direction) and let the axial displacements in the fiber and the matrix on the boundary of the unit cell ($r = D/2$) be denoted by u and v respectively. Then, by assuming that the difference in the axial displacements, $u - v$, is proportional to the shear stress at the matrix–fiber interface τ_0, or $d\sigma_f/dz$,

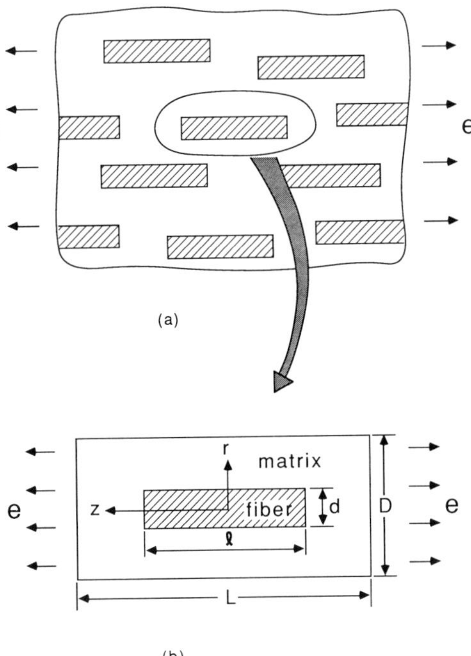

FIG. 2.3 Shear lag model for aligned short fiber composite; (**a**) representative short fiber, (**b**) unit cell for shear lag analysis.

one can obtain at z

$$\frac{d\sigma_f}{dz} = -\frac{4\tau_0}{d} = h(u - v) \tag{2.61}$$

where σ_f is the axial stress in the fiber and h is a constant which will be determined later. The first equation in eq. (2.61) was derived by considering the equilibrium of force along the z-direction. It is noted in eq. (2.61) that the positive direction of shear stress τ_0 is taken along the positive z-axis. In the fiber, a one-dimensional Hooke's law is valid

$$\sigma_f = E_f \frac{du}{dz} \tag{2.62}$$

where E_f is the axial Young's modulus of the fiber. The applied composite strain e is equal to dv/dz. Hence, from eqs. (2.61) and (2.62) we have the ordinary differential equation

$$\frac{d^2\sigma_f}{dz^2} = h\left(\frac{\sigma_f}{E_f} - e\right) \tag{2.63}$$

The general solution to eq. (2.63) is given by

$$\sigma_f = E_f e + C_1 \cosh \beta z + C_2 \sinh \beta z \tag{2.64}$$

where

$$\beta = \sqrt{\frac{h}{E_f}} \tag{2.65}$$

and C_1 and C_2 are unknown constants. Applying boundary conditions, $\sigma_f = \text{constant}\ (\sigma_0)$ at $z = l/2$ and $d\sigma_f/dz = 0$ at $z = 0$, we obtain the stress field in the fiber given by

$$\sigma_f = E_f e \left\{ 1 + \frac{\left(\dfrac{\sigma_0}{E_f e} - 1\right) \cosh \beta z}{\cosh\left(\dfrac{\beta l}{2}\right)} \right\} \tag{2.66}$$

It is noted in eq. (2.66) that $\sigma_0 = 0$ was used in the original derivation of Cox [14, 15], but σ_0 may not be zero for the case of strongly bonded fiber ends [16]. The average fiber stress σ_f is computed as

$$\begin{aligned} \bar{\sigma}_f &= \frac{2}{l} \int_0^{l/2} \sigma_f \, dz \\ &= E_f e \left\{ 1 + \frac{\left(\dfrac{\sigma_0}{E_f e} - 1\right) \tanh\left(\dfrac{\beta l}{2}\right)}{\left(\dfrac{\beta l}{2}\right)} \right\} \end{aligned} \tag{2.67}$$

Consider next the displacement along the z-direction at an arbitrary point ($r = r$) in the matrix, w, where $w(r = d/2) = u$, and $w(r = D/2) = v$. Force equilibrium at $r = d/2$ and arbitrary point ($r = r$) provides

$$2\pi r\tau = 2\pi\left(\frac{d}{2}\right)\tau_0 \tag{2.68}$$

The shear strain at $r = r$, γ is related to τ_0 as

$$\gamma = \frac{dw}{dr} = \frac{\tau}{G_m} = \frac{\tau_0}{2G_m}\frac{d}{r} \tag{2.69}$$

where τ is the shear stress in the matrix at $r = r$, and G_m is the shear modulus of the matrix. Integrating eq. (2.69) from $r = d/2$ to $r = D/2$, we obtain

$$v - u = \frac{\tau_0 d}{2G_m}\ln\left(\frac{D}{d}\right) \tag{2.70}$$

From eqs. (2.61) and (2.70), constant h is solved as

$$h = \frac{8G_m}{d^2 \ln(D/d)} \tag{2.71}$$

From eqs. (2.65) and (2.71), β is found as

$$\beta = \frac{2\sqrt{2}}{d}\sqrt{\frac{G_m/E_f}{\ln(D/d)}} \tag{2.72}$$

With β given by eq. (2.72) the average stress σ_f in the fiber can be calculated from eq. (2.67). To obtain the average stress in the composite along the loading (z) direction, σ_c can be estimated by using the law of mixtures, eq. (2.60), where $\bar{\sigma}_m$ and $\bar{\sigma}_f$ are interpreted as the average quantities in the relevant domain

$$\sigma_c = (1 - V_f)\bar{\sigma}_m + V_f\bar{\sigma}_f \tag{2.73}$$

where V_f is the volume fraction of fiber. For a given applied strain e, one can assume that

$$\bar{\sigma}_m = E_m e \tag{2.74}$$

$$\sigma_c = E_c e \tag{2.75}$$

A substitution of eqs. (2.67), (2.74) and (2.75) into (2.73) yields the Young's modulus of the composite E_c

$$E_c = (1 - V_f)E_m + V_f E_f\left\{1 - \frac{\tanh\left(\frac{\beta l}{2}\right)}{\frac{\beta l}{2}}\right\} \tag{2.76}$$

where $\sigma_0 = 0$ (i.e., no load transfer at fiber ends) was assumed. If $\sigma_0 = \sigma_m$ is assumed, then

$$E_c = (1 - V_f)E_m + V_f E_f \left\{ 1 + \frac{\left(\frac{E_m}{E_f} - 1\right)\tanh\left(\frac{\beta l}{2}\right)}{\frac{\beta l}{2}} \right\} \tag{2.77}$$

It should be noted in the above derivation that the definition of applied strain e is not consistent with the law of mixtures where the average strain in the fiber should be equal to e, which is not the case here, as seen from (eq. 2.67). Despite this inconsistency, the shear lag analysis in conjunction with the law of mixtures has been used extensively, mainly because of its simplicity. This shear lag model can also be applicable to prediction of flow stress during the plastic deformation of stress–strain curve of a metal matrix composite [16]. It is known that the shear lag model tends to give a poorer approximation than the other rigorous models such as Eshelby's model [17].

2.4.3 Laminated plate model

The basic element of laminate composites (or simply laminates) is a lamina which is a unidirectional composite with continuous fibers all being aligned in one direction. A laminate usually consists of laminae with different fiber orientations. Hence, the overall thermomechanical behavior of a laminate depends strongly upon the stacking sequence of laminae which have different fiber orientations. The laminate plate model provides an anaytical tool to predict such overall properties of the laminate. The details of the laminate plate model have been given by Jones [9]. The most popular laminated plate model stems from a special lamina, an orthotropic plate under plane stress condition. We will describe this type in some detail below.

Consider a unidirectional lamina where x_1, x_2 and x_3 coordinates are defined as shown in Fig. 2.4. This lamina is considered orthotropic and its constitutive relations are given by eq. (2.38) (elastic stress–strain relation) and eq. (2.46) (elastic strain–stress relation). The thickness of each lamina is small, thus the state of stress field in a lamina is considered plane stress, which gives rise to

$$\sigma_{33} = 0, \quad \sigma_{23} = 0, \quad \sigma_{31} = 0 \tag{2.78}$$

From eqs. (2.46) and (2.78), out of plane strain components are reduced to:

$$\begin{aligned} e_{33} &= -\frac{\nu_{13}}{E_1}\sigma_{11} - \frac{\nu_{23}}{E_2}\sigma_{22} \\ \gamma_{23} &= 0 \\ \gamma_{31} &= 0 \end{aligned} \tag{2.79}$$

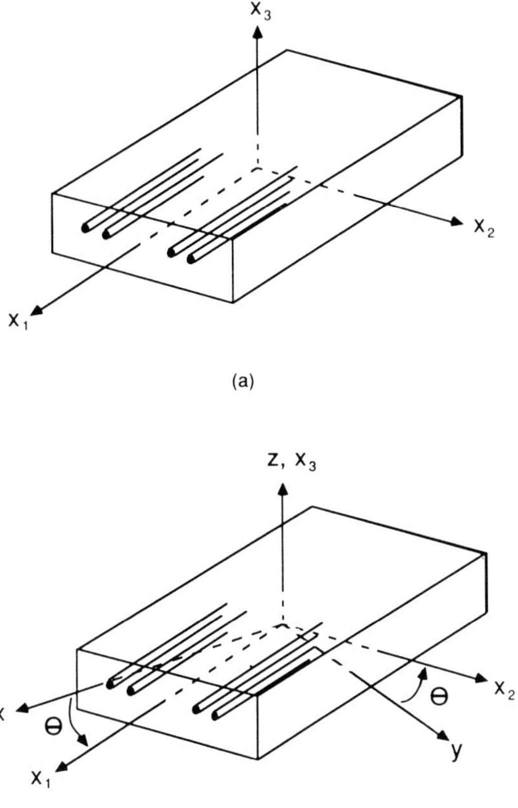

FIG. 2.4 Laminated plate model; (**a**) local coordinates (x_1, x_2) referred to a unidirectional lamina, (**b**) the same lamina referred to the global coordinates (x, y).

Hence, the main strain–stress relation for plane stress can be obtained from eq. (2.43) (or (2.46))

$$\begin{Bmatrix} e_{11} \\ e_{22} \\ \gamma_{12} \end{Bmatrix} = \begin{bmatrix} S_{11} & S_{12} & 0 \\ S_{12} & S_{22} & 0 \\ 0 & 0 & S_{66} \end{bmatrix} \begin{Bmatrix} \sigma_{11} \\ \sigma_{22} \\ \sigma_{12} \end{Bmatrix} \tag{2.80}$$

where

$$S_{11} = \frac{1}{E_1}$$

$$S_{12} = -\frac{\nu_{21}}{E_2} = -\frac{\nu_{12}}{E_1}$$

$$S_{22} = \frac{1}{E_2}$$

$$S_{66} = \frac{1}{G_{12}} \tag{2.81}$$

Now, let us look at a representative lamina in a laminate and denote the global coordinates system, x, y and z as shown in Fig. 2.4(b), where the local orthogonal coordinates x_1, x_2, and x_3, corresponding to the principal axes of the orthotropic lamina are defined by rotating the x_1–x_2 plane about the x_3 (or z-) axis by angle θ. Transformation of stress components from x_1, x_2, and x_3 coordinates to x, y, z coordinates can be made by following the transformation rule of second-order tensors [18] and it is given by

$$\begin{Bmatrix} \sigma_x \\ \sigma_y \\ \sigma_{xy} \end{Bmatrix} = [T]^{-1} \begin{Bmatrix} \sigma_1 \\ \sigma_2 \\ \sigma_{12} \end{Bmatrix} \tag{2.82}$$

where $[T]^{-1}$ is the inverse of $[T]$ matrix which is defined by

$$[T] = \begin{bmatrix} \cos^2\theta, & \sin^2\theta, & -2\sin\theta\cos\theta \\ \sin^2\theta, & \cos^2\theta, & 2\sin\theta\cos\theta \\ \sin\theta\cos\theta, & -\sin\theta\cos\theta, & \cos^2\theta - \sin^2\theta \end{bmatrix} \tag{2.83}$$

In eq. (2.82), a short notation of stress components are used, i.e., $\sigma_x = \sigma_{xx}$, $\sigma_1 = \sigma_{11}$, etc. Similarly, strain components in the x, y, z coordinates are related to those in the x_1, x_2, x_3 coordinates as

$$\begin{Bmatrix} e_x \\ e_y \\ \gamma_{xy}/2 \end{Bmatrix} = [T]^{-1} \begin{Bmatrix} e_1 \\ e_2 \\ \gamma_{12}/2 \end{Bmatrix} \tag{2.84}$$

It should be noted in eq. (2.84) that a factor of 1/2 associated with shear strains, γ_{xy} and γ_{12} is necessary to make them tensorial strains, i.e. e_{xy} and e_{12} (see eq. (2.35)). Upon substitution of eq. (2.82) and (2.84) into eq. (2.80), one can arrive at

$$\begin{Bmatrix} e_x \\ e_y \\ \gamma_{xy} \end{Bmatrix} = [\bar{S}] \begin{Bmatrix} \sigma_x \\ \sigma_y \\ \sigma_{xy} \end{Bmatrix} = \begin{bmatrix} \bar{S}_{11} & \bar{S}_{12} & \bar{S}_{16} \\ \bar{S}_{12} & \bar{S}_{22} & \bar{S}_{26} \\ \bar{S}_{16} & \bar{S}_{26} & \bar{S}_{66} \end{bmatrix} \begin{Bmatrix} \sigma_x \\ \sigma_y \\ \sigma_{xy} \end{Bmatrix} \tag{2.85}$$

where

$$\bar{S}_{11} = S_{11}\cos^4\theta + (2S_{12} + S_{66})\sin^2\theta\cos^2\theta + S_{22}\sin^4\theta$$

$$\bar{S}_{12} = S_{12}(\sin^4\theta + \cos^4\theta) + (S_{11} + S_{22} - S_{66})\sin^2\theta\cos^2\theta$$

$$\bar{S}_{22} = S_{11}\sin^4\theta + (2S_{12} + S_{66})\sin^2\theta\cos^2\theta + S_{22}\cos^4\theta$$

$$\bar{S}_{16} = (2S_{11} - S_{12} - S_{66})\sin\theta\cos^3\theta - (2S_{22} - 2S_{12} - S_{66})\sin^3\theta\cos\theta \tag{2.86}$$

$$\bar{S}_{26} = (2S_{11} - 2S_{12} - S_{66})\sin^3\theta\cos\theta - (2S_{22} - 2S_{12} - S_{66})\sin\theta\cos^3\theta$$

$$\bar{S}_{66} = 2(2S_{11} + 2S_{22} - 4S_{12} - S_{66})\sin^2\theta\cos^2\theta + S_{66}(\sin^4\theta + \cos^4\theta)$$

The stress–strain relationship of a lamina in the global coordinates can be constructed similarly, or inverted directly from eq. (2.85), and it is given by

$$\begin{Bmatrix} \sigma_x \\ \sigma_y \\ \sigma_{xy} \end{Bmatrix} = [\bar{Q}] \begin{Bmatrix} e_x \\ e_y \\ \gamma_{xy} \end{Bmatrix} = \begin{bmatrix} \bar{Q}_{11} & \bar{Q}_{12} & \bar{Q}_{16} \\ \bar{Q}_{12} & \bar{Q}_{22} & \bar{Q}_{26} \\ \bar{Q}_{16} & \bar{Q}_{26} & \bar{Q}_{66} \end{bmatrix} \begin{Bmatrix} e_x \\ e_y \\ \gamma_{xy} \end{Bmatrix} \tag{2.87}$$

where

$$\bar{Q}_{11} = Q_{11}\cos^4\theta + 2(Q_{12} + 2Q_{66})\sin^2\theta\cos^2\theta + Q_{22}\sin^4\theta$$

$$\bar{Q}_{12} = (Q_{11} + Q_{22} - 4Q_{66})\sin^2\theta\cos^2\theta + Q_{12}(\sin^4\theta + \cos^4\theta)$$

$$\bar{Q}_{22} = Q_{11}\sin^4\theta + 2(Q_{12} + 2Q_{66})\sin^2\theta\cos^2\theta + Q_{22}\cos^4\theta$$

$$\bar{Q}_{16} = (Q_{11} - Q_{12} - 2Q_{66})\sin\theta\cos^3\theta + (Q_{12} - Q_{22} + 2Q_{66})\sin^3\theta\cos\theta$$

$$\bar{Q}_{26} = (Q_{11} - Q_{12} - 2Q_{66})\sin^3\theta\cos\theta + (Q_{12} - Q_{22} + 2Q_{66})\sin\theta\cos^3\theta$$

$$\bar{Q}_{66} = (Q_{11} + Q_{22} - 2Q_{12} - 2Q_{66})\sin^2\theta\cos^2\theta + Q_{66}(\sin^4\theta + \cos^4\theta)$$

and where $[Q]$ is the stiffness matrix of the stress–strain relationship with plane stress, and is the inverse of the S_{ij} matrix as given in eq. (2.80).

If a laminate consists of n layered laminae, the macroscopic stress–strain relationship of the laminate can be obtained by summing up the contribution of each lamina multiplied by its thickness. The macroscopic behavior of the laminate, however, can be conveniently studied by looking at in-plane forces and bending moments which corresponds to the in-plane displacement and curvatures, respectively. The details of the macroscopic mechanical behavior of laminates are given in Ref. 9.

Although the above laminate plate model has been discussed in terms of the stress–strain (or strain–stress) relationship, it can be readily applicable to thermal properties, for example, coefficients of thermal expansion (α_{ij}) and thermal conductivity K_{ij}. Since both of them are second-order tensors, their behavior in the global coordinates basically follows the transformation rule defined by eqs. (2.82) and (2.83). For example, in the case of thermal conductivity of the mth lamina

$$\{\bar{K}_{ij}\}_m = [T]_m^{-1}\{K_{ij}\}_m \tag{2.88}$$

where the subscript m denotes the mth lamina and the top bar denotes the quantity referred to as the global coordinates. There, the macroscopic conductivity $\{K_{ij}\}_C$ of a laminate with n layered laminae can be estimated as

$$\{K_{ij}\}_c = \sum_{m=1}^{n} t_m \{\bar{K}_{ij}\}_m \tag{2.89}$$

where t_m is the thickness of the mth lamina.

2.4.4 Eshelby's models

Law of mixtures and shear lag models tend to give poor estimates for the stiffness and other mechanical properties of a composite where the aspect ratio of short fibers becomes small. The geometry of such a short fiber composite is three-dimensional, thus requiring a more rigorous mathematical treatment. Eshelby's model provides a relatively easy tool to solve three-dimensional problems in elasticity, and thus is suited for such a short fiber composite system. As special cases of short fiber, Eshelby's model is also applicable to particulate and continuous fiber composites. The details of Eshelby's model and various models derived from Eshelby's original model have been summarized by Mura [4].

Eshelby [19, 20] proposed a short-cut method to solve three-dimensional elasticity problems where inclusions or inhomogeneities of ellipsoidal shape are embedded in an infinite elastic body (or matrix). Let us first consider an inclusion problem where an ellipsoidal domain (denoted by Ω) is subjected to uniform non elastic strain (or eigenstrain [4]) e^*_{ij} (Fig. 2.5). Such an inclusion embedded in an infinite elastic body with elastic modulus tensor C_{ijkl} induces stress fields within and outside the inclusion (the inclusions in this case are labeled as Ω). Eshelby has proposed a solution procedure to obtain the stress

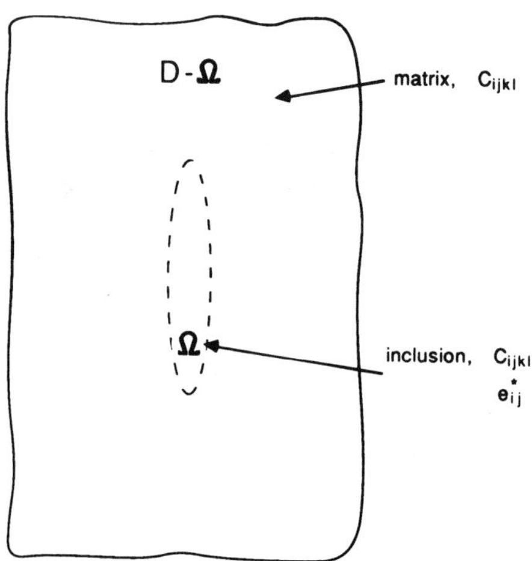

FIG. 2.5 Eshelby's model for ellipsoidal inclusion with eigenstrain e^*_{ij}.

field within Ω [19] and outside Ω [20]. The procedure to obtain the stress field outside Ω remains rigorous, but that for the stress field inside Ω is simplified (it is uniform for a given uniform eigenstrain e_{ij}^{*}). However, the results of the stress field inside Ω will provide vital information used to calculate the stress field just outside Ω and the total strain and potential energies of the entire body. The elastic modulus of a composite can be calculated from the total strain energy. Following Eshelby [19] the stress inside Ω is given by

$$\sigma_{ij} = C_{ijkl}(e_{kl} - e_{kl}^{*}) \tag{2.90}$$

where e_{kl} is the total (actual) strain and related to displacements by eq. (2.4) and also to eigenstrain e_{mn}^{*} by

$$e_{kl} = S_{klmn}e_{mn}^{*} \tag{2.91}$$

and where S_{klmn} is called 'Eshelby's tensor,' a function of the geometry of the ellipsoidal inclusion and the matrix Poisson's ratio when the matrix is isotropic, and its detailed expressions for various ellipsoidal shapes are given in Appendix C. S_{klmn} defined by eq. (2.91) should not be confused with compliance tensor S_{ijkl} discussed earlier. It is clear from eqs. (2.90) and (2.91) that the stress inside the inclusion Ω can be readily calculated algebraically for a given eigenstrain e_{ij}^{*}. Examples of eigenstrain are CTE mismatch strain defined by eq. (2.5) and plastic strain.

A practical and important outcome is an 'inhomogeneity problem' in which Ω is occupied by a material different from the surrounding matrix material. Examples of inhomogeneities are precipitates and fibers. When a composite consisting of a matrix with stiffness tensor C_{ijkl} and inhomogeneity or inhomogeneities with a stiffness tensor C_{ijkl}^{*} is subjected to the applied stress σ_{ij}^{0}, the actual (or total) stress field σ_{ij}^{t} in the inhomogeneity (Fig. 2.6(a)) is given by

$$\sigma_{ij}^{t} = \sigma_{ij}^{0} + \sigma_{ij} = C_{ijkl}^{*}(e_{kl}^{0} + e_{kl}) \tag{2.92}$$

where σ_{ij} and e_{kl} are the stress and strain disturbed by the existence of the inhomogeneity, respectively, and e_{kl}^{0} is related to the applied stress σ_{ij}^{0} by

$$\sigma_{ij}^{0} = C_{ijkl}e_{kl}^{0} \tag{2.93}$$

This inhomogeneity problem can be reduced to the inclusion problem as far as the disturbances of the stress and strain are concerned; namely, the inhomogeniety Ω with C_{ijkl}^{*} and unknown eigenstrain e_{kl}^{*} (Fig. 2.6(b)) if the following equality can be established.

$$C_{ijkl}^{*}(e_{kl}^{0} + e_{kl}) = C_{ijkl}(e_{kl}^{0} + e_{kl} - e_{kl}^{*}) \tag{2.94}$$

It is obivous from eqs. (2.93) and (2.92) that the stress disturbance σ_{ij} is given by eq. (2.90). Hence the inhomogeneity problem is reduced to the equivalent inclusion problem. Then, the strain disturbance e_{kl} is related to unknown

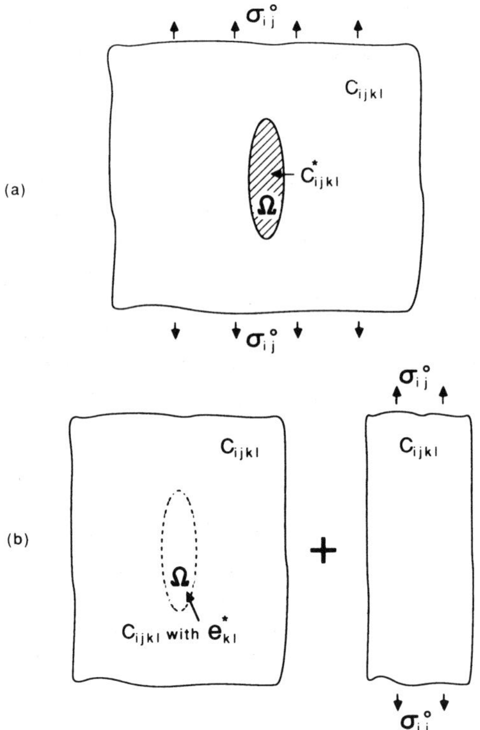

FIG. 2.6 Infinite body containing an inhomogeneous inclusion is subjected to the applied stress: (**a**) actual problem, (**b**) equivalent inclusion problem.

eigenstrain e^{*}_{mn} through eq. (2.91). Thus, eq. (2.94) is reduced to

$$C^{*}_{ijkl}(e^{0}_{kl} + S_{klmn}e^{*}_{mn}) = C_{ijkl}(e^{0}_{kl} + S_{klmn}e^{*}_{mn} - e^{*}_{kl}) \tag{2.95}$$

where the only unknown remaining to be determined is the eigenstrain e^{*}_{ij}. Once e^{*}_{ij} is obtained, the stress inside Ω can be obtained by eqs. (2.90) and (2.91).

If the eigenstrain e^{p}_{ij} is prescribed in the inhomogeneity, which is called the 'inhomogeneous inclusion problem,' the governing equation is similar to eq. (2.94), and can be established as

$$C^{*}_{ijkl}(e^{0}_{kl} + e_{kl} - e^{p}_{kl}) = C_{ijkl}(e^{0}_{kl} + e_{kl} - e^{*}_{kl}) \tag{2.96}$$

where e_{kl} is related to e^{*}_{kl} by eq. (2.91). Thus, the inhomogeneous inclusion problem is reduced to solving the unknown eigenstrain e^{*}_{kl} in eq. (2.96) where all the other quantities are known. An example of the inhomogeneous inclusion problem is the thermal stress induced in a fiber, which will be discussed in Section 3.5.

So far we have discussed the case of single inclusion inhomogeneity and inhomogeneous inclusion where the interaction between inhomogeneities need not be considered. In actual composites, however, the volume fractions (V_f) of fibers or particulates are finite, i.e., $0 < V_f < 1$, where the interaction between fibers (or inhomogeneities) becomes important. In this case the preceding formulations can still be used to make an approximate solution of the stress field if a term accounting for such an interaction is included. To this end, let us introduce the average stress disturbance in the matrix domain, $\langle \sigma_{ij} \rangle_m$, which is related to the average strain disturbance $\bar{e}_{kl}$ by

$$\langle \sigma_{ij} \rangle_m = C_{ijkl} \bar{e}_{kl} \tag{2.97}$$

Then, the formulation for finite V_f requires that $\bar{e}_{kl}$ be inserted within the brackets on both sides of the governing equations in the preceding formulations. For example, the inhomogeneity problem of eq. (2.94) will be modified to

$$C^*_{ijkl}(e^0_{kl} + \bar{e}_{kl} + e_{kl}) = C_{ijkl}(e^0_{kl} + \bar{e}_{kl} + e_{kl} - e^*_{kl}) \tag{2.98}$$

Thus, the formula for the stress disturbance is reduced to

$$\sigma_{ij} = C_{ijkl}(\bar{e}_{kl} + e_{kl} - e^*_{kl}) \tag{2.99}$$

Since the integral of σ_{ij} over the entire composite domain D vanishes [4], we have

$$\bar{e}_{kl} + V_f(e_{kl} - e^*_{kl}) = 0 \tag{2.100}$$

By combining eqs (2.91), (2.98) and (2.100), we can solve for e^*_{kl}, then the stress field is given by eq. (2.99).

One of the advantages of using Eshelby's model is that the strain and potential energies of a composite can be easily calculated once eigenstrain e^*_{ij} is solved. Eshelby showed that the elastic strain energy W of a composite with inhomogeneous inclusions is given by

$$W = \frac{1}{2}\int_D \sigma_{ij} e^0_{ij} \, dv + \frac{1}{2}\int_\Omega \sigma^0_{ij} e^*_{ij} \, dv - \frac{1}{2}\int_\Omega \sigma_{ij} e^p_{ij} \, dv \tag{2.101}$$

where Ω refers to the region of all the inhomogeneities or all the inclusions. For Ω being inhomogeneity ($e^p_{ij} = 0$), then W is given by

$$W = \frac{1}{2}\int_D \sigma^0_{ij} e^0_{ij} \, dv + \frac{1}{2}\int_\Omega \sigma^0_{ij} e^*_{ij} \, dv \tag{2.102}$$

Equation (2.102) is the basic formula from which the effective elastic moduli of a composite can be calculated. To this end the left-hand terms are set equal to the total strain energy expressed in terms of the applied stress σ^0_{ij} and the compliance tensor of the composite, $\bar{C}^{-1}_{ijkl}$. Noting that e^*_{ij} is constant in ellipsoidal fibers with a volume fraction V_f, we can show that

$$\frac{1}{2}\bar{C}^{-1}_{ijkl}\sigma^0_{ij}\sigma^0_{kl} = \frac{1}{2}C^{-1}_{ijkl}\sigma^0_{ij}\sigma^0_{kl} + \frac{1}{2}V_f\sigma^0_{ij}e^*_{ij} \tag{2.103}$$

Substitution of e^*_{ij} (found from eq. (2.98)) into eq. (2.103) will provide the compliance (or alternatively the stiffness) of the composite.

Several energy principles associated with eigenstrains have been proposed by Eshelby and his followers, and they are well documented in a book by Mura [4], where a number of practical examples are also given. Eshelby's model has also been applied to elastic–plastic bodies. It is also found to be applicable to the problem of steady-state heat conduction in composites [21]. This is due to the analogy between elasticity and steady-state heat conduction in which stress σ_{ij}, strain e_{ij}, and tiffness tensors, C_{ijkl} are analogous to heat flux q_i, temperature gradient $T_{,j}$ and thermal conductivity K_{ij}, respectively. For the inhomogeneity problem in a composite with finite V_f, the governing equations are eqs. (2.98), (2.100) and (2.91), while the corresponding governing equation for the steady state heat conduction in a composite are given by

$$\begin{aligned} q^0_i + q_i &= k_{ij}(T_{,j}{}^0 + \bar{T}_{,j} + T_{,j} - T_{,j}{}^*) \\ &= k^*_{ij}(T_{,j}{}^0 + \bar{T}_{,j} + T_{,j}) \end{aligned} \tag{2.104}$$

$$\bar{T}_{,j} + f(T_{,j} - T_{,j}{}^*) = 0 \tag{2.105}$$

$$T_{,i} = S_{ij}T_{,j}{}^* \tag{2.106}$$

where S_{ij} is Eshelby's tensor for heat conduction problems. The details are given in Appendix D.

2.4.5 Other models

In addition to the preceding composite models, several rigorous models have been proposed. Among these, self-consistent models and the models based on variational principles are the most popular.

A self-consistent model which was originally proposed by Hershey [22] and Kröner [23] is valid particularly for cases where the volume fraction of fibers or particulates is comparatively larger. Self-consistent model was applied extensively to composites by Budiansky and Wu [24–26], Hill [27, 28], and Hutchinson [29]; it can be best discussed within the framework of Eshelby's fundamental inclusion problem. Namely, in eq. (2.95), the matrix stiffness tensor C_{ijkl} must be replaced by those of the composite $\bar{C}_{ijkl}$,

$$C^*_{ijkl}(e^0_{kl} + S_{klmn}e^*_{mn}) = \bar{C}_{ijkl}(e^0_{kl} + S_{klmn}e^*_{mn} - e^*_{kl}) \tag{2.107}$$

It should be noted here that Eshelby's tensor S_{klmn} also contain unknowns; for example, Poisson's ratio of the composite $\bar{\nu}$ if the composite is isotropic.

Once eigenstrain e^*_{kl} are obtained as a function of known quantities and unknown $\bar{C}_{ijkl}$, then one can use the average stress σ_{ij} and strain e_{ij} in the composite to calculate the composite stiffness tensor $\bar{C}_{ijkl}$, or the energy equations associated with eigenstrain e^*_{ij}. In most cases, however, any self-consistent model requires rigorous numerical calculations such as the iterative method.

The composite models based on variational principles are aimed at providing the upper and lower bounds on the thermomechanical properties. Hashin and Shtrikman considered a new composite model based on variational principles, which is valid for anisotropic and nonhomogeneous elastic bodies [30]. The variational principles of Hashin and Shtrikman [30] can be discussed in terms of Eshelby's method [4], i.e., eq. (2.94) or eq. (2.95). First let us define stress σ^*_{ij} as

$$\sigma^*_{ij} = C_{ijkl} e^*_{kl} \tag{2.108}$$

where C_{ijkl} is the matrix stiffness tensor and e^*_{kl} is the eigenstrain. The strain disturbance e_{kl} is related to the eigenstrain by eq. (2.91) for an ellipsoidal inhomogeneity, or in general, to σ^*_{mn} by integral operator Γ_{klmn} [4]

$$e_{kl} = \Gamma_{klmn} \sigma^*_{mn} \tag{2.109}$$

Then, eq. (2.94) can be rewritten as

$$(C^*_{ijkl} - C_{ijkl})^{-1} \sigma^*_{kl} + e^*_{ij} + \Gamma_{ijkl} \sigma^*_{kl} = 0. \tag{2.110}$$

Hashin and Shtrikman considered the following functional U

$$U = \frac{1}{2}\{\sigma^*_{ij}, (C^*_{ijkl} - C_{ijkl})^{-1} \sigma^*_{kl}\} + \frac{1}{2}\{\sigma^*_{ij}, \Gamma_{ijkl} \sigma^*_{kl}\} + \{\sigma^*_{ij}, e^o_{ij}\} \tag{2.111}$$

where

$$\{f, g\} \equiv \int_D f(\mathbf{x}) g(\mathbf{x})\, dv \tag{2.112}$$

and where D is the entire domain of the composite. By taking a variation on U with σ^*_{ij} as a variable, one arrives at the stationary equation given by (2.110) and the stationary value of U, U_s is given by

$$\begin{aligned} U_s &= \frac{1}{2}\{\sigma^*_{ij}, e^o_{ij}\} \\ &= \frac{1}{2}\int_D \sigma^*_{ij} e^o_{ij}\, dv \end{aligned} \tag{2.113}$$

U_s corresponds to the second term in the total strain energy of the composite, W defined by eq. (2.102). Willis [31] proved that U_s becomes minimum for a positive definite $(C^{*}_{ijkl} - C_{ijkl})$ and maximum for a negative definite $(C^{*}_{ijkl} - C_{ijkl})$. In the former case, by using eqs. (2.102) and (2.113) and noting that W also can be expressed as $1/2\int C_{ijkl} e_{ij} e^{o}_{kl}\, dv$, one can obtain the following inequality:

$$\frac{1}{2}\int_{D} \bar{C}_{ijkl} e^{o}_{ij} e^{o}_{kl}\, dv - \frac{1}{2}\int_{D} C_{ijkl} e^{o}_{ij} e^{o}_{kl}\, dv$$

$$\leqslant \frac{1}{2}\int (C^{*}_{ijkl} - C_{ijkl})\sigma^{*}_{ij}\sigma^{*}_{kl}\, dv + \frac{1}{2}\int_{D} \Gamma_{ijkl}\sigma^{*}_{ij}\sigma^{*}_{kl}\, dv + \int_{D} \sigma^{*}_{ij} e^{*}_{ij}\, dv \quad (2.114)$$

In the latter case a similar inequality will be derived, which is used to obtain the lower bound on C_{ijkl} [30]. Once σ^{*}_{ij} are solved from the stationary

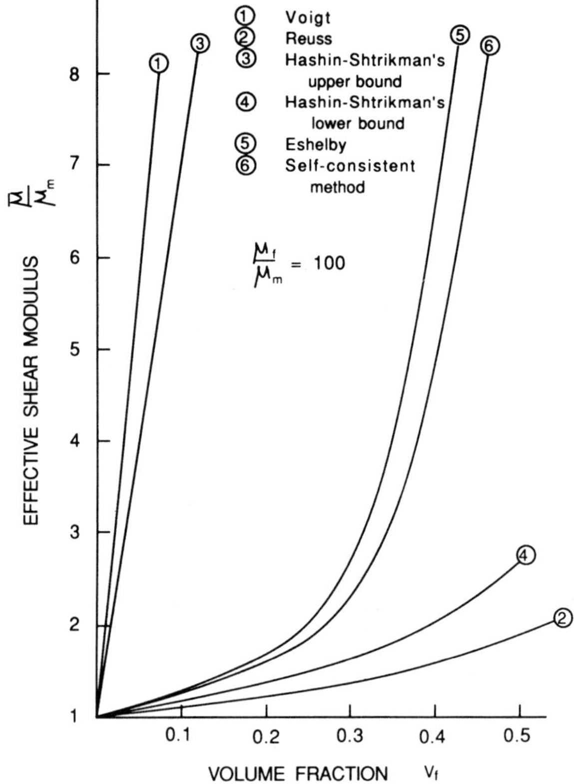

FIG. 2.7 Effective shear modulus ($\bar{\mu}$) of particle composite vs particle volume fraction V_f, predicted by various models where the ratio of the fiber shear modulus (μ_f) to the matrix's (μ_m) was kept 100 [4].

equation, eq. (2.110), one can obtain the upper and lower bounds on C_{ijkl} by choosing two extreme cases of C_{ijkl}, i.e., 0 and ∞.

Mura [4] has computed the effective shear and bulk moduli of a composite with spherical reinforcement by using the several models discussed above, and the results of the ratio of the composite shear modulus $\bar{\mu}$ to the matrix shear modulus μ_m are shown as a function of the volume fraction of spherical filler V_f in Fig. 2.7.

In order to simplify the otherwise rigorous solution procedure, one can often assume a periodicity of reinforcement where continuous or short fibers (including spheres) of the identical size and shape are distributed periodically. This assumption of periodicity has been used extensively for prediction of the thermomechanical properties of a continuous unidirectional composite system, stress–stress curve [31, 32], and thermal properties (CTE, conductivity) [33–36].

2.5 Problems

(1) Derive the Hooke's law for volumetric expansion from eq. (2.2) and show the law in terms of bulk modulus K, volumetric CTE ξ, and volumetric stress (σ) and strain (e).
(2) Derive eq. (2.15) from eq. (2.12) and also express J'_2 in terms of principal stresses σ_I, σ_{II}, and σ_{III}.
(3) What is the relationship between the average dislocation density $\bar{\rho}$ and the average plastic strain $\bar{\varepsilon}_p$?
(4) Discuss the validity of the power law creep, eq. (2.21), as compared with the exponential law, eq. (2.23). What would be the mechanisms corresponding to these creep laws?
(5) Write the governing equation relating explicitly to steady-state heat conduction in an orthotropic plate.
(6) Derive $C_{44} = (C_{11} - C_{12})/2$ for isotropic elastic material where C_{ij} is defined in eq. (2.37).
(7) In order to obtain the engineering constants for orthotropic material as defined by eq. (2.46), what types of mechanical tests would you conduct? Show the process of obtaining the constants in details.
(8) Discuss the cases where the law of mixtures model provides accurate predictions and also where it fails in the case of the stiffness of a composite.
(9) What would be the cases where Eshelby's model would fail?
(10) Can Eshelby's model be applied to prediction of the physical properties other than thermal conductivity?

References

1. Caratheodory, C., *Math. Annalan*, Vol. 67, 1909, pp. 355–386.
2. Boley, B. A. and Weiner, J. H., *Theory of Thermal Stress*, John Wiley & Sons, 1960.
3. Kroner, E., *Kontinuumstheorie der Versetzungen und Eigenspannungen*, Springer-Verlag, 1958.
4. Mura, T., *Micromechanics of Defects in Solids*, 2nd edn, Martinis Nijhoff, 1987.
5. Hill, R., *The Mathematical Theory of Plasticity*, Oxford University Press, 1960.
6. Frost, H. J. and Ashby, M. F., *Deformation-Mechanism Maps*, Pergamon Press, 1982.
7. Mukherjee, A. K., Bird, J. E. and Dorn, J. E., *Trans. ASM*, Vol. 62, 1969, pp. 155–179.
8. Carslaw, H. S. and Jaeger, J. C., *Conduction of Heat in Solids*, 2nd edn, Oxford University Press, 1959.
9. Jones, R. M., *Mechanics of Composite Materials*, McGraw-Hill, 1975.
10. Wooster, W. A., A *Textbook on Crystal Physics*, Cambridge University Press, 1938.
11. Cristensen, R. M., *Mechanics of Composite Materials*, John Wiley, 1979.

12. Voight, W., *Wied. Ann.*, Vol. 38, 1889, pp. 573–587.
13. Hill, R., *Proc. Phys. Soc.*, Vol. A65, 1952, pp. 349–354.
14. Cox, H. L., Br, *J. Appl. Phys.*, Vol. 3, 1952, p. 72.
15. Kelly, A., *Strong Solids*, 2nd edn, Oxford University Press, 1973, Chapter 5.
16. Nardone, V. C. and Prewo, K. M., *Scripta Metall.*, Vol. 20, 1986, pp. 43–48.
17. Taya, M. and Arsenault, R. J., *Scripta Metall.*, Vol. 21, 1987, pp. 349–354.
18. Borisenko, A. I. and Tarapov, I. E., *Vector and Tensor Analysis with Applications*, Prentice Hall, 1968.
19. Eshelby, J. D., *Proc. Roy. Soc.*, Vol. 241, 1957, pp. 376–396.
20. Eshelby, J. D., *Proc. Roy. Soc.*, Vol. A252, 1959, pp. 561–569.
21. Hatta, H. and Taya, M., *Int. J. Eng. Sci.*, Vol. 24, 1986, pp. 1159–1172.
22. Hershey, A. V., *J. Appl. Mech.*, Vol. 21, 1954, pp. 236–240.
23. Kröner, I. E., *Kontinuumstheorie der Verstzungen und Eigenspannungen*, Springer-Verlag, 1958.
24. Budiansky, B. and Wu, T. T., *Proc. 4th US National Congress Applied Mech.*, 1962, pp. 1175–1185.
25. Budiansky, B., *J. Mech. Phys. Solids*, Vol. 13, 1965, pp. 223–227.
26. Wu, T. T., *Int. J. Solids, Structs.*, Vol. 2, 1966, pp. 1–8.
27. Hill, R., *J. Mech. Phys. Solids*, Vol. 13, 1965, pp. 189–198.
28. Hill, R., *J. Mech. Phys. Solids*, Vol. 13, 1965, pp. 213–222.
29. Hutchinson, J. W., *Proc. Roy. Soc. Lond.*, Vol. A319, 1970, pp. 247–272.
30. Hashin, Z. and Shtrikman, S., *J. Mech. Phys. Solids*, Vol. 10, 1962, pp. 335–343.
31. Willis, J. R., *J. Mech. Phys. Solids*, Vol. 25, 1977, pp. 185–202.
32. Accorsi, M. L. and Nemat-Nasser, S., *Mechanics of Materials*, Vol. 5, 1986, pp. 209–220.
33. Aboudi, J., *Int. J. Eng. Sci.*, Vol. 22, 1984, pp. 439–449.
34. Lord Rayleigh, *Phil. Mag.*, Vol. 34, 1882, pp. 481–502.
35. Springer, G. S. and Tsai, S., *J. Comp. Mater.*, Vol. 1, 1967, pp. 166–173.
36. Gurtman, G. A., Rice, M. H. and Maewal, A., *Thermomechanical Analysis of Graphite/Metal Matrix Composites*, DARPA Final Report (No. 3877), SSS-R-81-4862.

CHAPTER 3

Basic Mechanical Behavior

3.1 Introduction

The primary reason for using metal matrix composites for aerospace and high-temperature applications is to utilize their high specific mechanical properties and their mechanical behavior in severe environments. The mechanical behavior of metal matrix composites has been considered to be the most important property, and therefore it has been studied reasonably well compared with other properties, such as corrosion resistance. The mechanical behavior can be divided approximately into two groups: the basic mechanical behavior at room temperature and that in severe environments which resemble actual use environments. In this chapter the basic mechanical behavior of metal matrix composites will be discussed, and the mechanical behavior in use environments will be discussed in Chapter 4.

A stress–strain curve of a metal matrix composite can conveniently represent its basic mechanical behavior: stiffness, yield, flow and fracture stresses. The stress–strain curve of a typical continuous fiber metal matrix composite consists at most of three stages, the first, second and third stages [1, 2]. In the first stage both matrix and fiber deform elastically; in the second stage the matrix deforms plastically while the fiber remains elastic; and then both matrix and fiber deform plastically in the third stage of the stress–strain curve. This is schematically shown in Fig. 3.1 where the stress–strain curves of both matrix metal and fiber (most likely, in this case, refractory metal fiber) are also shown. The existence of the second and third stages of the stress–strain curve of a metal matrix composite depends on the types of the matrix metal and fiber, and also on the environmental condition under which the mechanical tests to obtain the stress–strain curves are conducted. Most of metal matrix composite systems, however, appear to possess the first and second stages.

In the case of short fiber metal matrix composites, the stress–strain curves do not have three distinct stages. The stress–strain curves of these composites can be divided into two classes. In the first class there is a distinct linear region in the initial portion of the stress–strain curve corresponding to stage I in the continuous fiber composites, followed by a parabolic stress–strain curve, corresponding to stage III in the continuous fiber composites. The second class of fiber metal matrix composites have

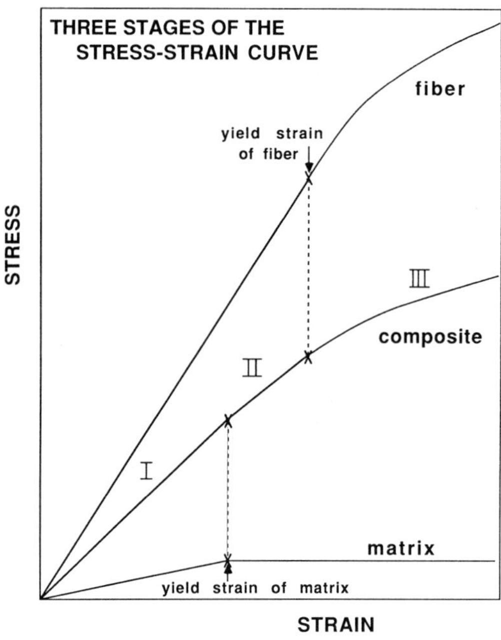

FIG. 3.1 Stress–strain curves of typical matrix metal, fiber and metal matrix composite with three stages, stages I, II and III [1, 2].

stress–strain curves which have no detectable initial linear region; their stress–strain curves are parabolic in shape over the entire stress–strain range.

Though stress–strain curves provide most of the basic mechanical behavior, the toughness of a metal matrix composite is considered to be different from its strength or fracture stress, thus the study on the toughness has been treated differently from the strength or fracture. In addition to the above basic mechanical bahavior, thermal residual stresses in the metal matrix composites will also be discussed in this chapter, for the thermal residual stresses are known to strongly influence some of the basic mechanical properties.

In the remainder of this chapter some of the sections will contain two parts; the first devoted to continuum considerations and the second based on microstructural considerations.

3.2 Stiffness

Stiffness in engineering terminology represents the elastic constants of a metal matrix composite. These elastic constants are often anisotropic due to the geometric and material properties of reinforcing phase (see Section 2.2 in Chapter 2). Though stiffness is one of the mechanical properties of a metal matrix composite that has been studied extensively, only some of the anisotropic elastic constants have been well documented [1–8]; these are the longitudinal (E_L) and transverse Young's moduli (E_T), where the longitudinal and transverse directions are taken along the fiber axis and perpendicular to it, respectively. Thus E_L becomes equal to E_T for particulate (or particle) reinforced metal matrix composites.

The mechanism for the stiffening during the Stage I deformation, i.e., elastic part of the stress–strain curve (Fig. 3.1), is that the applied stress σ_c is shared by both the matrix and fiber with the stress in each phase proportional to its stiffness and the stress in the matrix is transferred to the fiber across the matrix–fiber interfaces. The simplest, yet reasonably accurate, model to predict the stiffness of a composite is a law of mixtures and it is expressed as

$$E_c = V_f E_f + V_m E_m \tag{3.1}$$

where E_i is the stiffness (for example, Young's modulus) of the ith phase and $i = \mathrm{f}$, m and c denote fiber, matrix and composite, respectively, and V_f and V_m are the volume fractions of the fiber and matrix, respectively. Equation (3.1) is considered to be accurate in predicting E_L (the longitudinal modulus) of continuous fiber metal matrix composites, as illustrated in Fig. 3.2, where the experimental values E_L of continuous tungsten fiber/copper matrix composites are plotted as a function of V_f as filled circles and the prediction by

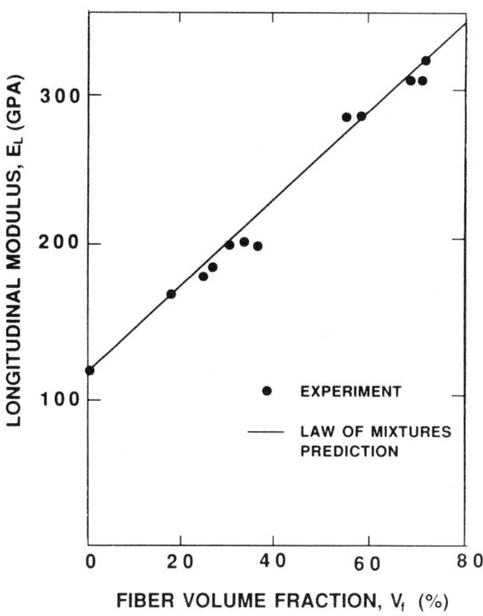

FIG. 3.2 Dynamic modulus of elasticity of a continuous tungsten fiber/copper composite where the experimental results are shown by filled circles and the prediction based on eq. (3.1) by solid line [1].

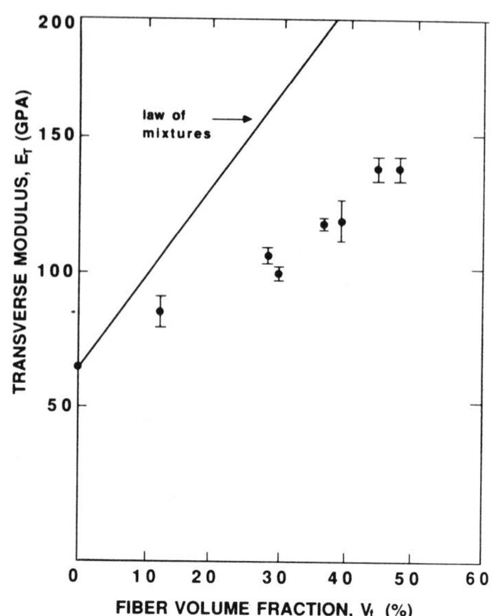

FIG. 3.3 Transverse modulus (E_T) of a BORSIC fiber/6061 Al composite as a function of fiber volume fraction (V_f) [4].

eq. (3.1) is shown by a solid line [1]. The law of mixtures prediction is also reasonably accurate for aligned short fiber metal matrix composites, but it gives rise to poor estimates of E_L of misoriented short fiber metal matrix composites and E_T (the transverse modulus) of continuous and short fiber metal matrix composites as shown in Fig. 3.3, where E_T of continuous boron fiber/aluminum matrix composites are plotted as filled circles with error band and the prediction by eq. (3.1) by a solid line [4]. It is clear from Fig. 3.3 that the experimentally measured E_T is far below the value predicted by a law of mixtures.

When the prediction based on a law of mixtures (eq. 3.1) is poor, then more rigorous models (see Section 2.3) can be used to better predict the stiffness of a metal matrix composite. This is often the case with short fiber metal matrix composites whose elastic properties depend not only on V_f and E_f/E_m (see eq. 3.1), but also on fiber aspect ratio (l/d) and other constituent parameters. Figure 3.4 [9] shows the prediction of the longitudinal modulus of an aligned 20% V_f SiC whisker/Al alloy composite (E_c) normalized by the matrix Young's modulus (E_m) by two models—the Eshelby type model (solid curve) [10–12], and the shear-lag type model [13]. In the same figure the experimental results [14] are also shown by open circles. It follows from Fig. 3.4 that the Eshelby type model can provide the best estimate of E_c, followed by the shear-lag type model, and both predictions approach the law of mixtures values as fiber aspect ratio l/d increases.

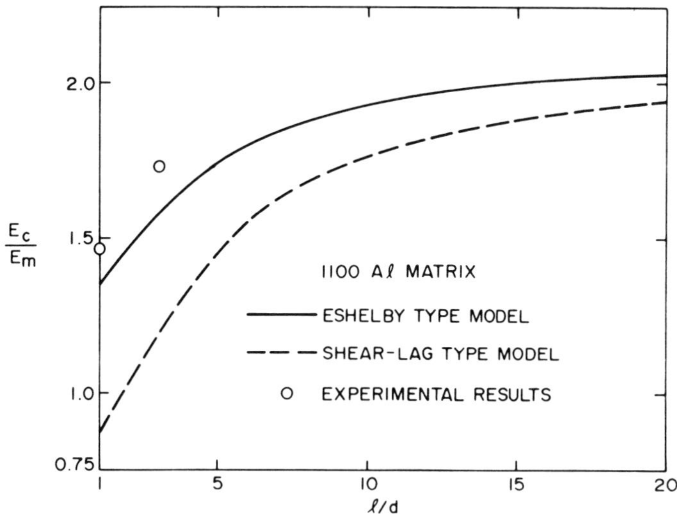

FIG. 3.4 Modulus (E_c) of short fiber metal matrix composites normalized by the matrix's modulus (E_m) as a function of fiber aspect ratio (l/d) [9].

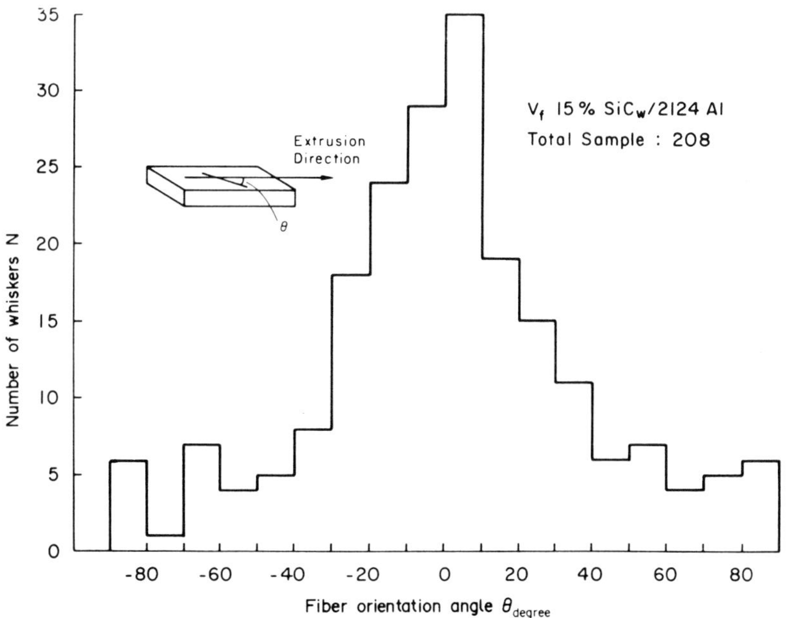

FIG. 3.5 Distribution of orientation angle θ of SiC whiskers in 15% V_f SiC whisker/2124 Al composite where θ is measured with respect to extrusion direction [15].

In addition to its dependence on V_f, E_f/E_m and l/d, the stiffness of a metal matrix composite is strongly dependent on fiber misorientation (this is particularly true for a short fiber composite system), and micro-damage that exists in the matrix and fiber phases as well as at the matrix–fiber interfaces. An example of fiber misorientation is shown in Fig. 3.5 where the experimentally measured density of orientation angle θ of SiC whiskers in a 15% V_f SiC whisker/2124 Al matrix composite, $\rho(\theta)$ is shown [15]. The angle θ is measured with respect to a particular direction (in this case along the extrusion direction). Takao et al. have developed an Eshelby type model to predict E_L of such a misoriented short fiber composite [16] where two types of distribution function are considered;

$$\begin{aligned} &\text{uniform type:} \quad \rho(\theta) = \rho_0 \text{ (constant)} \\ &\text{cosine type:} \quad \rho(\theta) = \rho_0 \cos a\theta \end{aligned} \tag{3.2}$$

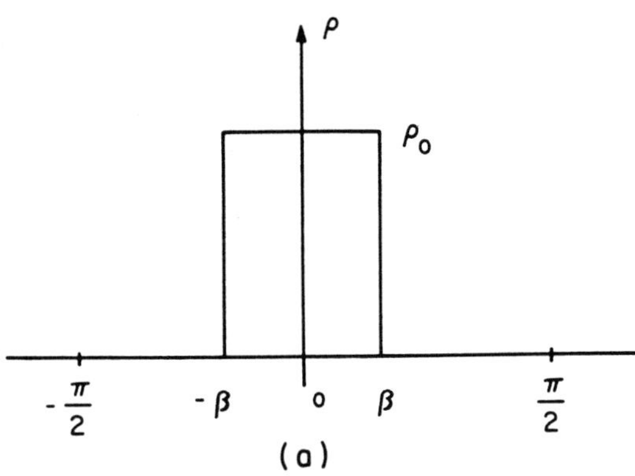

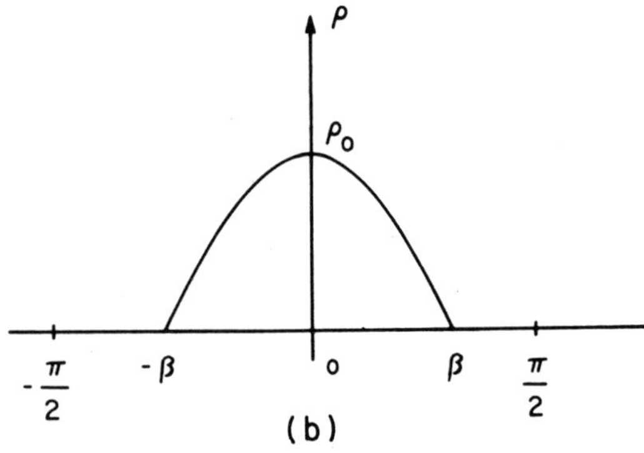

FIG. 3.6 Distribution functions $\rho(\theta)$ for the fiber orientation angle θ, uniform type (a) and cosine type (b).

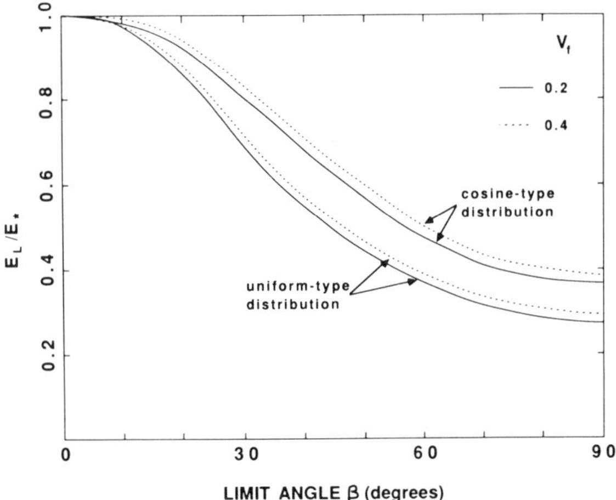

FIG. 3.7 Longitudinal modulus (E_L) of a misoriented short fiber composite normalized by that of an aligned short fiber composite (E_*) as a function of limit angle α for two types of density distribution function, uniform and cosine type [16].

where $|\theta| \leq \beta$ and β denotes the limit angle for these distribution functions (see Fig. 3.6). It should be noted that the cases of $\beta = 0$ and $\pi/2$ in the uniform type correspond to aligned and completely random short fiber composites, respectively. Typical results of E_L of various misoriented short fiber composites normalized by Young's modulus of an aligned short fiber metal matrix composite (E_*) are shown as a function of β in Fig. 3.7 where the analytical results based on the uniform and cosine types are shown as solid and dashed curves, respectively [16]. Figure 3.7 clearly illustrates strong dependence of E_L on β.

The length of short fibers in a short fiber metal matrix composite is often non-uniform due to the processing route that the composite had gone through. Arsenault [17] has examined the non-uniformity of fiber length (or the fiber aspect ratio) of SiC whiskers and the results of the percentage of fiber volume fraction V_i (%) at various fiber aspect ratios $(l/d)_i$ are shown in Fig. 3.8. In various models the fiber aspect ratio has been assumed to be constant to facilitate the computations, despite its non-uniformity. Takao and Taya have recently studied the effect of this non-uniformity of fiber aspect ratio and found that the use of the mean value of (l/d) for prediction of the stiffness can be justified provided the intensity distribution of fiber orientation angle is not widely scattered [18].

Micro-damage (voids, cracks and the interface debonding, etc.) is inevitably induced in a metal matrix composite either by the processing or thermomechanical loading to which the metal matrix composite is subject in its use environment. Whatever the source of the micro-damage may be, its

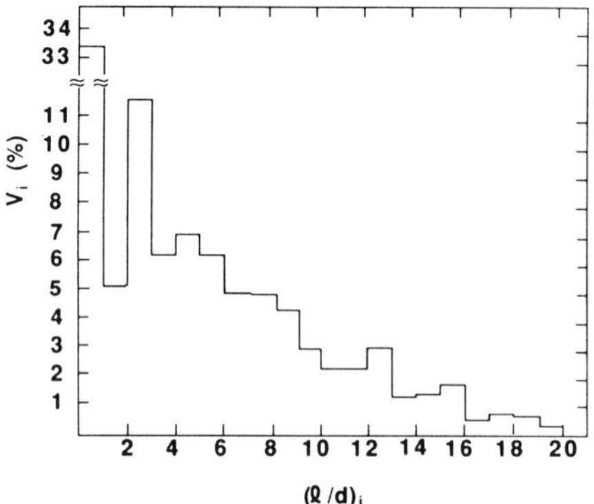

FIG. 3.8 A histogram of the percentage of fiber volume fraction V_i (%) for fiber aspect ratio (l/d) [17].

effect on the mechanical behavior of the composite is known to be from modest to significant. Taya and his co-workers have studied analytically the longitudinal Young's modulus weakened by various types of micro-damage, cracks in the matrix [19] and fiber-end cracks [12, 20] and found that these types of micro-damage are influential in reducing the stiffness of a composite.

3.3 Yield and Flow Stresses

The yield and flow stresses are defined in terms of the stress of various levels of plastic strain. If the yield stress is defined as the stress at a plastic strain of 0.2% (which is commonly used), then many continuous fiber metal matrix composites do not reach their yield stress before they fracture. However, it is possible to choose some other definition of the yield stress, and the yield stress may occur in stage II or III. The flow stress in general always defines stresses at plastic strains greater than the plastic strain chosen for the yield stress. The yield stress of metal matrix composites is always greater than that of unreinforced metal. The yield and flow stresses can be explained easily by using a law of mixtures type model, particularly for continuous fiber composites i.e., the composite stress σ_c can be predicted as

$$\sigma_c = V_f\sigma_f + V_m\sigma_m$$

or (3.3)

$$\sigma_c - \sigma_m = V_f(\sigma_f - \sigma_m)$$

where σ_i is the stress (yield or flow) of the ith phase with $i = \text{m}$ (matrix), f (fiber) and c (composite). It can be proved from eq. (3.3) that $\sigma_c > \sigma_m$ as long as $\sigma_f > \sigma_m$, which is generally the case, as seen from Fig. 3.1, where for a fixed strain e, $\sigma_f > \sigma_m$. In fact, the second and third stages of the stress–strain curve

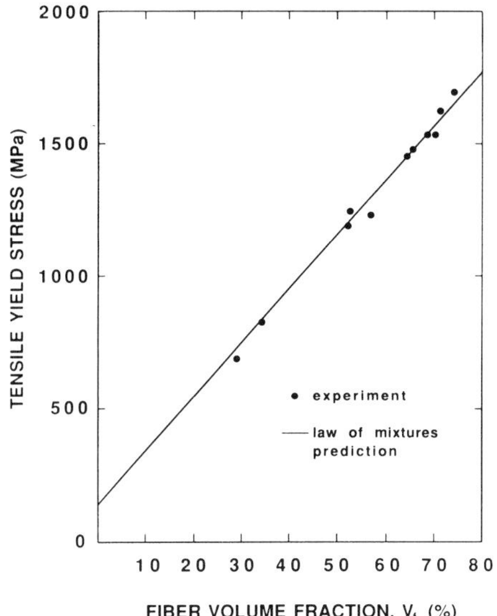

FIG. 3.9 Tensile yield stress of a continuous tungsten fiber/copper composite as a function of fiber volume fraction, V_f (%).

of a metal matrix composite can be approximated by eq. (3.3). This approximation is reasonably accurate for the longitudinal behavior of a continuous fiber metal matrix composite. The fact that yield stress increases with fiber volume fraction is well illustrated in Fig. 3.9, where the flow stress of 0.2% offset strain of continuous tungsten fiber/copper matrix composite is plotted as filled circles against fiber volume fraction [1].

The strengthening mechanism of short (discontinuous) fiber metal matrix composites has also been studied, both expeimentally [1, 2, 5] and analytically [2, 8, 9, 21–28]. The important findings in the above studies are:

1. Yield and flow stresses increase with the fiber volume fraction and fiber aspect ratio.
2. Yield and flow stresses are strongly dependent upon the fiber orientation with respect to loading direction.
3. Yield stress in compression is larger than that in tension due to the residual stress in a metal matrix composite which was caused by the mismatch in coefficients of thermal expansion (CTEs) between the matrix metal and fiber, presumably during the fabrication process.

The dependence of the yield stress of short fiber metal matrix composites on fiber aspect ratio is shown in Fig. 3.10(a), where the experimental results of the 0.2% offset yield stresses (σ_{yc}) normalized by the yield stress of the unreinforced matrix metals (σ_{ym}) are plotted as various symbols against fiber

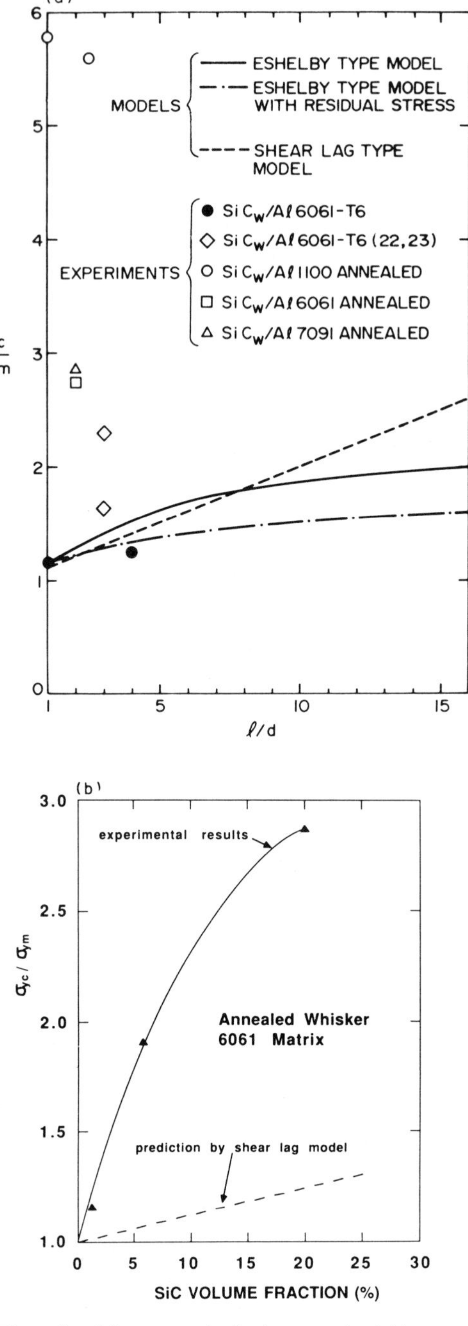

FIG. 3.10 (a) The ratio of the composite (σ_{yc}) to matrix yield stress (σ_{ym}) of 20% V_f SiCw/Al composite as a function of fiber aspect ratio (l/d): prediction by several models vs. experimental results. (b) The ratio of yield stresses of the composite (σ_{yc}) to matrix metal (σ_{ym}) vs. volume fraction of SiC whisker.

aspect ratio (l/d) [9]. In the same figure the analytical results based on several models are also shown by various curves; solid (Eshelby type model) [10], dash (shear lag type model) [8] and dash–dot (the Eshelby type model modified to account for the residual stress) [22]. Although the predictions by these models agree with the experimental results of T6-treated SiC/Al composites, they tend to underestimate for non-heat-treated SiC/Al composites (open symbols). The shear lag model predicts a linear increase of the σ_{yc}/σ_{ym} with increasing volume fraction as shown in Fig. 3.10b (dashed line); however, the experimental results (filled triangles connected by solid line) do not agree with the predicted linear increase (Fig. 3.10b).

So far we have discussed the strengthening mechanism based on a continuum model, which may be a crude approximation to the actual behavior of a metal matrix composite. This is particularly true for a short fiber metal matrix composite where non-continuum type models such as a dislocation model may provide a better analytical tool for describing the non-homogeneous deformation in the matrix. In the following, the strengthening mechanism based on one of non-continuum models, i.e., a dislocation model, will be discussed.

3.3.1 Strengthening mechanisms based on dislocations

The framework of a strengthening mechanism for short fiber metal matrix composites is slowly evolving to account for the strengthening due to the addition of a discontinuous reinforcement to a metal matrix. However, it should be kept in mind that the number of detailed investigations of these composites is rather limited. The basic strengthening mechanisms are:

- high dislocation densities due to dislocation generation as a result of differences in coefficients of thermal expansion
- small subgrain size as a result of the generation of a high dislocation density
- residual elastic stresses
- differences in texture
- classical composite strengthening (load transfer)
- dispersion strengthening

This subsection will be divided into several parts, beginning with a discussion of dislocation densities and ending with a consideration of dispersion strengthening, followed by an overall summary.

The fact that dislocations could be generated from the interface of a misfitting particle and a ductile matrix was studied by Hedyes and Mitchell [23] in the late 1950s and by Ashby and Johnson [24] in the late 1960s. In composite systems the generation of dislocations due to misfit strains was also studied by several workers. In the investigation of the copper tungsten system, Chawla and Metzeger [29], using an etch pitting technique, observed that the dislocation density is much higher at the copper/tungsten interface than in the bulk matrix, although the difference in coefficient of thermal expansion (CTE) is only 4:1. Weatherly [25] stated that in the silica copper system, multiple dislocation tangles were actually observed around silica due to the difference in CTE. Similar results were also reported by Ashby, Gelles and Tanner [26], and in the Ni/W eutectic composite system by Williams and Garmong [30].

Based on various experimental facts, Arsenault and Fisher [31] proposed that the increased strength observed in SiC/Al composites could be accounted for by a high dislocation density in the aluminum matrix. An actual experimental simulation of dislocation generation in the SiC/Al system, by analyzing the generation of slip lines around a SiC cylinder in an aluminum disk due to thermal cycling, has been demonstrated by Flom and Arsenault [27]. However, dislocation generation due to factors other than CTE differences is also possible. One other possibility is dislocation generation due to plastic deformation during materials processing, and possibly trapping these dislocations by SiC particles during annealing. If large differences (10:1) in CTE are the primary cause for high dislocation density, dislocation generation should be seen after cooling from annealing temperatures in an in-situ high-voltage electron microscope (HVEM) experiment, as suggested by Arsenault and Fisher [31]. An extensive investigation was undertaken by Vogelsang et al. [28] involving in-situ high-voltage electron transmission microscopy of the heating and cooling of SiC/Al composites. It was conclusively shown that a very high dislocation density could be produced in the Al matrix upon cooling the composite from 773 to 300 K.

Though an excellent agreement has been achieved with the dislocation generation model due to CTE mismatch, other investigators are still active in the classical mechanics approach. Several investigators, such as Hoffman [32], Garmong [33], Dvorak and his co-workers [34, 35], Mehan [36], Koss and Copley [37] and most recently Nardone and Prewo [8], have studied different systems in explaining the strengthening phenomenon due to the existence of reinforcement by analyzing the stress–strain field. It is assumed in these models that the plastic strain in the matrix is uniform and the metal matrix behaves following the plastic behavior of the unreinforced metal. Although these models can predict the flow stress of continuous fiber metal matrix composites qualitatively, they are not capable of predicting accurately the observed strengthening of short fiber metal matrix composites. Arsenault and Shi [38, 39] have recently proposed a simple model of non-continuum type for prismatic punching. This model will be described below.

The assumption was made that both SiC and aluminum were elastically isotropic and the SiC reinforcement was assumed to be parallelepiped particles and prismatic punching was assumed to occur equally on all faces of the particles, as shown in Fig. 3.11. Consider the reinforcement particles with

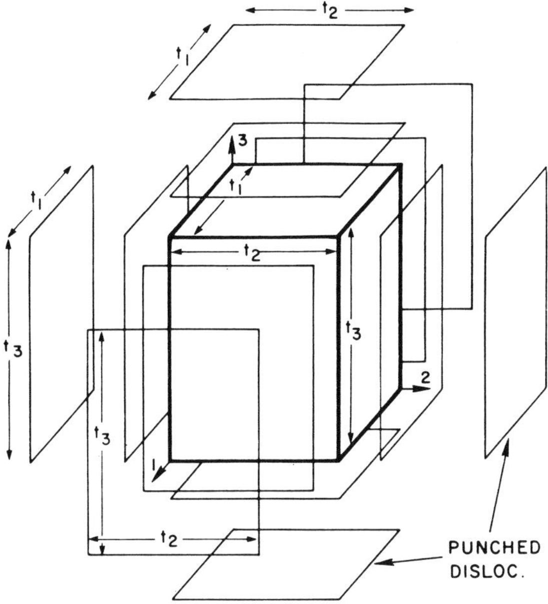

FIG. 3.11 Schematic diagram of the particle and several punched dislocations.

dimensions of height t_1, width t_2 and thickness t_3; the misfit strain due to the difference in the CTE between aluminum and SiC is

$$\varepsilon = \frac{\Delta t_i}{t_i} = \frac{\Delta \mathrm{CTE} \cdot \Delta T}{2} \tag{3.4}$$

where ΔT is the temperature difference, and total number of dislocation loops may be expressed as

$$N_i = t_{ii} \frac{\varepsilon}{b} \tag{3.5}$$

where N_i is the number of prismatic loops punched in the ith dimension, ε is the misfit strain, b is the magnitude of the Burger's vector, t_{ii} is a specific case defined as the contribution tensor and t_{ij}, the contribution of the particle height in the jth dimension in punching a dislocation loop in the ith direction where in this case i is equal to j.

It is believed that, by taking account of dislocation back stresses and dislocation interaction, t_{ii} can be written in a general form

$$t_{ii} = f(t_i, k_p, k_m, \nu_m, \nu_p) \tag{3.6}$$

where k_p and k_m are the bulk modulus of the particle and the matrix respectively, ν_p and ν_m are Poisson's ratio for the particle and the matrix respectively, and t_i is the actual dimension in the ith dimension of the particle.

With the help of the contribution tensor, the total length of the dislocation loops punched out in the ith direction is given by

$$l_i = \frac{2\varepsilon}{b} t_{ij}\delta_{ij}t_{ik}(1 - \delta_{ik}) \tag{3.7}$$

where δ_{ij} and δ_{ik} are Kronecker deltas. The Einstein suffix notation is used here.

In order to compare with the experimental data further simplifications will be made. Note the fact [28] that the speed of the dislocation motion was at least as fast as the rate of the dislocation generation. This would result from relatively low friction stress and therefore result in relatively low back stress. These two factors would then be neglected in the subsequent derivations. It is further assumed that the misfit strain is so relaxed that the residual elastic strain results from misfit of the magnitude which less than one Burgers vector, and the interaction stresses among different particles and those between punched dislocations from one particle with the dislocation punched out by another platelet are neglected.

Based on these considerations, the proceeding formulations can be further simplified. On the assumption of rigid expansion

$$t_{ii} = t_i \tag{3.8}$$

Then

$$N_i = \left[t_i \frac{\varepsilon}{b} \right] \quad (i = 1, 2, 3) \tag{3.9a}$$

where N_i is integer, and the square brackets represent the closest smaller integer taken from the real number enclosed. Since N_i is a large number

$$N_i = t_i \frac{\varepsilon}{b} = \frac{\Delta t_i}{b} \tag{3.9b}$$

and it is also reasonable to assume

$$t_{ij} = t_{kj} = t_j$$

Therefore eq. (3.7) can be expanded in the simple form

$$\begin{aligned} l_1 &= \frac{2\varepsilon}{b} t_1(t_2 + t_3) \\ l_2 &= \frac{2\varepsilon}{b} t_2(t_1 + t_3) \\ l_3 &= \frac{2\varepsilon}{b} t_3(t_1 + t_2) \end{aligned} \tag{3.10}$$

which may also be directly extracted from Fig. 3.11. Thus the total length of the dislocations generated by one particle is

$$L = \Sigma l_i = l_1 + l_2 + l_3 \tag{3.11}$$

Without considering the dislocation back stresses and dislocation–particle interaction, the arrangement of the particles will not affect the dislocation density ρ due to the difference between the CTEs of aluminum and SiC. The number of particles in unit volume is

$$n = \frac{A_v}{t_1 t_2 t_3} \tag{3.12}$$

where n is the number of particles in unit volume and A_v is the volume fraction of particles. Therefore, the total length L_L of the dislocations generated by all particles will be

$$\begin{aligned} L_L &= nL \\ &= (l_1 + l_2 + l_3)\frac{A_v}{t_1 t_2 t_3} \end{aligned} \tag{3.13}$$

and the dislocation density generated by the particles in the matrix is

$$\begin{aligned} \rho &= \frac{L_L}{1 - A_v} \\ &= \frac{4A_v\varepsilon}{b(1 - A_v)}\left(\frac{1}{t_1} + \frac{1}{t_2} + \frac{1}{t_3}\right) \end{aligned} \tag{3.14a}$$

Rearranging eq. (3.14a), the following is obtained:

$$\rho = \frac{2A_v\,\varepsilon}{b(1 - A_v)}\frac{S}{V} = ks \tag{3.14b}$$

where k is a constant, S is the total surface area of the particle, s is the total surface area per unit volume of SiC that is the particle-specific surface area, and V is the volume of the particle. In general, the smaller the size or the larger the surface area per unit volume of reinforcement material, the higher a dislocation density will be produced.

Reconsidering eq. (3.14b) and being aware of eq. (3.12), eq. (3.14b) will become

$$\rho = \frac{2\varepsilon A_n}{b} S = cS \tag{3.14c}$$

where A_n is the number of particles per unit matrix volume. If c is treated as a constant, and it is assumed that the original dislocation density in the matrix ρ_0 is very small compared to the density which was generated by CTE mismatch,

$$\rho_{total}(S) = \rho_0 + cS \tag{3.15}$$

This has a form of linear approximation of Taylor's series expansion of $\rho_{total}(S)$.

Let us consider the cases of composites with a whisker of aspect ratio ($R = D/d$) of 0.5, and a platelet of $R = 2$, where D is the diameter and d is the thickness (length) of the particle. The dislocation density for both cases are

$$\rho_w = \frac{10A_v\varepsilon}{b(1-A_v)}\frac{1}{t} \tag{3.16}$$

$$\rho_p = \frac{8A_v\varepsilon}{b(1-A_v)}\frac{1}{t} \tag{3.17}$$

with t representing the smallest dimension, while ρ_w and ρ_p stand for the dislocation densities in the whisker and platelet composite, respectively. It can be seen that the prismatic dislocation density is higher in a whisker composite than in a platelet composite of the same volume fraction (Fig. 3.12).

In general, the dislocation density ρ due to punching can be written

$$\rho = \frac{BA_v\varepsilon}{b(1-A_v)}\frac{1}{t} \tag{3.18}$$

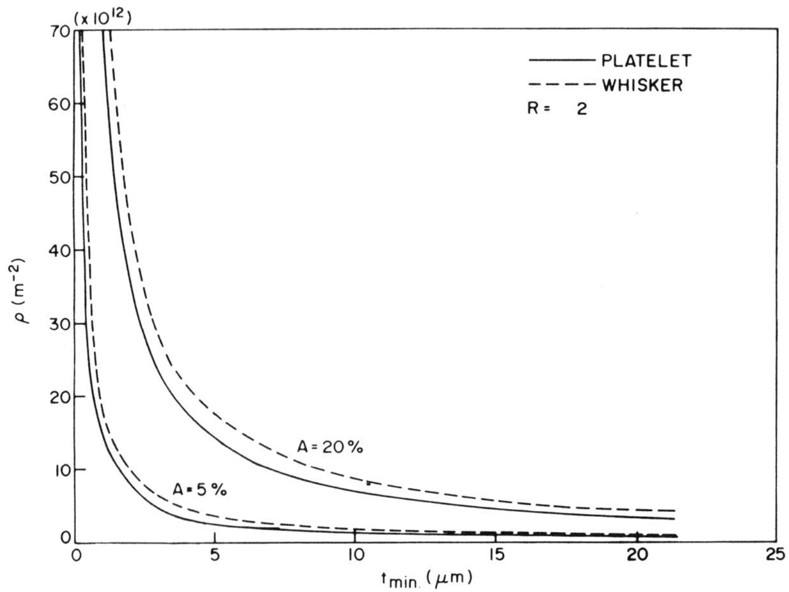

FIG. 3.12 Calculated dislocation density ρ due to prismatic punching as a function of minimum particle thickness t for both whiskers (dashed lines) and platelets (solid lines) of the same volume fraction A_v.

where B is a geometric constant which is theoretically between 4 ($R = \infty$) and 12 ($R = 1$).

If the increase of dislocation density is believed to have contributed to strengthening, the following equation can be used:

$$\Delta\sigma = \beta\mu b\rho^{1/2} \tag{3.19}$$

where $\Delta\sigma$ is the increase in yield strength, μ is the shear modulus of the matrix and β is a geometric constant. Hansen [40] obtained a β value of 1.25 for aluminum, which will be used here.

$$\Delta\sigma = \beta\mu b\left(\frac{A_v}{1 - A_v}\frac{B\varepsilon}{b}\right)^{1/2}\left(\frac{1}{t}\right)^{1/2} \tag{3.20}$$

Figure 3.13 shows the decrease of yield strength due to an increase in platelet diameter since the diameter is easier to measure experimentally. Two experimental data points were included for comparison [38]. It is not surprising that, when the diameter of the platelet is increased, the effect of strengthening decreased drastically. It is also noticeable that the experimental data are well fitted to the theoretical curve when the diameter of the platelet is moderate. Yet, when particle size is large, the experimental results do not agree as well as that when the particle size is small.

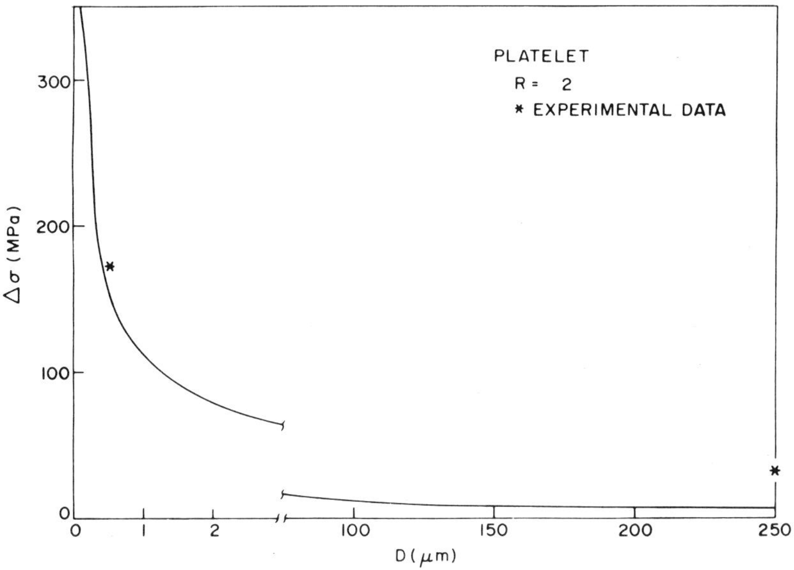

FIG. 3.13 Calculated increase in the yield stress of the composite over that of the matrix material on account of the dislocations due to CTE mismatch between Al and SiC where $\Delta\sigma_y = \sigma_{yc} - \sigma_{ym}$.

In the above discussion the strengthening depends on particle size: however, an explicit picture of how such a material property changes with the morphology of the particle is also needed. Upon manipulating eq. (3.14a), the following equation can be obtained

$$\rho = \frac{4A_v\varepsilon}{b(1-A_v)}(R)^{2/3}\left(1+\frac{2}{R}\right)\left(\frac{1}{V}\right)^{1/3} \tag{3.21}$$

It is noted here that when $R<1$ it represents whisker morphology, when $R>1$ it is a representation of platelet morphology, and when $R=1$ it is a equi-axed particle which is an approximation of a spherical composite. On subsequent substitution into eq. (3.19), it becomes

$$\Delta\sigma = 2\alpha\mu\left(\frac{bA_v\varepsilon}{1-A_v}\right)^{1/2}(R)^{1/3}\left(1+\frac{2}{R}\right)^{1/2}\left(\frac{1}{V}\right)^{1/6} \tag{3.22}$$

Equi-volume plots of ρ and $\Delta\sigma$ as a function of R are shown in Figs. 3.14 and 3.15, respectively, where the volume of the particle is assumed to be 1 μm^3. Equation (3.22) predicts that a spherical reinforcement is not recommended as far as increase of strengthening is concerned. For the same volume fraction the tensile strength of a composite with spherical reinforcement will be the weakest. This is for the case of the same volume fraction of reinforcement, and the reinforcement has the same volume, i.e., the volume of each sphere is the same as that of a single whisker or plate.

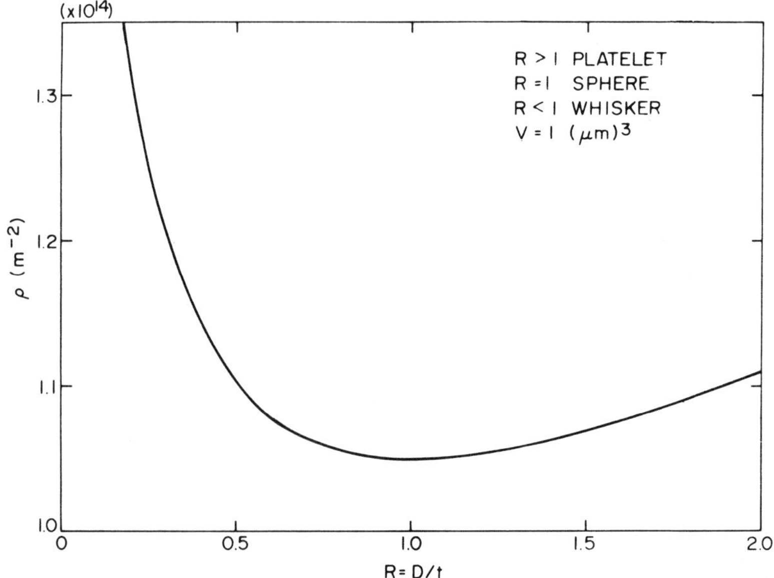

FIG. 3.14 Calculated dislocation density ρ vs. fiber aspect ratio R.

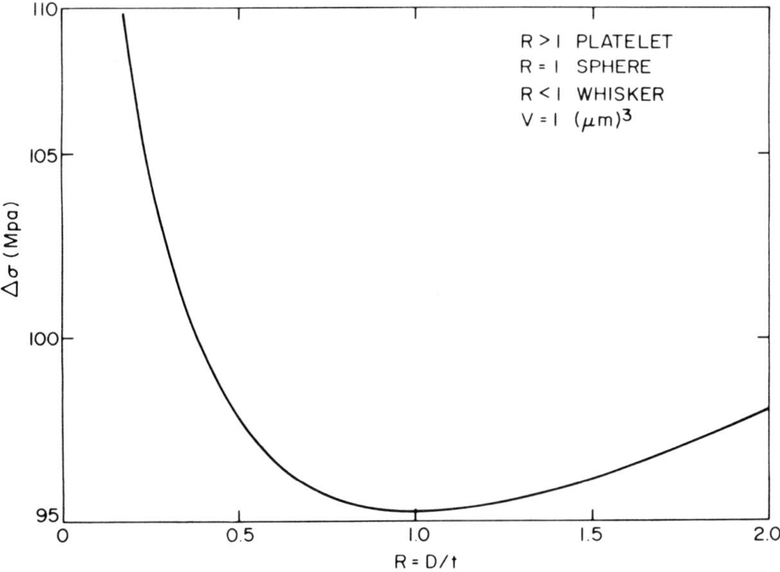

FIG. 3.15 Predicted increase of yield stress vs. fiber aspect ratio R, based on eq. (3.22).

Experimentally producing equi-volume particles of different morphology is not a trivial task. The equation was checked with experimental data [41] using different volume parameters, as shown in Table 3.1. It is worthwhile mentioning that, although the increase of strengthening in the spherical composite is higher than that in the whisker sample, it does not contradict the result in Fig. 3.15, since the volume parameter is different. If the average size of the particles was measured for the above-mentioned composites it is in the moderate size range, as can be calculated from Table 3.1. This once again proves the statement made before, that experimental data are well fitted to the theoretical curve when the size of the particle is moderate ($0.5 \sim 2\ \mu m$).

Table 3.1 *Increase of strengthening of different composites* (Al alloy 1100 matrix)*

Particle parameter	Increase in strength $\Delta\sigma$ (MPa)	
	Theoretical prediction (MPa)	Experimental data (MPa)
$R = 1$; $V = 0.125\ \mu m^3$	135	166
$R = 0.25$; $V = 0.5\ \mu m^3$	117	152

* Additional experimental data are shown in Fig. 3.13.

The idea that the high dislocation density in the matrix is due to a large difference in coefficient of thermal expansion, and can account for the increase of strength, as proposed by Arsenault and Fisher, has been proven to a first order by a simple prismatic punching theoretical model. The prismatic punching which was adopted in this theoretical model is a more efficient way to relax the strain field than any other dislocation generation mechanisms, and it predicts a dislocation density of $2.4 \times 10^{13}\ \mathrm{m}^{-2}$ for an average size platelet of 5 μm, which is quite in agreement with what was being observed ($2.0 \times 10^{13}\ \mathrm{m}^{-2}$) [28]. Therefore, the theoretical model gives a reasonable lower bound of dislocation density at least for moderate particle sizes.

It is also noted in Fig. 3.13 and Table 3.1 that the experimental data [38, 39] agree with theoretical prediction quite well when the particle size is small, but they are several factors off when the particle size is large. One of the reasons is believed to be that moderate particles have a large edge length over surface area ratio; that means more of the dislocations generated were from the particle edge than from the interface area, which is classified as a secondary dislocation generation, and it is well known that prismatic punching is associated solely with the edge length of particle. As a result, the actual dislocation density is close to what the theoretical model predicted; therefore, there is a good agreement in increase of yield strength. On the other hand, dislocations that were generated from 250 μm platelet are, in a large part, secondary dislocations from the particle matrix interface, as can be seen in Fig. 3.16. That is to say that the dislocation density predicted by prismatic punching is much lower than the actual one. If fiber aspect ratio remains constant, the size has no effect in the classical strengthing model.

An attempt was made to determine the upper bound of the dislocation density due to differences in CTE. The results of the investigation indicated in all cases that the dislocation density would approach infinity. At this time, all that can be stated is that the upper bound of the dislocation density is very large.

Taya and Mori [42] have recently studied the punching of the surface dislocations in a short fiber metal matrix composite and found that the punching distance is a function of fiber aspect ratio; namely, the larger the fiber aspect ratios, the less favorable the punching becomes, and at and beyond a threshold fiber aspect ratio the punching becomes impossible. Then, at fiber aspect ratios larger than the threshold value, the other mechanisms must be operative to reduce the high stress concentration at the matrix–fiber interface that was caused by CTE mismatch. However, experimentally it has been observed in continuous filament composites that dislocation generation does occur due to CTE differences. Other possible mechanisms are plastic deformation of fiber (this is possible only for refractory metal fiber), fiber fracture, and the debonding of the matrix–fiber interfaces. The above difficulty of punching in a long fiber metal matrix

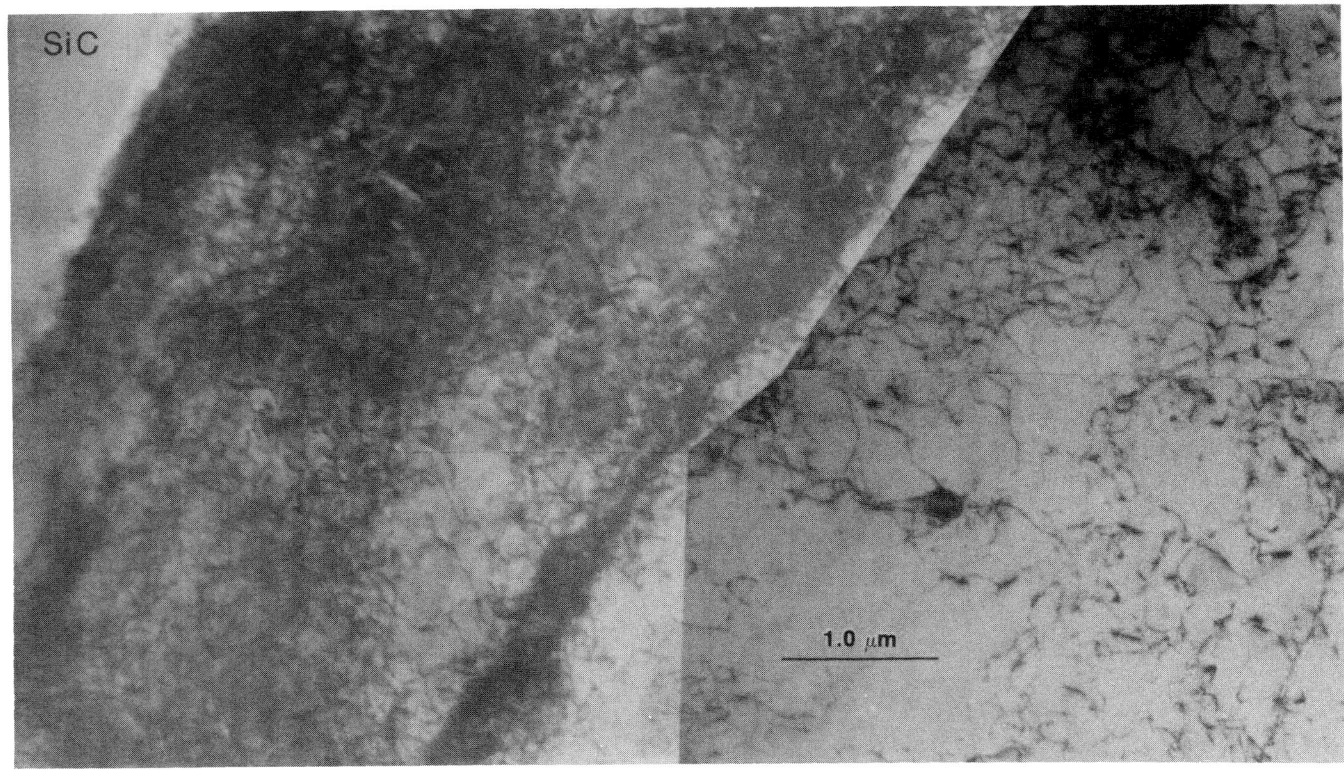

FIG. 3.16 A TEM photo taken from a bulk 20% V_f SiC/Al composite which was annealed for 12 hours at 810 K and furnace cooled.

composite, however, can be overcome by alloying the matrix metal which will increase the friction stress of the matrix, thus retarding the relaxation of internal stress.

The modeling of the formation of subgrains about a discontinuous reinforcement has not been undertaken, although there have been several observations [27, 28, 38] of subgrains about the reinforcement and the formation (Fig. 3.16). The subgrain boundaries are quite complex, which means that more than two slip systems were involved. One of the additional slip systems could result in geometrically necessary dislocations [24], which are necessary to prevent void nucleation about the particle during the production of misfit dislocation.

If we now consider the proportional limit of the composite, an interesting correlation can be obtained. The proportional limits of some composites [17, 43] are approximately equal to σ_{ym}. This correlation gives rise to two important points. First, the modulus (E_c) predicted by the Eshelby type model, which is in agreement with the experimental data, is for the initial portion of the stress–strain curve, i.e., for stresses up to the proportional limit. Therefore, the Eshelby type model does operate up to the proportional limit, but the predicted moduli are smaller than the experimental data. Secondly, the increase in stress with strain in the stress region from the proportional limit to σ_{yc} is caused by an exhaustion phenomenon. In this stress–strain region, dislocation motion occurs in the lower dislocation density regions within the matrix. The increase in stress between the proportional limit and the σ_{yc} is not caused by general work-hardening, for there is no general increase in dislocation density [44]. Examination of composite samples indicates there is only an increase in dislocation density in the region of the fracture, i.e., within 100 μm of the fracture surface. The dislocation density in the remainder of the sample is identical to that of the undeformed sample. However, only limited deformation can occur because of the small volume of matrix containing a low density of dislocations. In order to have macro-deformation, i.e., to reach 0.2% offset strain, additional dislocation motion must occur in the higher stress regions of the matrix which are often quite localized in the specimen. Therefore, the macro-yielding of the composite is controlled by the inhomogeneous matrix which is a mixture of the high–low dislocations regions. There are several remaining possible considerations. It has been clearly shown that there is tensile residual stress in the matrix which results in higher yield stress in compression than in tension [43].

Classical composite strengthening mechanisms have already been shown to be inadequate. Also, dispersion strengthening is not capable of predicting the strengthening observed [45], and it was shown that addition of the reinforcement has no effect on the texture. In other words if the matrix alloy is processed in the same way as the composite, both have the same texture. Therefore, texture differences do not contribute to strengthening [45].

In summary, it has been clearly demonstrated that the classical continuum model of composite strengthening cannot account for the observed strengthening in discontinuous metal matrix composites. A model based on a high dislocation and small subgrain sizes due to difference in CTE can easily predict the observed strengthening of discontinuous metal matrix composite system.

3.4 Fracture and Toughness

Fracture stress and fracture toughness have often been treated as a combined terminology, 'strength.' Despite the fact that fracture stress and fracture toughness possess common features in their micromechanical process leading to the final fracture, they should be treated separately. In fact they often conflict with each other; for example, the metal matrix composite with large fracture stress tends to have smaller fracture toughness (K_{IC}). Thus, in this section, fracture stress and fracture toughness will be discussed separately.

3.4.1 Fracture stress (strength)

Fracture stress is sometimes defined as strength (we will call it strength hereafter). It has been studied extensively and has been summarized in several books on composites [46–48]. Of the previous experimental data and the analytical models proposed for the strength of composites, those relevant to metal matrix composites will be discussed here. The fracture mode which results in the fracture depends on the morphology of the reinforcement (i.e., fiber morphology) and also its orientation to the loading axis. Figure 3.17 illustrates three fiber morphologies subjected to several possible loading axes where (a), (b) and (c) denote continuous fibers which are unidirectional, unidirectionally short and misoriented short fiber metal matrix composites, respectively, and σ_i with $i =$ L(longitudinal, solid line in the figure), T (transverse, dashline) and θ (off-axis, dash–dot line), denotes the applied stress along the ith direction. It is noted in Fig. 3.17(c) that the longitudinal loading cannot be uniquely defined for a misoriented metal matrix composite, though in practice σ_L is taken along the majority of fiber axes. The strength of the above three fiber morphologies will be discussed in the following.

Continuous fiber metal matrix composite

The strength of a continuous fiber metal matrix composite, like its stiffness, is anisotropic, which is strongly dependent upon the loading direction (Fig. 3.17(a)). The strengths that have been studied are the longitudinal (σ_L), transverse (σ_T) and off-axis (σ_θ) stress, which wil be discussed separately in the following.

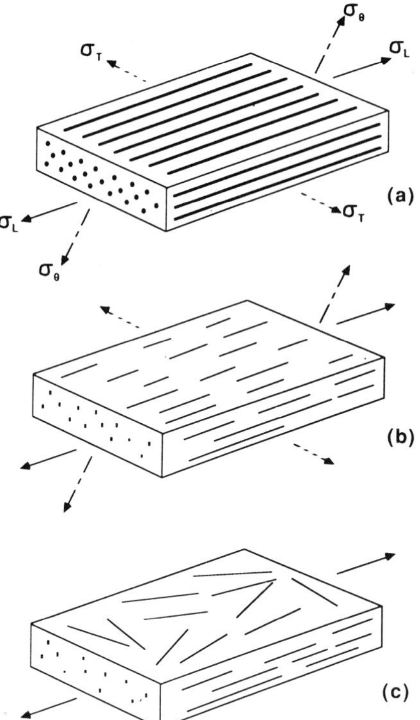

FIG. 3.17 Three types of fiber reinforcing morphology: (a) continuous aligned fiber, (b) aligned short fiber and (c) misoriented short fiber metal matrix composites under three different applied stresses along the longitudinal σ_L (solid line), transverse σ_T (dashed line) and off-axis σ_θ (dash–dot line).

Longitudinal tensile strength σ_L

Longitudinal (tensile) strength of a continuous fiber metal matrix composite can be expressed following a law of mixtures model:

$$\sigma_L = V_f \sigma_{fb} + (1 - V_f)\sigma_{mb} \tag{3.23}$$

where σ_{fb} and σ_{mb} are the strength of the fiber and matrix metal, respectively. Equation (3.23) is valid only if the longitudinal failure strain of the matrix metal and fiber is the same [46], which is less likely to be the case for most of the metal matrix composite systems, where the failure strain of fiber is much less than that of the matrix metal. Then, the law of mixtures equation given by eq. (3.23) should be modified as

$$\sigma_L = V_f \sigma_{fb} + (1 - V_f)\sigma'_m \tag{3.24}$$

where σ'_m is the stress in the matrix when the applied strain reaches that of the

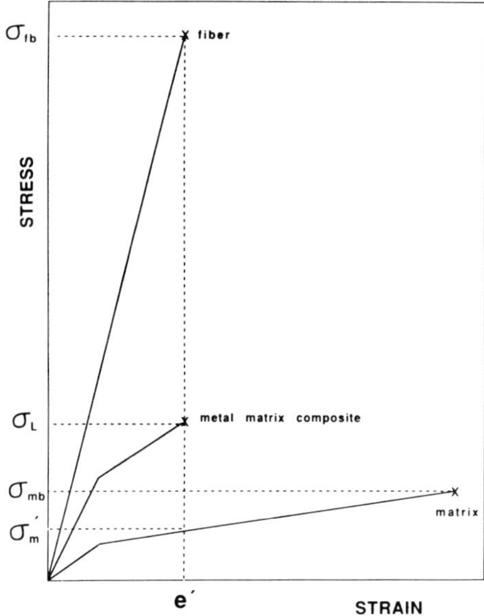

FIG. 3.18 Stress–strain curve of a continuous fiber metal matrix composite where ductile matrix and brittle fiber are used and their stress–strain curves and the definition of variables used in eq. (3.24) are also shown.

failure strain of the fiber (e'). This is illustrated in Fig. 3.18, where the stress–strain curves of typical fiber (brittle), matrix metal (ductile) and metal matrix composite are shown together with the definition of σ_L, σ_{fb}, σ_{mb}, σ'_m, and e' that are used in eqs. (3.23) and (3.24). If the volume fraction of fibers is low, then it is possible that the predicted strength of the composite at the failure strain of the fiber is less than the fracture strength of the matrix, as shown in Fig. 3.19. If this is the case, then the longitudinal fracture is controlled by the fracture of the matrix. This is illustrated in Fig. 3.19. At the applied strain e' a continuous fiber is broken into pieces, but the applied load can be supported by the work-hardening matrix metal whose fracture stress σ_{mb} is larger than σ'_m. Thus the applied stress can increase until the matrix stress reaches σ_{mb}. Hence the longitudinal fracture stress of the composite, σ_L, is given by

$$\sigma_L = (1 - V_f)\sigma_{mb} \tag{3.25}$$

σ_L predicted by the above three equations are shown in Fig. 3.20 where V'_f is defined as a fiber volume fraction of the intersection point of eqs. (3.24) and (3.25). In actual continuous fiber metal matrix composites σ_L are expected to follow the lines PQR indicated in the figure. In fact, the data of σ_L on continuous tungsten fiber/copper composites (filled circles in Fig. 3.20) appear to agree well will the line PQR [46]. It follows from Fig. 3.20 that σ_L can

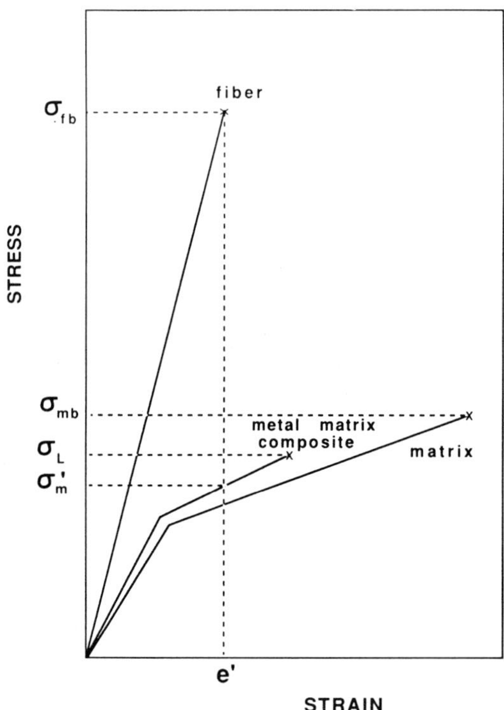

FIG. 3.19 Stress–strain curve of a continuous fiber metal matrix composite is schematically shown along with the stress–strain curves of brittle fiber and work-hardening matrix metal (see also eq. (3.25)).

be predicted well by eq. (3.24) for $0 \leqslant V_f \leqslant V'_f$ and eq. (3.23) for $V'_f \leqslant V_f \leqslant 1$. In most cases, however, V_f is much larger than V'_f. Thus eq. (3.23) will become the key equation to predict σ_L.

Though the experimental results of σ_L of continuous tungsten fiber/copper composite agree with the prediction of the law of mixtures based on eqs. (3.23) and (3.24), this may not be the case with other metal matrix composites. Figure 3.21 shows such an example where the experimental data on σ_L of a continuous SiC fiber/commercially high-purity aluminum (CPAl) composite are plotted as open circles, those of a continuous SiC fiber/casting grade aluminum (A384) composite as filled circles [49], and the law of mixtures prediction based on eq. (3.23) is shown by a solid line [49]. σ_{fb} used in eqs. (3.23) is the mean value of the strength of the fibers which were extracted from the composites, thus leading to the more accurate prediction by the law of mixtures equation. If the mean value of σ_{fb} of as-supplied fibers is used, the law of mixtures prediction shown by the dashed line overestimate the experimental results significantly. The experimental data of σ_L of a continuous SiC fiber/A384 matrix composite (filled circles) appear to follow the law of mixtures prediction given by eq. (3.25) where the strength of fiber does not contribute to the composite strength σ_L; instead σ_L is controlled solely by the

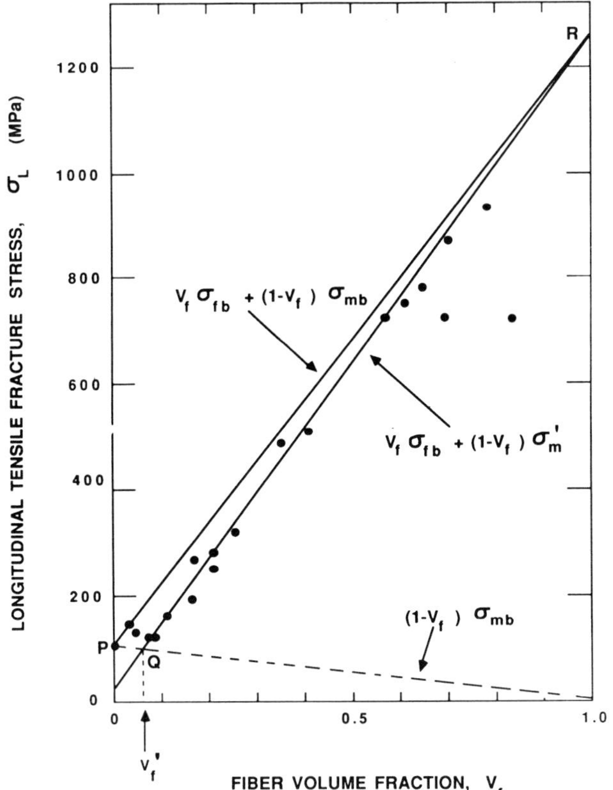

FIG. 3.20 Longitudinal tensile strength σ_L of a continuous tungsten fiber/copper composite as a function of fiber volume fraction V_f: experimental results (filled circles) [46] vs. prediction.

matrix strength σ_{mb}. The casting type aluminum alloy (A384) upon high processing temperature is known to react with SiC fiber, resulting in a reaction product of relatively large thickness which is usually brittle. According to the Metcalfe–Klein model (see Section 4.1), the SiC fibers with such thick reaction zone fracture at a strain much smaller than the fracture strain of as-received SiC fiber. Thus, at low levels of strain the reacted fibers fracture and then the increased load will be supported by the matrix. This is why σ_L of SiC/A384 composite appears to follow eq. (3.25).

As seen above, the law of mixtures model is sometimes too crude to predict the strength of a composite. Rosen [50] and Zweben [51] proposed more accurate models for prediction of σ_L, which are based on the statistical distribution of flaws or imperfections that result in fiber failure under applied stress. In the following we will review these models briefly.

Before discussing the composite strength, we must begin first by examining the fiber strength σ_{fb} in a statistical manner as suggested by Coleman [52].

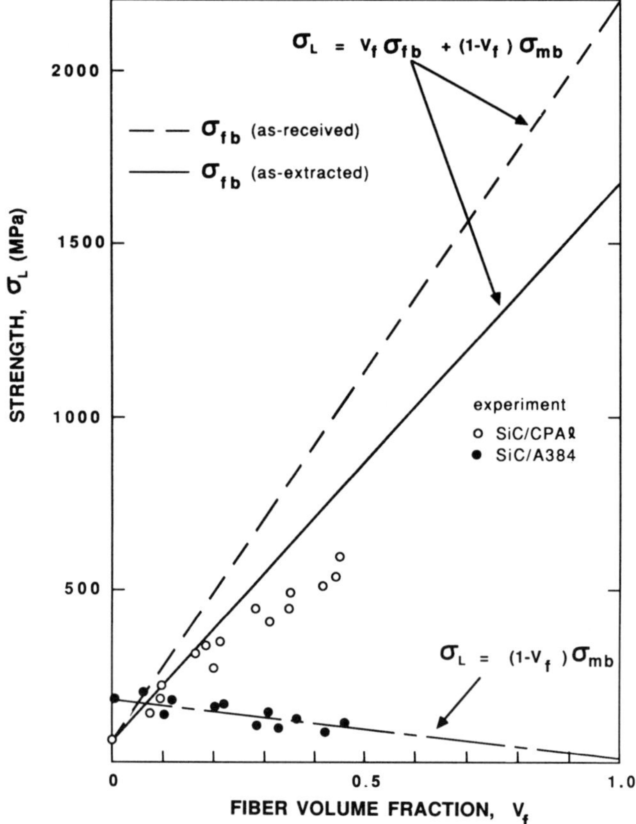

FIG. 3.21 Longitudinal tensile strength σ_L of as-received continuous SiC fiber-/commercially high-purity aluminum (open circles) and A384 aluminum (filled circles) composites as a function of fiber volume fraction V_f, [49] vs. prediction based on the law of mixtures equations (eqs. (3.23) and (3.25)) where σ_{fb} is the mean value of as-received fibers (dashed lines) and that of as-extracted fibers from the composite (solid line).

σ_{fb} is assumed to obey the Weibull distribution:

$$f(\sigma_{fb}) = l\alpha\beta\sigma_{fb}^{\beta-1}\exp(-l\alpha\sigma_{fb}^{\beta}) \tag{3.26}$$

where α and β are constants that can be determined from tests on monofilament fibers, l is the fiber length with the corresponding cumulative distribution function

$$F(\sigma_{fb}) = \int_0^{\sigma_{fb}} f(\sigma_{fb})\mathrm{d}\sigma_{fb} \tag{3.27}$$

The mean value of the fiber strength, $\bar{\sigma}_{fb}$ is then given by

$$\bar{\sigma}_{fb} = (l\alpha)^{-1/\beta}\Gamma(1+1/\beta) \tag{3.28}$$

where Γ is the gamma function. α and β in eqs. (3.26) and (3.28) are also related to the standard deviation s and coefficient of variation C as [46].

$$s=(\alpha l)^{-1/\beta}\{\Gamma(1+2/\beta)-\Gamma^2(1+1/\beta)\}^{1/2} \tag{3.29}$$

$$C=\frac{s}{\bar{\sigma}_{fb}}=\frac{\{\Gamma(1+2/\beta)-\Gamma^2(1+1/\beta)\}^{1/2}}{\Gamma(1+1/\beta)} \tag{3.30}$$

Instead of tests on monofilament fibers, we now consider tests on bundles, each of which contains N fibers. Then, the distribution of the average fiber stress at bundle failure, $\bar{\sigma}_B$, approaches a normal distribution with expectation (or mean values) $\bar{\sigma}_B$ [53],

$$\bar{\sigma}_B=\rho_{max}[1-F(\sigma_{max})] \tag{3.31}$$

where σ_{max} can be determined from

$$\left.\frac{dP(\sigma)}{d\sigma}\right\}_{\sigma=\sigma_{max}}=\frac{d}{d\sigma}\quad\{\sigma[1-F(\sigma)\}_{\sigma=\sigma_{max}}=0 \tag{3.32}$$

where $P(\sigma)$ is the load carried by the bundle, and the bundle strength is characterized by the distribution function.

$$\omega(\sigma_B)=\frac{1}{\psi_B(2\pi)^{1/2}}\exp\left[-\frac{1}{2}\frac{(\sigma_B-\bar{\sigma}_B)}{\psi_B}^2\right] \tag{3.33}$$

where ψ is the standard deviation. With the associated cumulative function:

$$\Omega(\sigma_B)=\int_0^{\sigma_B}\omega(\sigma_B)d\sigma_B \tag{3.34}$$

σ_{max} and σ_B can be obtained from eqs. (3.26), (3.27), (3.31), and (3.32) as

$$\sigma_{max}=(l\alpha\beta)^{-1/\beta} \tag{3.35}$$

$$\bar{\sigma}_B=(l\alpha\beta e)^{-1/\beta} \tag{3.36}$$

Coleman [52] showed that $\bar{\sigma}_B\leqslant\sigma_{fb}$ where the equality holds for no dispersion in the fiber breaking stress.

Rosen considered that in a continuous fiber composite (lamina) consists of N fibers of length l which is divided into M elements with the length l_c and at applied stress, σ_L, the weaker fibers are broken while the others remain intact as shown in Fig. 3.22(a). This composite is converted to a chain of M links (or bundle) as shown in Fig. 3.22(b). He assumed that when the weakest bundle (or link) in Fig. 3.22(b) is subject to its breaking stress σ_B given by eq. (3.36) with l replaced by l_c, then the composite failure would occur. Hence, using eq. (3.32), the fiber breaking stress in the composite at failure, σ_L^* is given by

$$\sigma_L^*=(l_c\alpha\beta e)^{-1/\beta} \tag{3.37}$$

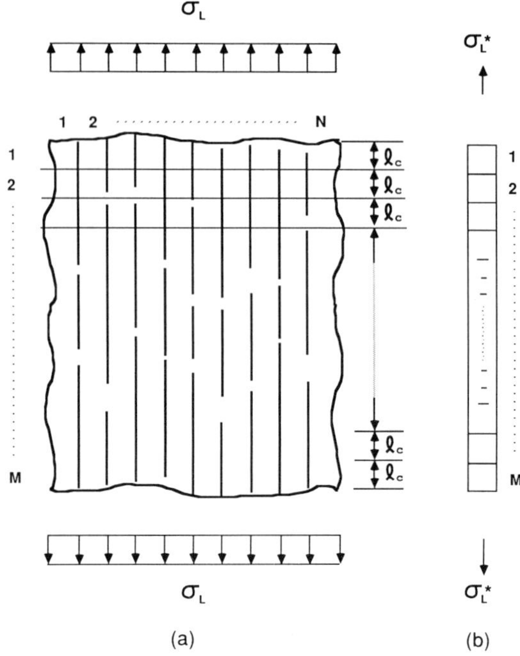

FIG. 3.22 (a) Continuous fiber composite with N fibers and M divisions of length l_c which is equal to an ineffective fiber length, (b) which is converted to a chain of M links [50].

Then, the corresponding strength of the composite, σ_L, is given by

$$\sigma_L = V_f \sigma_L^* = V_f (l_c \alpha \beta e)^{-1/\beta} \tag{3.38}$$

In eqs. (3.37) and (3.38), l_c is taken as the length of a fiber over which the fiber stress is less than $\phi\, \sigma_{f0}$ where σ_{f0} is the uniform fiber stress undisturbed by the ends of a broken fiber, e is the base of natural logarithm, and ϕ is a parameter with $0 = \phi < 1$. l_c is called the ineffective length or critical fiber length, and it is determined by the shear lag stress analysis which depends on several geometrical and material parameters including the strength of the matrix–fiber interface τ_i. l_c is given by

$$l_c = \frac{\sigma_{fb} d}{2\tau_i} \tag{3.39}$$

where d and τ_i are the diameter of a short fiber and the matrix–fiber interfacial shear strength.

In the Rosen model the effect of load concentration due to a broken fiber-end was neglected. Zweben [51] considered this effect in his cumulative fracture-propagation model. He assumed that a composite consists of M times N elements (see Fig. 3.22(a)) and the breaking stress of an element with length l_c can be characterized by a cumulative distribution function $F(\sigma)$. In a composite with NM elements, the number that can be expected to fracture when subjected to σ is

$$E_1 = MNF(\sigma) \tag{3.40}$$

If E_1 elements are broken, then there exist two elements adjacent to each of those broken elements (breaks), which will be subjected to an increased load $K_1\sigma$ where K_1 is called the load concentration factor and is equal to the ratio of the maximum stress of the adjacent fibers to the average fiber stress σ. K_1 can be determined by the shear lag analysis used by Hedgepeth [54]. Then, the probability that one of the two adjacent fibers will break due to this load concentration is

$$P_{2/1} = 2[F(K_1\sigma) - F(\sigma)] - 2[F(K_1\sigma) - F(\sigma)]^2 \tag{3.41}$$

and the probability that both adjacent fibers will break simultaneously is

$$P_{3/1} = [F(K_1\sigma) - F(\sigma)]^2 \tag{3.42}$$

In eqs. (3.41) and (3.42), $P_{i/j}$ is the probability of having i fibers broken given that j fibers are already broken. Propagation of broken elements in a two-dimensional plane with MN elements can be estimated by following the above rule. However, this process of computation seems to be quite time-consuming.

Zweben's model has been extended by Fukuda [55], Kimpara and Ozaki [56], and Ochiai and Osamura [57]. Fukada computed the load concentration factors of two idealized cases; (1) the infinite model where two adjacent fibers are broken and they are surrounded by unbroken fibers, and (2) the repeating model where two broken fibers and unbroken fibers are located alternately [55], the results can be used to predict σ_L of continuous fiber composites including hybrid composites. Kimpara and Ozaki constructed a more realistic model which can simulate the dynamic fracture process in a continuous fiber composite [56]. In this model the effect of debonded interface, dynamic stress wave propagation and matrix's load-bearing capacity also are considered. Ochiai and Osamura [57] have extended Zweben's model to the case of continuous fiber metal matrix composites with more realistic interfaces, i.e., debonded interface with some friction stress and interface with reaction layer. The accuracy of the fracture stress σ_L predicted by these models is expected to increase with the number of fibers N, thus requiring a computer with higher speed and larger storage capacity.

Transverse tensile strength σ_T

The transverse (tensile) strength σ_T is known to depend strongly on the strength of the matrix–fiber interface relative to the transverse fiber strength, the matrix strength, and also fiber volume fraction. Thus, σ_T is usually much smaller than σ_L of the same metal matrix composite. Prewo and Kreider [4] have studied experimentally the transverse tensile fracture of boron fiber/

aluminum composites. Typical stress–strain curves of as-fabricated (F-condition) and T-6 treated boron fiber/6061 Al composites are shown in Fig. 3.23 where the F-conditioned curve is markedly different from the T-6 treated case, and the '×' marks denote the fracture points. Although the fracture stress of the T-6 treated composite is larger than that of the unreinforced matrix metal, Fig. 3.23 indicates that T-6 heat treatment resulted in the stronger matrix–fiber interfacial bonding relative to the matrix and fiber, thus leading to a relatively flat fracture path, while in the F-conditioned composite the interfacial bonding is the weakest, and hence the fracture process is a mixture of the interfacial debonding and the matrix fracture. Dependence of σ_T on fiber volume fraction V_f is shown in Fig. 3.24, where the experimental results of σ_T of a BORSIC fiber/2024 Al composite are shown by open symbols and the solid curves denote the prediction based on a simple model. The simple model assumes that the fibers are replaced by holes, thus the transverse strength is controlled by the matrix fracture strength times the reduced sectional area. However, this model (Fig. 3.25(b)) is considered to be an oversimplification of the actual fracture mode (Fig. 3.25(a)) where the fracture path is presumably a mixture of interfacial debonding, fiber splitting, and the matrix fracture. In order to better simulate the actual fracture mode, one must know the statistical data of the variation of the interfacial bond strength for all the matrix–fiber interfaces in the actual continuous fiber metal matrix composites and of the matrix and transverse fiber strengths. Kyono et al. [58] have attempted to simulate the transverse tensile stress–strain curve of a boron fiber/1100 aluminum composite by using a model based on the Eshelby method. The Kyono et al. model assumes that a composite subjected to transverse loading, σ_T, has a mixture of debonded interfaces and completely bonded interfaces, and the interfacial bond strength (τ_i) follows a three-point Weibull distribution.

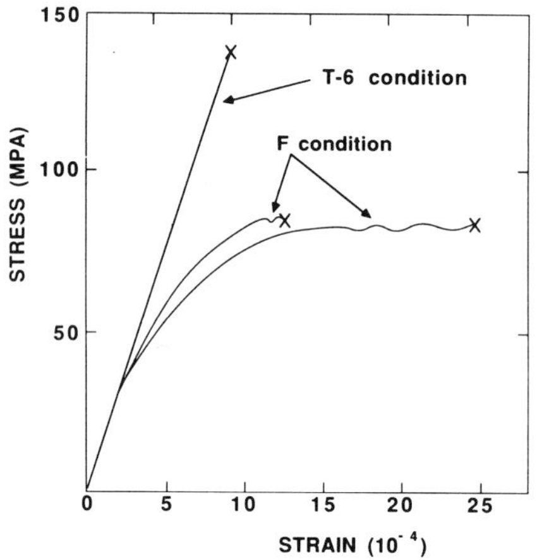

FIG. 3.23 Transverse tensile stress–strain curves of a T-6 conditioned and F-conditioned BORSIC fiber/6061 Al composite [4].

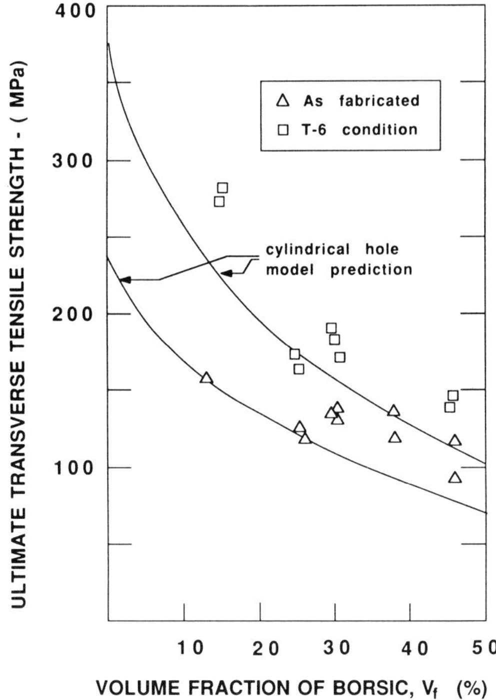

FIG. 3.24 Transverse tensile strength σ_T of a continuous BORSIC fiber/2024 aluminum composite as a function of fiber volume fraction V_f [4].

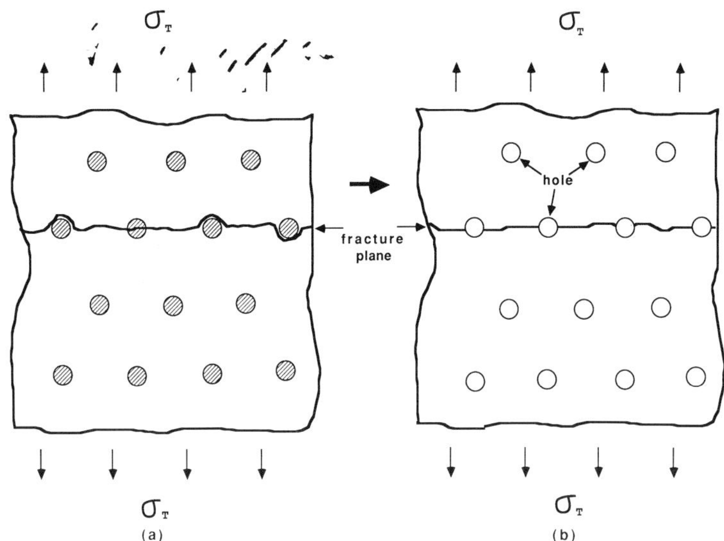

FIG. 3.25 Fracture path for an actual continuous fiber metal matrix composite subjected to transverse loading σ_T (a), which is converted to a simple model where fibers in the crack plane are replaced by cylindrical holes (b) [4].

Off-axis tensile strength

When a continuous fiber composite (lamina) is subjected to the applied stress (σ_θ) whose direction makes angle θ with respect to the fiber axis, the fracture of the composite is controlled by three possible fracture modes, longitudinal fracture (σ_L), transverse fracture (σ_T) and shear fracture (τ). The first two modes have been discussed above, while the shear fracture mode is caused by the applied shear stress τ along the perpendicular to the fiber axis. τ can be the matrix shear strength (τ_m) or the interfacial shear strength (τ_i), and σ_T can be equal to the matrix tensile strength (σ_{mb}). Jackson and Cratchley [59] studied the off-axis tensile fracture of metal matrix composites both experimentally and analytically. The applied stress σ_θ can be related to the above three fracture stresses:

$$\sigma_\theta = \begin{cases} \dfrac{\sigma_L}{\cos^2\theta} & \text{for longitudinal tensile failure} \\[2ex] \dfrac{\tau}{\sin\theta\cdot\cos\theta} & \text{for shear failure} \\[2ex] \dfrac{\sigma_T}{\sin^2\theta} & \text{for transverse failure} \end{cases} \tag{3.43}$$

The predicted σ_θ is plotted as a function of angle θ (loading direction with respect to the fiber axis) in Fig. 3.26 where the experimental results of the off-axis fracture stress σ_θ of 50% V_f silica fiber/aluminum matrix composite are shown by filled circles. It follows from Fig. 3.26 that the prediction by eq. (3.43) agrees well with the experiment and the off-axis strength σ_θ drops

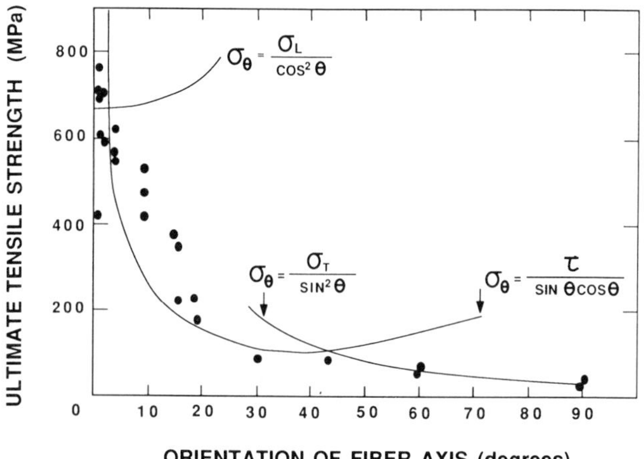

FIG. 3.26 Off-axis tensile fracture stress σ_θ of a unidirectionally aligned 50% V_f silica/aluminum composite as a function of off-axis orientation where the solid curves denote the prediction by eq. (3.43) [59].

significantly for $\theta \geqslant \theta_c$ where θ_c is the angle corresponding to the intersection of the two curves, eq. (3.43), and in the present example, θ_c is about 3°. The sharp drop in the strength σ_θ beyond θ_c can be avoided by using a laminated composite plate with $\pm\theta$ orientation, leading to the smooth change of σ_θ with respect to the alternate lamination angle θ. Figure 3.27 illustrates such a case [59]. The reason for maintaining the relatively high strength of the laminated composite with $\pm\theta$ over $0 \leqslant \theta \leqslant 30°$ is that the shear stress τ in each ply is cancelled out, retarding the shear type failure (eq. (3.43)).

The above fracture model (eq. 3.43) appears to be valid for a single ply (or lamina) and does not account for the interaction between the three fracture modes which will become enhanced for laminated composites. Modifying the Von Mises yield criterion for plane stress anisotropic plate, Azzi and Tsai [60] proposed a failure criterion (which is called 'Tsai–Hill criterion') to predict the state of two-dimensional stress that causes the failure of composite plates, particularly laminated composites. The Tsai–Hill criterion is given by

$$\left(\frac{\sigma_x}{\sigma_L}\right)^2 + \left(\frac{\sigma_y}{\sigma_T}\right)^2 - \left(\frac{\sigma_x \sigma_y}{\sigma_L^2}\right) + \left(\frac{\sigma_{xy}}{\tau}\right)^2 = 1 \tag{3.44}$$

where σ_L, σ_T and τ are the fracture stresses of a composite plate in the longitudinal, transverse and longitudinal shear directions, respectively. Fracture of the composite plate occurs when the state of stress (σ_x, σ_y, τ_{xy}) in the composite plate satisfies eq. (3.44), whereas the composite does not fracture when the left-hand side value of eq. (3.44) is less than unity.

Short fiber metal matrix composites

The short fibers in actual short fiber metal matrix composites are often misoriented and have various lengths due mainly to processing. Figures 3.5 and 3.8 show the intensity of fiber misorientation [15], and that of various fiber aspect ratios of SiC whisker/aluminum composites [17], respectively. Thus, the fracture stress of an actual short fiber metal matrix composite has not been discussed fully, due to this complicated morphology of short fibers, although a number of experimental results of the fracture stress of the composites have been reported, for examples see Refs 6, 8, 61, 62, 63.

In order to study the effect of various material and geometrical parameters on the strength of short fiber metal matrix composites, use of an idealized (or model) short fiber metal matrix composite is preferred. Kelly [64] and Kelly and Tyson [65] proposed such a model to predict the strength of an aligned short fiber metal matrix composite. The model is based on the shear lag analysis discussed in Section 2.3. The fiber stress builds up from zero at the fiber ends to the maximum stress at the center of the fiber. This high stress in the center portion of the fiber should be bounded by the fiber breaking stress σ_{fb}. Hence, the correct distribution of the fiber stress is given by Fig. 3.28

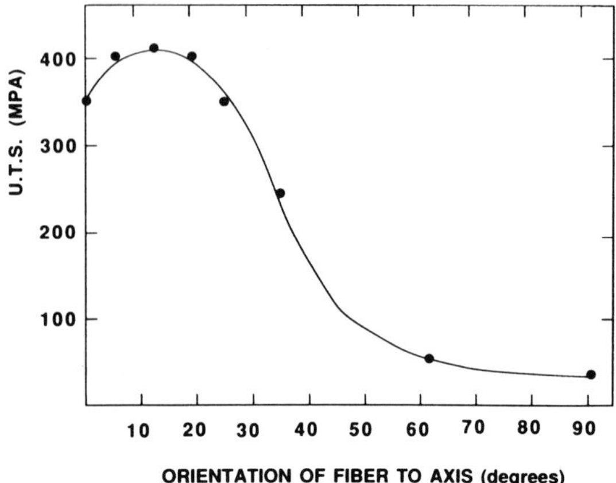

FIG. 3.27 Tensile fracture stress σ_θ of a stainless-steel fiber/aluminum laminated composite plate with $\pm\theta$ plies as a function of the lamination angle θ [59].

where the dash–dot line is a linearization of the build-up portion over the length $l_c/2$, and l_c is the critical fiber length that is defined by eq. (3.39). The physical meaning of the critical fiber length is that for the fiber length $l < l_c$, the fiber is not loaded to its full capacity (σ_{fb}), while for $l \geqslant l_c$, the central portion of a short fiber bears the maximum stress (σ_{fb}). Based on the simplified stress distribution (the dash–dot line for the build-up portion and solid line for the center portion in Fig. 3.28) and a law of mixtures, Kelly and Tyson [65] proposed the following formulae to predict the longitudinal strength (σ_L) of an aligned short fiber metal matrix composite.

$$\sigma_L = \begin{cases} V_f \sigma_{fb}\left(1 - \dfrac{l_c}{2l}\right) + V_m \sigma'_m & \text{for } l \geqslant l_c \\ V_f \sigma_{fb} \dfrac{l}{2l_c} + V_m \sigma_m & \text{for } l < l_c \end{cases} \tag{3.45}$$

where σ_m is the strength of the matrix metal and σ'_m is the flow stress of the matrix at the failure strain of the fiber. The predictions based on the formulae were compared with the experimental results of short tungsten fiber/copper and short molybdenum fiber/copper composites, resulting in a good agreement. In using eqs. (3.45), one must know *a priori* the critical fiber length l_c given by eq. (3.39), which requires the knowledge of the matrix–fiber interfacial strength τ_i. Kelly and Tyson estimated the critical length l_c by using the experimental results and eqs. (3.45). Then, they calculated τ_i by using eq. (3.39) and found that the value of τ_i so estimated is close to the yield stress of the matrix metal (single crystal copper). They also conducted fractography analysis to examine the fiber fractures to pull-outs, whose results were found to support the aforementioned estimate. Thus, the accuracy of the prediction by eqs. (3.39) and (3.45) depends on that of τ_i (or l_c). To this end several experimental methods have been proposed. The most popular method is the

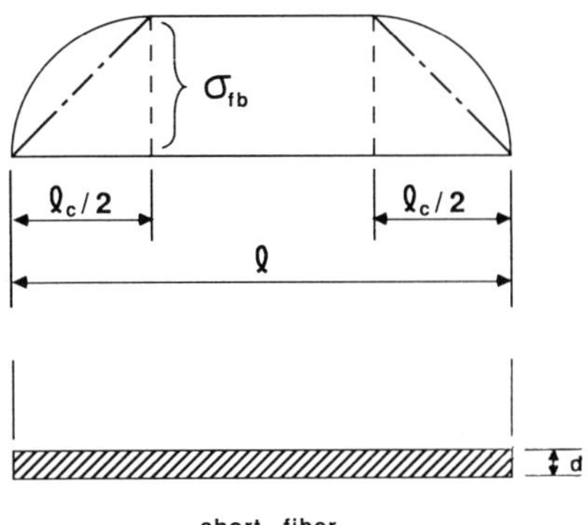

FIG. 3.28 Distribution of the axial stress in a short fiber (solid line) with build-up region of length $l_c/2$ which is linearized by dash–dot lines.

fiber pull-out test, which is shown schematically in Fig. 3.29(a) [66]. A monofilament fiber of diameter d is embedded in the matrix metal with the embedded fiber length l_e, and it is then subjected to stress σ_{p0}. The results of the fiber pull-out tests will be plotted in terms of the pull-out stress σ_{p0} vs. embedded length l_e as demonstrated in Fig. 3.29(b), which indicates two regions, the build-up linear region and the flat region. $l_c/2$ corresponds to the transition point between these regions. This fiber pull-out test appears to be a straightforward method, but it would not be so when the fiber diameter is small, resulting in the difficulty of fabricating such a test specimen.

In actual short fiber metal matrix composites, fiber lengths vary even though all the fibers are aligned. In this case the Kelly–Tyson model can still be applicable. Namely, fibers are separated to two groups, one with length $l<l_c$ and the other with $l \geqslant l_c$. The strength (σ_L) of an aligned short fiber metal matrix composite with variable fiber lengths can then be predicted by

$$\sigma_L = \sum^{l_i<l_c} \frac{(V_f)_i \sigma_{bf} l_i}{2l_c} + \sum^{l_i \geqslant l_c} (V_f)_i \sigma_{bf}\left(l - \frac{l_c}{2l_i}\right) + V_{m1}\sigma'_m + V_{m2}\sigma_m \tag{3.46}$$

where $(V_f)_i$ is the volume fraction of the fibers with length l_i, V_{m1} and V_{m2} are the volume fractions of the fibers with $l<l_c$ and those with $l \geqslant l_c$, respectively. σ'_m has been defined in eq. (3.24). The relation between l_i and $(V_f)_i$ should be obtained from the fibers extracted from the as-fabricated composite, for the majority of short fibers of initially larger length will be broken to shorter lengths during processing.

When short fibers are misoriented, the above model is not valid except that its prediction can serve as an upper bound on σ_c. The only existing model

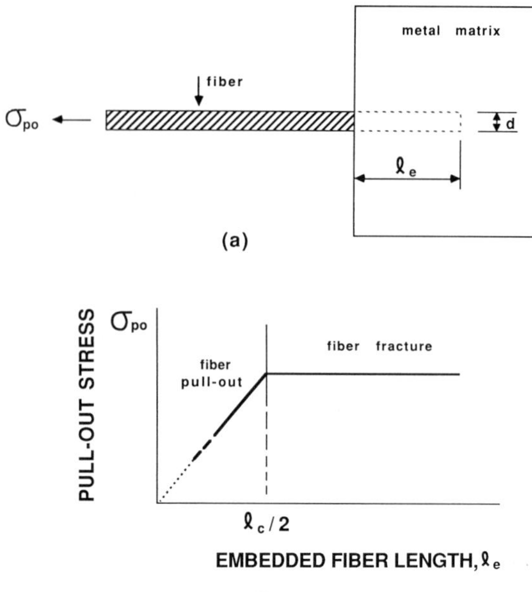

FIG. 3.29 (a) Fiber pull-out test by embedding a monofilament fiber in the matrix metal. (b) The results of pull-out stress σ_{po} are plotted as a function of the embedded length l_e, indicating two regions, corresponding to fiber pull-outs and fiber fractures.

that can account for the fiber misorientation is the Tsai–Hill equation (eq. (3.44)), which can be modified as [47]

$$\sigma_\theta = \left[\frac{\cos^4 \theta}{\sigma_L^2} + \left(\frac{1}{\tau^2} - \frac{1}{\sigma_L^2} \right) \sin^2 \theta \cos^2 \theta + \frac{\sin^4 \theta}{\sigma_T^2} \right]^{-1/2} \tag{3.47}$$

where the misorientation of short fibers with respect to the loading axis is assumed in-plane (Fig. 3.30(a)), and characterized by its density distribution function $\rho(\theta)$ (Fig. 3.30(b)). Hence, the strength of the composite σ_c is obtained as

$$\sigma_c = \int_{-\pi/2}^{\pi/2} \sigma_\theta \rho(\theta) \mathrm{d}\theta \tag{3.48}$$

where

$$\int_{-\pi/2}^{\pi/2} \rho(\theta) \mathrm{d}\theta = 1 \tag{3.49}$$

In order to apply the modified Tsai–Hill equations (eqs. (3.47) and (3.48)), one must know *a priori* three fracture stresses (σ_L, σ_T and τ) and density

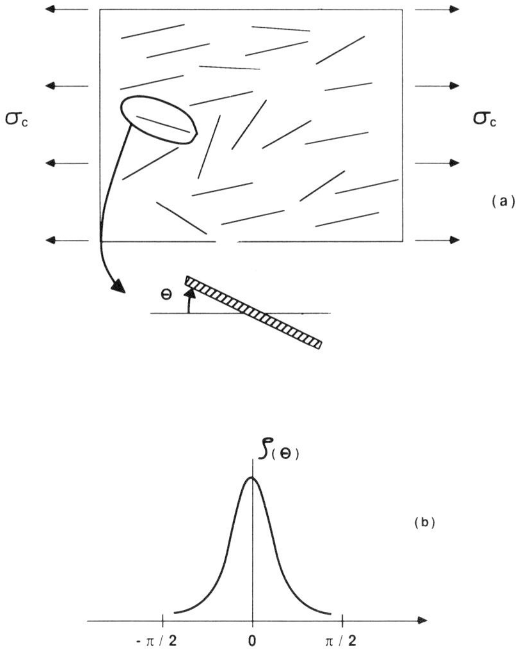

FIG. 3.30 (a) A two-dimensionally misoriented short fiber composite (b) with its density distribution function $\rho(\theta)$ where fiber misorientation θ is measured with respect to the loading axis.

distribution function $\rho(\theta)$. To obtain σ_L, σ_T and τ is a quite elaborate process since three independent fracture tests must be conducted on aligned short fiber composites (see Fig. 3.31). Two extreme cases of $\rho(\theta)$ are worth noting, $\delta(\theta)$ (Dirac delta function) and uniform density distribution $\rho(\theta)=\rho_0$, $|\theta|\leqslant\pi/2$. The former and latter cases correspond to aligned short fiber and in-plane random short fiber metal matrix composites, respectively. When the composite plate becomes thicker, it is considered a three-dimensionally (3D) misoriented short fiber metal matrix composite. If this 3D composite is fabricated by extrusion it will become transversally isotropic. The corresponding density distribution function $\rho(\theta)$ can then simulated by eq. (3.2). The strength σ_c of such a 3D misoriented short fiber metal matrix composite can still be predicted by eqs. (3.47) and (3.48) except that σ_L, σ_T and τ are now interpreted as the longitudinal strength (along the extrusion direction) (Fig. 3.31(a)), the transverse strength (perpendicular to the extrusion direction), (Fig. 3.31(b)), and longitudinal shear strength (Fig. 3.31(c)) of an aligned short fiber composite, respectively. Again the tests to obtain σ_L, σ_T and τ are usually difficult, mainly due to the difficulty in preparing such aligned short fiber metal matrix specimens. At present, no dislocation models have been proposed for the fracture strength of a short fiber metal matrix composite.

In discontinuous composites where the strengthening cannot be predicted

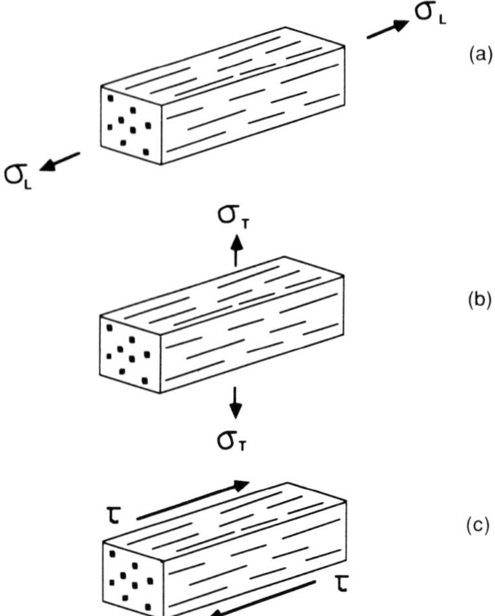

FIG. 3.31 Tensile strength of an aligned short fiber metal matrix composite, (a) longitudinal strength σ_L, (b) transverse strength σ_T and (c) longitudinal shear strength (τ).

by any of the law of mixtures formulation, i.e., in cases where the strengthening is due to high dislocation density, the prediction of the fracture strength is difficult. If we consider a standard equation for the stress–strain relationship

$$\sigma = K\varepsilon_p^n$$

where K is a constant and n is a work-hardening parameter. Fracture occurs when $\varepsilon_p = n$. This is a simple and straightforward equation, but in the case of discontinuous composites there are local regions of higher ε_p than others, due to local variations of reinforcement. Therefore, fracture will occur at ε_p which have no correlation with the ε_p to fracture of the matrix. The fracture of the reinforcement (provided the reinforcement is small $< 10\ \mu$m) does not play a part in the fracture process.

3.4.2 Fracture toughness

As discussed in the previous subsection, the strength of a metal matrix composite is simply the fracture stress of the metal matrix composite, while fracture toughness is the ability of the metal matrix composite to resist the propagation of a crack that exists within the composite. If a composite is

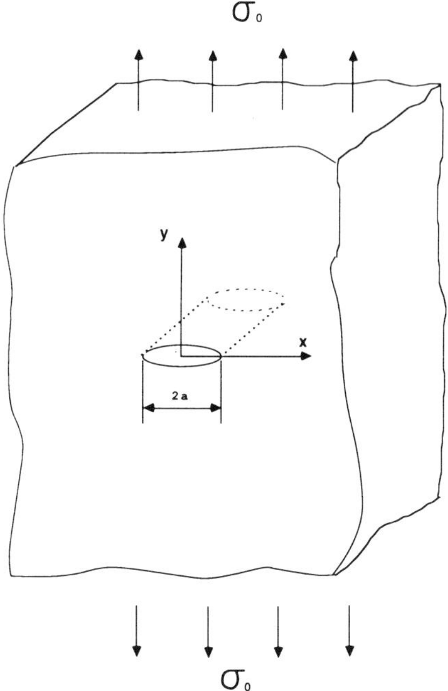

FIG. 3.32 A two-dimensional crack of length $2a$ is located in an infinite homogeneous body subjected to applied stress σ_0.

linear elastic up to the final fracture, then the strength (σ_0) and fracture toughness (K_{IC}) of the composite can be directly related to each other, i.e.,

$$K_{IC} = \sigma_0(\pi a)^{1/2} \tag{3.50}$$

where a is one-half of the length of an existing two-dimensional crack (see Fig. 3.32). It was assumed in deriving eq. (3.50) that the composite is isotropic and homogeneous, the current crack will propagate along the same crack plane (self-similar propagation), and the width of the composite body is much larger than the crack size. K_{IC} in eq. (3.50) is called the critical stress intensity factor for mode I and it is considered a material constant and also the most popular way of defining 'fracture toughness.' If σ_0 does not reach the composite fracture stress, then an equation similar to eq. (3.50) still holds,

$$K_I = \sigma_0(\pi a)^{1/2} \tag{3.51}$$

where K_I is called stress intensity factor for mode I loading. Thus for a metal matrix composite which is isotropic and homogeneous, and contains a two-dimensional crack of size $2a$, the fracture criterion based on linear elastic fracture mechanics (LEFM) can be rewritten as

$$K_I = K_{IC} \tag{3.52}$$

The left side value (K_I) is a function of applied stress σ_0, crack size $2a$, matrix elastic constants and the geometrical factors, and it can increase with the applied stress and with crack size, whereas the right side value (K_{IC}) represents 'fracture toughness for mode I,' thus a material constant. When a composite is no longer linear elastic, then the fracture toughness can be conveniently defined by energy release rate G and the fracture criterion is then given by

$$G_I = G_{IC} \tag{3.53}$$

Again the interpretation of the left and right side terms in eq. (3.53) is basically the same as in eq. (3.52). G_I is related to K_I if a composite is linear elastic and isotropic:

$$G_I = \frac{K_I^2}{E_c} \tag{3.54}$$

where E_c is the modulus of the composite. In the above equations the case of a plane stress condition was considered. For a plane strain case, E_c in eq. (3.39) must be replaced by $E_c/(1 - \nu_c^2)$ where ν_c is the Poisson's ratio of the composite. G_{IC} is also related to fracture surface energy γ by

$$G_{IC} = 2\gamma \tag{3.55}$$

where γ is defined per-unit crack surface area.

Determination of fracture toughness (the right side of eqs. (3.52) and (3.53)) is straightforward if we use one of the standard experimental methods [67], for example, the three-point bent test with a center-notched specimen (Fig. 3.33(a)) and the compact tension test with a side-notched specimen (Fig. 3.33(b)). A typical force (P)–displacement (Δ) curve that can be obtained by these standard tests is schematically shown in Fig. 3.34. Then K_{IC} is expressed by

$$K_{IC} = \begin{cases} \dfrac{PS}{BW^{3/2}} Y_b\left(\dfrac{a}{w}\right) & \text{for the three-point bend test} \\[2ex] \dfrac{P}{BW^{1/2}} Y_c\left(\dfrac{a}{w}\right) & \text{for the compact tension test} \end{cases} \tag{3.56}$$

where B, W and a are thickness, width and crack length as defined in Fig. 3.33, S is the full span length for the three-point bend test (Fig. 3.33(a)), and $Y_i(a/w)$ is a function of only non-dimensional crack size (a/w) (its detailed expressions are given in any standard textbook in fracture mechanics, for example, Ref. 67), and where $i = b$ (three-point bend test) and c (compact tension test). Once a P–Δ curve is obtained, one can calculate the instantaneous value of G_i (energy release rate) from the curve which can be converted to stress intensity factor (K_R) for plane stress

$$K_R = (G_i E_C)^{1/2} \tag{3.57}$$

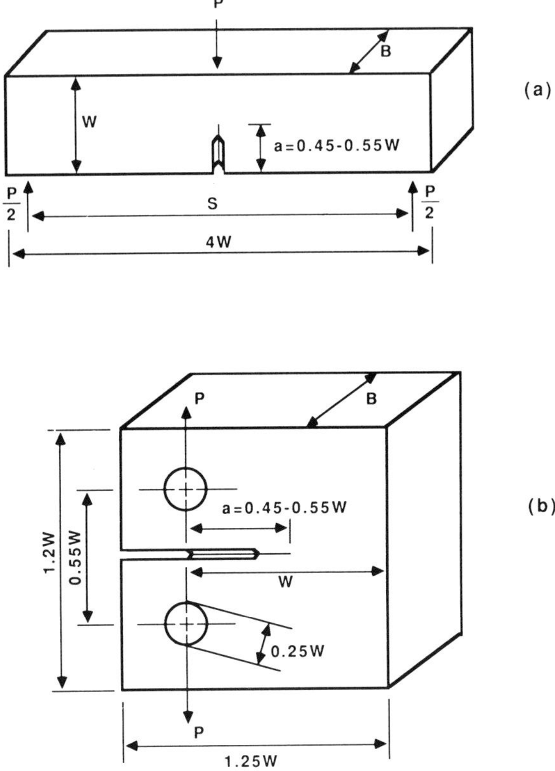

FIG. 3.33 Standard tests to measure fracture toughness K_{IC} (or K_{IQ}), (a) three-point bend test with a center-notched specimen and (b) compact tension test with a side-notched specimen.

If the values of K_R are constant over crack length a, then K_R is called the stress intensity factor for steady-state crack growth, and it can be considered as a material constant (fracture toughness). A K_R a relation can also be obtained from one of the standard tests (eq. 3.56). An example of a non-uniform K_R is shown in Fig. 3.35 where the value of K_R was calculated [68] from the P–Δ curve of a continuous carbon fiber/6061 aluminum composite [59]. The above non-uniformity in K_R which is indicative of the departure from the fracture criterion based on LEFM, can be attributed to the micro-mechanical fracture modes taking place at the crack-tip region which is characteristic of the continuous fiber metal matrix composite. Wright et al. [69] reported similar results on a continuous boron fiber/6061 aluminum composite. Namely, the critical stress intensity factor K_{IC} measured by the compact tension test (see Fig. 3.33(b)) and center-notched three-point bend specimens (see Fig. 3.33(a)) were highly dependent on the dimensions of the

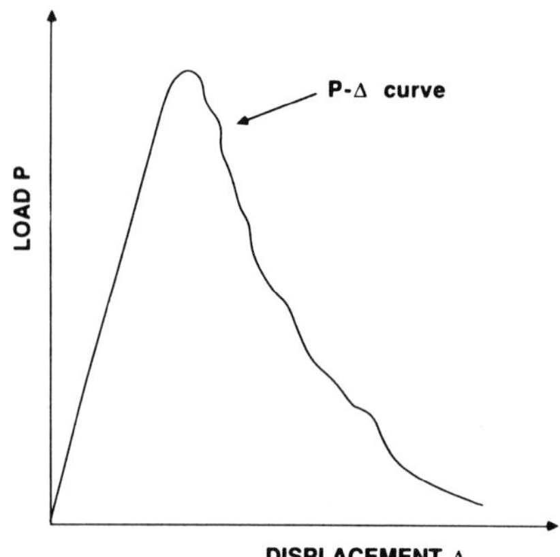

FIG. 3.34 A typical load (P)–displacement (Δ) curve that can be obtained by a standard testing.

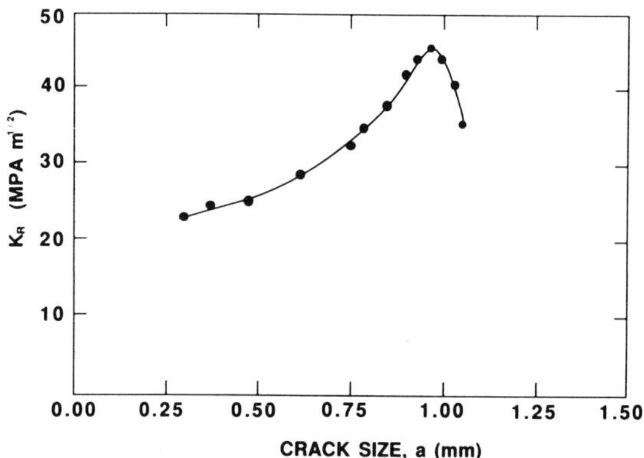

FIG. 3.35 The stress intensity factor for steady-state crack growth K_R vs. crack size a for a continuous carbon fiber/Al 6061 composite [68].

specimens. Kanninen et al. [70] made a critical review on the applicability of LEFM type models to the failure of a composite, and concluded that a new model must be constructed to account for the anisotropic and inhomogeneous fracture characteristic of a composite, particularly a continuous fiber composite. Kanninen et al. have also proposed such a new model in which a continuous fiber composite with a crack is converted to a local heterogeneous region embedded in a linear elastic anisotropic continuum (Fig. 3.36), although they did not derive any specific formula. Recently, Shibata et al. [71] proposed a model that can account for crack bridging by short fibers with friction at the matrix-fiber interface. This model can predict several possible modes of crack propagation which depends on the applied stress, interfacial friction, current crack size, and fiber volume fraction.

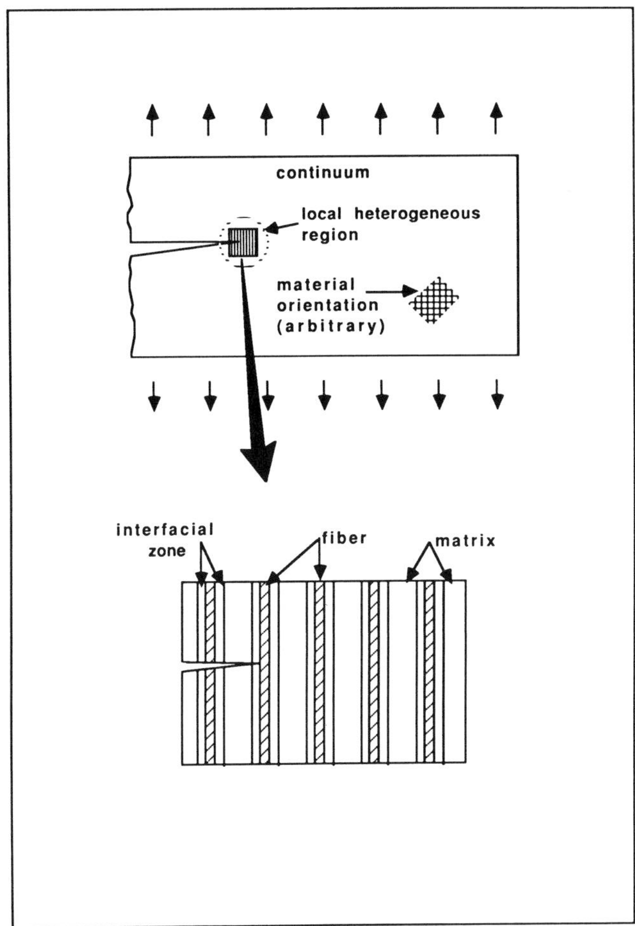

FIG. 3.36 The analytical model of Kanninen et al. [70] for a continuous fiber composite with a local heterogeneous region embedded in a linear elastic anisotropic continuum.

Earlier efforts to estimate the energy dissipation due to the local heterogeneous region at the crack-tip were made mainly by British researchers. Kelly [66] examined analytically the fracture process at a local heterogeneous region and gave explicit formulae to compute the work of fracture per fiber due to fiber pull out (W_{po}). He discussed two cases separately, i.e., aligned short fiber composite and continuous fiber composite.

In the case of an aligned short fiber composite, the work done in pulling out a fiber over the distance x is given by

$$W_{po} = \int_0^x \pi d\tau_i x dx = \frac{\pi}{2}\tau_i d x^2 \qquad (3.58)$$

where τ_i is the matrix–fiber interfacial shear strength and is assumed to be constant, and d is the fiber diameter. For these fibers with length (l) less than or equal to the critical fiber length (l_c), the average work done per fiber is

$$\bar{W}_{po} = \frac{2}{l}\int_0^{l/2} W_{po}\,dx = \frac{\pi d\tau_i l^2}{24} \quad \text{for } l \leqslant l_c \tag{3.59}$$

On the other hand, for those fibers with $l \geqslant l_c$, only a fraction (l_c/l) will be pulled out, contributing to the average work done per fiber which is then given by

$$\bar{W}_{po} = \left(\frac{l_c}{l}\right)\frac{\pi d\tau_i}{24}\,l_c^2 = \frac{\pi d\tau_i l_c^3}{24l} \quad \text{for } l \geqslant l_c \tag{3.60}$$

W_{po} can be termed 'pull-out energy' per fiber, which can be converted to pull-out work per unit fracture surface area. To this end, we multiply eqs. (3.59) and (3.60) by $V_f/\pi(d^2/4)$. Hence, pull-out energy per unit fracture area $\bar{W}_{po}$ is given by

$$\bar{W}_{po} = \begin{cases} \dfrac{V_f\tau_i l^2}{6d} & \text{for } l \leqslant l_c \\ \dfrac{V_f\tau_i l_c^3}{6dl} & \text{for } l \geqslant l_c \end{cases} \tag{3.61}$$

The dependence of pull-out energy (either per fiber or per unit area) on fiber length l is schematically illustrated in Fig. 3.37. Namely, for $l \leqslant l_c$, pull-out energy increases with l, while for $l \geqslant l_c$ it decreases with l. The pull-out energy becomes the maximum at $l = l_c$.

For $l \leqslant l_c$ the fiber stress increases linearly from the fiber end to the maximum value (σ_{max}) at the center when τ_i is assumed to be constant. Thus, σ_{max} is given by

$$\sigma_{max} = \frac{2\tau_i l}{d} \quad \text{for } l \leqslant l_c \tag{3.62}$$

while for $l \geqslant l_c$ σ_{max} is equal to σ_{fb} which is related to τ_i as defined by eq. (3.39). By using eqs. (3.62) and (3.34), $\bar{W}_{po}$ can be rewritten as

$$\bar{W}_{po} = \begin{cases} \dfrac{1}{12}V_f\sigma_{max}l & \text{for } l \leqslant l_c \\ \dfrac{1}{12}V_f\sigma_{fb}\dfrac{l_c^2}{l} & \text{for } l \geqslant l_c \end{cases} \tag{3.63}$$

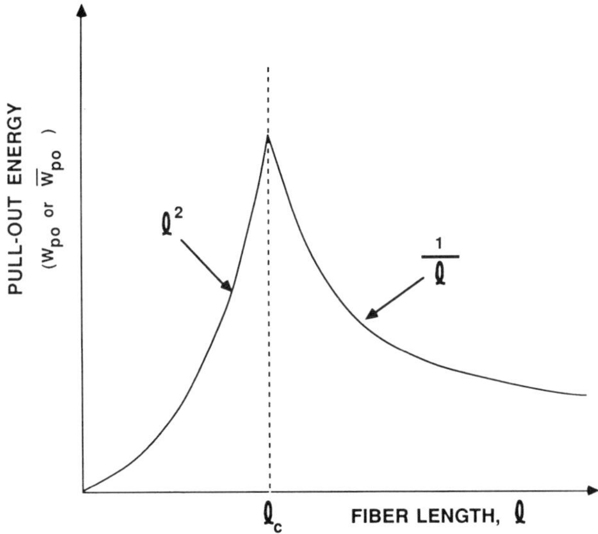

FIG. 3.37 Pull-out energy ($W_{\rm po}$ or $\bar{W}_{\rm po}$) as a function of fiber length l [66].

The maximum pull-out energy per unit area occurs when $l = l_c$

$$(\bar{W}_{\rm po})_{\max} = \frac{1}{12} V_{\rm f}\sigma_{\rm fb} l_{\rm c} \tag{3.64}$$

or

$$= \frac{1}{24}\frac{V_{\rm f}\sigma_{\rm fb}^2 d}{\tau_{\rm i}}$$

It follows from eq. (3.64) that pull-out energy increases with $\sigma_{\rm fb}$ and d, but it decreases with $\tau_{\rm i}$.

The above formulation must be modified in order to calculate the fiber pull-out energy for continuous fiber composite system. To this end, the strength of continuous fibers was assumed to be $\sigma_{\rm fb}$ everywhere except for weak points which have the strength $\sigma_{\rm fb}^*(<\sigma_{\rm fb})$ [72]. When such continuous fibers bridge a straight crack plane (Fig. 3.38), two cases of fracture process may be considered; when the average spacing between flaws $(l) < l_c^*$ (Fig. 3.38(a)) and when $l > l_c^*$ (Fig. 3.38(b)), where l_c^* is defined by

$$l_c^* = \frac{(\sigma_{\rm fb} - \sigma_{\rm fb}^*)}{\sigma_{\rm fb}} l_{\rm c} \tag{3.65}$$

where $\sigma_{\rm fb}$ and $\sigma_{\rm fb}^*$ are the strength of the unflawed portion and flawed points (weak points) in the fibers, and l_c is the critical fiber length defined by eq. (3.39). For $l < l_c^*$, all fibers will break at weak points lying within a distance $l_c^*/2$ on either side of the crack and will pull out, resulting in the

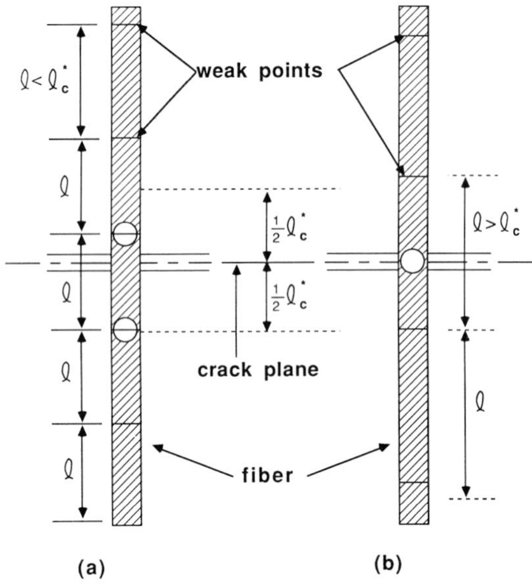

FIG. 3.38 Flawed continuous fibers (with the average flaw sparing l) bridging a plane crack in (a) the case of $l < l_c^*$, and (b) $l > l_c^*$ where open circle marks denote the locations of fiber fractures [66].

average pull-out energy per fiber

$$W_{po} = \frac{\pi d \tau_i l^2}{24} \tag{3.66}$$

On the other hand, for $l > l_c^*$, a fraction (l_c^*/l) of the fibers will break at weak points and pull out, and the remaining fraction $(1 - l_c^*/l)$ will break in the crack plane and will not pull out. In this case, W_{po} is given by

$$W_{po} = \frac{\pi d \tau_i (l_c^*)^3}{24l} \tag{3.67}$$

Equations (3.66) and (3.67) are similar to eqs. (3.59) and (3.60), respectively. The only difference between the above two sets of equations is that l_c in eq. (3.60) is replaced by l_c^* in eq. (3.67). It is obvious from eqs. (3.65)–(3.67) that l_c^* is reduced to l_c for $\sigma_{fb}^* = 0$; i.e., discontinuous (short) fibers and that for non-flawed continuous fibers ($\sigma_{fb}^* = \sigma_{fb}$, thus $l_c^* = 0$), all the fibers will break at the crack plane, but they will not pull out, i.e., $W_{po} = 0$. The average pull-out energy per unit area, $\bar{W}_{po}$ can be obtained by multiplying eqs. (3.66) and (3.67) by $V_f/(\pi d^2/4)$, leading to

$$\bar{W}_{po} = \begin{cases} \dfrac{V_f \tau_i l^2}{6d} & \text{for } l < l_c^* \\[2ex] \dfrac{(1-\eta)^3 V_f \tau_i l_c^3}{6dl} & \text{for } l > l_c^* \end{cases} \tag{3.68}$$

where l_c is the critical fiber length defined by eq. (3.39) and η is defined by

$$\eta = \frac{\sigma_{fb}^*}{\sigma_{fb}} \tag{3.69}$$

It is noted in eq. (3.68) that the average pull-out energy per unit fracture area $\bar{W}_{po}$ is equal to $2\gamma_{po}$ where γ_{po} is the surface fracture energy by fiber pull-out. The maximum $\bar{W}_{po}$ occurs when $\eta = 0$ and $l = l_c^*$ and it is given by

$$\begin{aligned} (\bar{W}_{po})_{max} &= \frac{V_f \tau_i l_c^2}{6d} \\ &= \frac{V_f d \sigma_{fb}^2}{24\tau_i} \end{aligned} \tag{3.70}$$

where eq. (3.39) was used. Equation (3.70) coincides with eq. (3.64). Thus, it can be concluded that the fiber pull-out energy of a continuous fiber composite is usually less than that of an aligned short fiber composite, although the stiffness (E_L) and the strength (σ_L) of the former composite is much larger than those of the latter composite.

In order to obtain the fracture toughness (G_{IC}) of a composite, one must add the fracture toughness (G_{Ilocal}) contributed by the fracture process in a local heterogeneous region near the crack-tip region such as by fiber pull-outs, to the macroscopic fracture toughness (G_{Imacro}) of an anisotropic, but homogeneous body with the composite properties. If the fiber pull-out is the only local contribution, then we have

$$G_{Ilocal} = \bar{W}_{po} \tag{3.71}$$

In the case of a continuous fiber metal matrix composite, fibers bridging the plane crack will do work (plastic work) on the matrix before they fracture. Piggott [73] accounted for such a plastic work by assuming lateral constraint, i.e., the transverse strains being zero and arrive at the fracture toughness per unit area, $\bar{W}_p$, as

$$\bar{W}_p = \frac{\pi d^3 \sigma_{fb}^3}{48\tau_y E_{fr}} \tag{3.72}$$

where

$$E_{fr} = \frac{(1 - \nu_f)E_f}{(1 + \nu_f)(1 - 2\nu_f)} \tag{3.73}$$

In eqs. (3.72) and (3.73), τ_y is the yield stress of the matrix metal, E_f and ν_f are the modulus and Poisson's ratio of the fiber. Ishikawa and Taya [74] made a more accurate prediction of $\bar{W}_p$ by considering the crack opening displacement δ_0. The $\bar{W}_p$ predicted by Ishikawa and Taya is given by

$$\bar{W}_p = \pi d^2(\sigma_{fb} - \sigma_f)\delta_0/2 \tag{3.74}$$

where σ_f is the average fiber stress away from the crack plane and δ_0 is the crack opening displacement at the fiber location and its detailed expression is given in Ref. 74. Thus, G_{Ilocal}, which account for both fiber pull-out energy and plastic work, is

$$G_{\text{Ilocal}} = \bar{W}_{\text{po}} + \bar{W}_{\text{p}} \tag{3.75}$$

The macroscopic fracture toughness of a continuous fiber metal matrix composite G_{Imacro} can be calculated following the classical LEFM model if one can assume isostrain condition (Fig. 3.39(a)) [74]. The stiffness and stress of the composite are given by the law of mixtures equations, i.e., eqs. (3.1) and (3.3), respectively (Fig. 3.39(b)). G_{Imacro} can be defined as

$$G_{\text{Imacro}} = \lim_{\delta a \to 0} \frac{1}{\delta a} \int_0^{\delta a} \sigma_y^c u_y^c \, dr \tag{3.76}$$

where σ_y^c and u_y^c are the normal stress and displacement along the y-axis at a crack tip (Fig. 3.39(b)), and they can be approximated by the law of mixtures

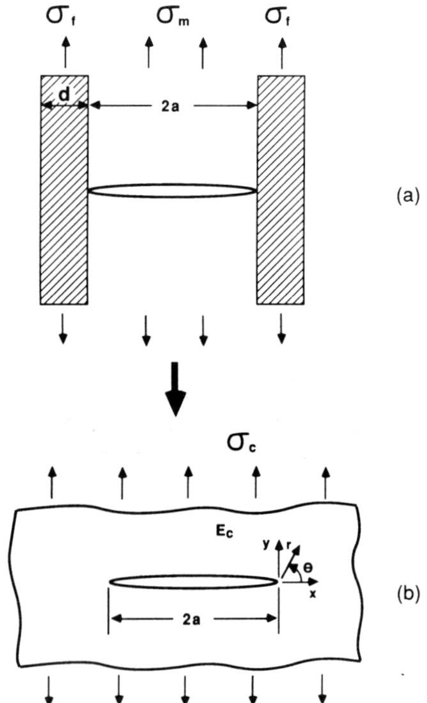

FIG. 3.39 (a) A continuous fiber composite converted to (b) homogeneous material [74, 79].

equations, i.e.,

$$\begin{aligned} \sigma_y^c &= V_m \sigma_y^m + V_f \sigma_y^f \\ u_y^c &= V_m u_y^m + V_f u_y^f \end{aligned} \tag{3.77}$$

where V_m and V_f are the volume fractions of matrix and fiber, respectively, and σ_y^i and u_y^i are the normal stress and displacement along the y-axis at a crack tip when a two-dimensional crack of length $2a$ is embedded in an infinite homogeneous ith phase medium subjected to the far-field applied stress σ_i. σ_c^i, σ_y^i, and u_y^i are related to each other.

$$\begin{aligned} u_y^i &= \frac{4}{\sqrt{2}} \frac{(1-v_i^2)}{E_i} \sqrt{ra}\,\sigma_i \\ \sigma_y^i &= \sigma_i \sqrt{\frac{a}{2r}} \end{aligned} \tag{3.78}$$

and where i denotes c (composite), m (matrix), or f (fiber) phases.

From eqs. (3.76)–(3.78), G_{Imacro} can be obtained as

$$G_{\text{Imacro}} = \{V_m(E_m G_{\text{Im}}/E_c)^{1/2} + V_f(E_f G_{\text{If}}/E_c)^{1/2}\}^2 \tag{3.79}$$

where G_{Ij} is the energy release rate of a plane strain crack of size $2a$ embedded in an infinite body of the j-phase material with $j = \text{m}$ or f, and defined by

$$G_{\text{Ij}} = \lim_{\delta a \to 0} \frac{1}{\delta a} \int_0^{\delta a} \sigma_y^j u_y^j \, dr \tag{3.80}$$

If Poisson's ratios of three phases are equal, i.e., $v_c = v_m = v_f$, then eq. (3.79) is further reduced to

$$G_{\text{Imacro}} = \pi(1 - v_c^2)\sigma_c^2 a/E_c \tag{3.81}$$

where σ_c and E_c are already defined by eqs. (3.1) and (3.3), respectively. The macroscopic energy release rate given by eq. (3.81) is just the standard formula for a plane strain crack embedded in an infinite homogeneous body with E_c and v_c subjected to the applied stress σ_c.

Based on the above discussion, the fracture toughness (energy release rate) G_{IC} of a continuous fiber metal matrix composite is given by

$$G_{\text{IC}} = G_{\text{Imacro}} + \bar{W}_{po} + \bar{W}_p \tag{3.82}$$

If one wants to obtain the fracture toughness in terms of K_{IC}, one can use eq. (3.82) to obtain

$$K_{\text{IC}} = \left\{\frac{E_c G_{\text{IC}}}{(1 - v_c^2)}\right\}^{1/2} \tag{3.83}$$

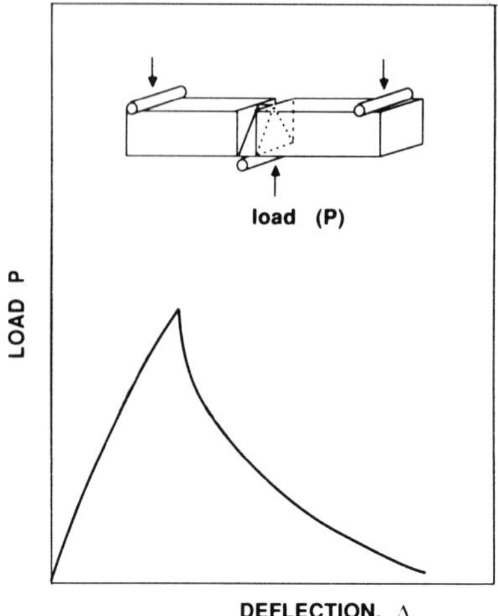

FIG. 3.40 A three-point bend test for a chevron-notched specimen and its result, load (P)–displacement (Δ) curve.

The formula obtained by Piggott [73] belongs to this type where $G_{IC} = G_{Imacro} + \bar{W}_p$.

It is sometimes convenient to evaluate the total fracture energy of a composite as opposed to its fracture toughness (G_{IC} or K_{IC}). One of the most popular methods of obtaining experimentally the total fracture energy (W_F) is to conduct a three-point bend test with a chevron notched specimen (see Fig. 3.40). The result of this test is obtained in terms of force (P)–displacement (Δ) curve. The total fracture energy W_F is equal to the area underneath the P–Δ curve. Then the fracture surface energy γ_F, which is called 'work of fracture,' is obtained as

$$\gamma_F = \frac{W_F}{2A} \tag{3.84}$$

where A is the area of the triangular ligament. The concept of γ_F was originally proposed for monolithic ceramics [75] and later applied to other materials [76], including metal matrix composites [77–81]. Taya and his co-workers [74, 79–81] obtained the formula to predict γ_F by integrating eq. (3.82) from the initial (a_0) to the final size (a_f) of a penny-shaped crack. It is assumed in this model that a continuous fiber metal matrix composite contains a penny-shaped crack of radius a_0 initially, which grows along the same crack plane to the final penny-shaped crack of radius a_f, leading to final fracture. The penny-shaped crack at the intermediate stage is shown schematically in Fig. 3.41. The formula to predict γ_F of a continuous fiber metal

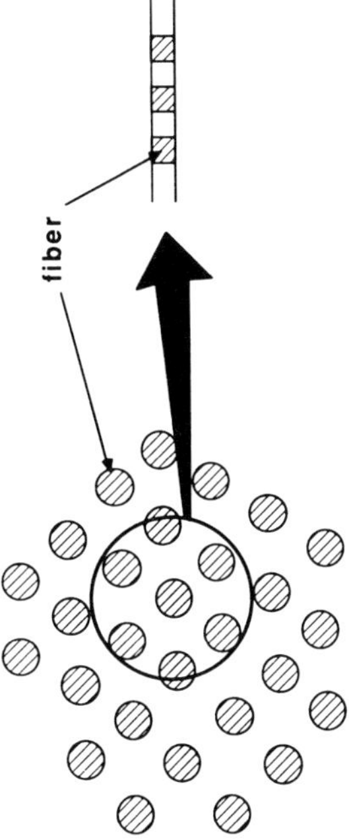

FIG. 3.41 Penny-shaped crack located perpendicular to the fiber axis in a continuous fiber metal matrix composite [74, 79].

matrix composite is given in Ref. 81.

$$\gamma_F = \frac{1}{2}\left[\frac{2\pi(1-\nu_c^2)\sigma_c a_f}{3E_c}(1+\alpha+\beta)+(1-\eta)^2\frac{V_f d\sigma_{fb}^2}{24\tau}\right] \tag{3.85}$$

where

$$\alpha = \frac{2(1-\eta^2)(\sigma_{fb}+\sigma_{fb}^*-2\sigma_f)(\sigma_{fb}+\sigma_{fb}^*)E_c}{\pi\sigma_c^2 E_m}\cdot \left\{\frac{4}{5}\left(\frac{\sigma_c}{\sigma_{fb}+\sigma_{fb}^*}\right)\left(\frac{a_0}{a_f}\right)^{1/2}-\frac{1}{2\pi}\left(\frac{d}{a_f}\right)\right\} \tag{3.86}$$

$$\beta = \frac{8\eta(\sigma_{fb}-\sigma_f)\sigma_{fb}E_c}{\pi\sigma_c^2 E_m}\left\{\frac{2}{5}\left(\frac{\sigma_c}{\sigma_{fb}}\right)\left(\frac{a_0}{a_f}\right)^{1/2}-\frac{1}{2\pi}\left(\frac{d}{a_f}\right)\right\}$$

where all the variables have already been defined in this chapter and σ_f can be calculated from eq. (3.3) with σ_c interpreted as the strength of the composite. τ in eq. (3.85) is interpreted as the interfacial shear strength (τ_i), which may be set equal to the shear yield stress of the matrix metal for some types of metal matrix composites. γ_f of a continuous graphite fiber/6061 aluminum composite [79] and a boron fiber/1100 composite [81] have been obtained experimentally, which are in reasonably good agreement with the prediction by eq. (3.85).

So far, the fracture toughness of only continuous fiber metal matrix composites has been discussed, except for the fiber pull-out energy of aligned short fiber metal matrix composites. Unlike the case of continuous fiber metal matrix composites, the fracture toughness based on linear elastic fracture mechanics (LEFM), i.e., G_{Imacro} or K_{IC}, has been successfully applied to the case of short fiber and particulate metal matrix composites, although a limited amount of work [82–88] has been centered on SiC whisker, or SiC particulate/aluminum alloy composites. It was found in these studies that the fracture toughness (K_{IC} or K_{IQ}) of SiC whisker and SiC particulate aluminum

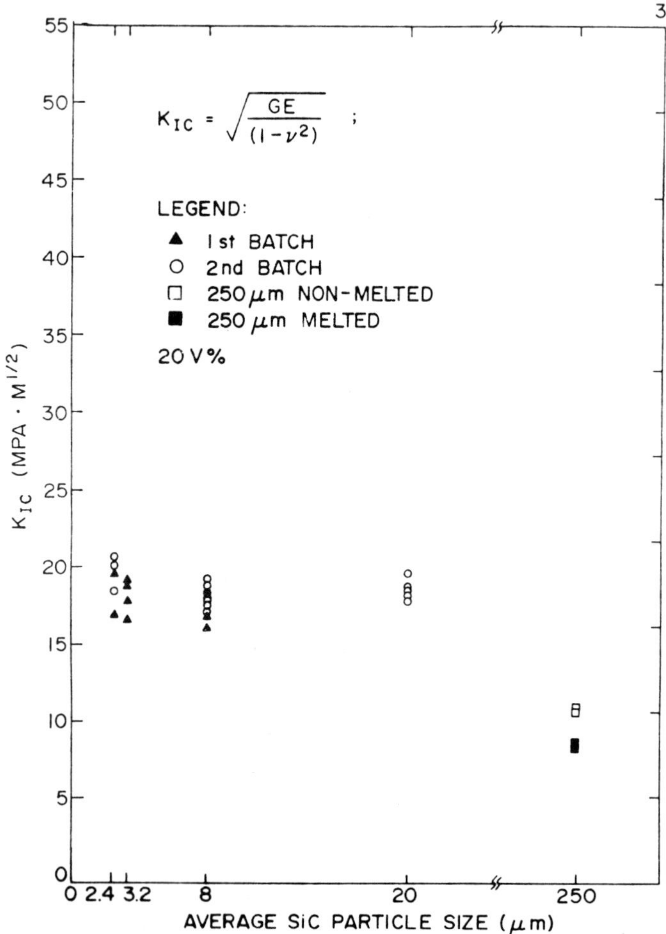

FIG. 3.42 The fracture toughness (K_{IC}) of a SiC particle/aluminum composite vs. average SiC particle size [87].

composites is still in the range 12–20 Mpa · $m^{1/2}$ and the size distribution and orientation of SiC whiskers (SiC_w) and particles (SiW_p) control the more of the fracture path, thus influencing the fracture toughness. The dependence of fracture toughness (K_{IC}) on SiC particles is demonstrated in Fig. 3.42 [87]. It follows from Fig. 3.42 that the K_{IC}'s for smaller SiC particles in the size range 2–20 μm, remain almost constant, but for the larger SiC particles of average size 250 μm, it drops to about one-half of that of the composite with smaller SiC particles. A scanning electron microscope (SEM) examination of the vicinity of a typical crack tip in a 20% V_f SiCp/aluminum composite did not reveal the presence of the voids below and above the fracture path, and instead indicated a series of short cracks in the matrix in front of the crack (Fig. 3.43). It is obvious from Fig. 3.43 that the crack path is extremely narrowly banded, and the damage zone is located mainly in front of the crack. In the narrow crack zone the crack path appears to be attracted toward SiC particles adjacent to the fracture plane. In the case of SiCw/Al 6061 matrix composite, Crowe et al. [84] found that the fracture toughness is dependent on whisker orientation. Namely, K_{IC} measured for the crack perpendicular to the axis of the majority of SiC whiskers is larger than that for the crack parallel to the whisker axis. The crack path in the former case is expected to encounter SiC whiskers more often, and this would increase the total fracture surface area, contributing to the higher toughness, compared with the latter case.

During recent decades there has been a considerable number of investigations of fracture characteristics of two-phase alloys where the second phase is a discrete precipitate or particulate [88–104] and the matrix is ductile. Discontinuous SiC in an Al matrix would appear to be a materials system which should be easily analyzed in terms of those existing theories. The review by Schwalbe [100] and more recent publications by Bates [102], Gerberich [103] and Firrao and Roberti [104] are very informative on this subject.

In ductile fracture characterized by void nucleation and growth (VNG), the spacing between the void nucleating particles is generally considered to be a critical microstructural parameter that, taken together with the tensile properties, controls the toughness of a given material [100–104]. This can be represented by the following simplified expression [104]:

$$K_{IC} = [\sigma_Y \cdot \varepsilon_F^* \cdot E \cdot f(n)]^{1/2} \cdot s^{1/2} \qquad (3.87)$$

where K_{IC} is the plane strain fracture toughness, σ_Y is the yield stress (depending on the analysis, this could be the yield stress of the matrix or the yield stress of the composite), ε_F^* is the maximum strain acting at the crack tip, E is the Young's modulus, $f(n)$ is some function of the strain-hardening exponent, and s is the average inclusion spacing in the matrix. The experimental data collected by various investigators are usually given in terms of

FIG. 3.43 SEM photo showing the region ahead of the crack tip in a compact tension tested SiC particle/aluminum composite [87].

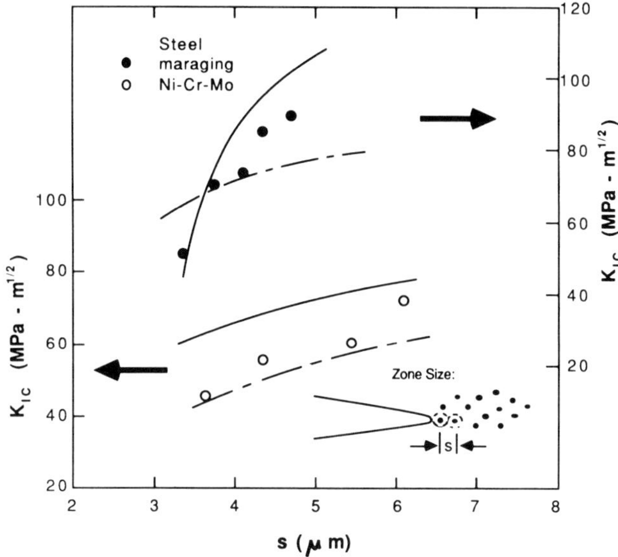

FIG. 3.44 Fracture toughness as a function of process zone size where the process zone size is equated with the average spacing (s) between void nucleating particles [103]. Various forms of eq. (3.87) are employed by Gerberich [103] to predict the fracture toughness values (solid and dashed lines) which are compared with the experimental data reported by Shwalbe [100].

the volume fraction of the second-phase particles (V_f), or less often in terms of s (see Fig. 3.44).

The dimple morphology of the fracture surfaces is the most common observation made by a number of researchers studying the fracture process in discontinuous SiC/Al composites [82, 105–110]. This indicates that the VNG mechanism may be active in these composites; therefore corresponding theories generalized by eq. (3.87) can be employed to describe the fracture process in SiC/Al composites.

The adverse influence of the increase of the volume fraction of SiC reinforcement on fracture toughness can be established rather well when one combines the data on fracture toughness testing available in the literature, as shown in Fig. 3.45.

A rather strict geometric consideration requires that all three microstructural variables V_f, s, and particle size (d) be related. There are several expressions available in the literature relating to V_f, s, and d [88, 101, 111]. Since there is no substantial difference between these expressions, we select one derived by LeRoy et al. [101] for the equiaxial particles:

$$s = 0.77 \cdot d \cdot V_f^{-1/2} \tag{3.88}$$

The data in Fig. 3.45 can now be replotted as a function of the average

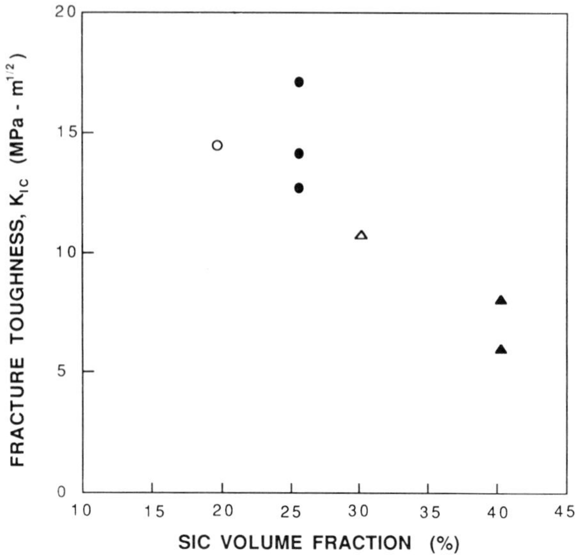

FIG. 3.45 Fracture toughness of SiC/Al composite as a function of volume fraction of SiC particle.

center-to-center SiC particle spacing (s) provided that the SiC particle size remains constant, which is generally the case in commercial composites. Figures 3.46 shows the dependence of K_{IC} on s, which was plotted using the data in Fig. 3.45 and eq. (3.88), assuming that the average SiC particle is about 5 μm [107]. Notice that, in this case, s changed by varying V_f.

Mathematically, a similar result can be achieved by keeping V_f constant and letting particles change their size (d). In accordance with eq. (3.88), by increasing d, s is proportionally increasing. Thus, a plot similar to one shown in Fig. 3.46 should result if the size of SiC particles is increased and their volume fraction is kept constant (see Fig. 3.47).

It therefore, appears that the toughness of SiC/Al composites can be improved when the size of the SiC particles is increased, provided that at least one of the two following assumptions is met:

1. voids are nucleated by SiC particles, and/or
2. the response of the SiC/Al system to the change of s remains the same regardless of whether this change is caused by varying the volume fraction or size of SiC particles.

An attempt to find a reference in the literature where these issues have been addressed in a systematic fashion did not give a positive result.

As can be seen from Fig. 3.42, K_{IC} is independent of interparticle spacing at a constant volume fracture. This is contrary to the predicted result of

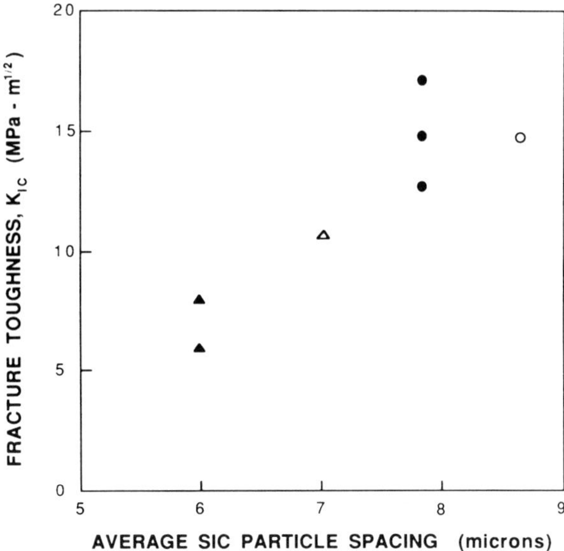

FIG. 3.46 Fracture toughness as a function of the average SiC particle spacing s. The experimental data of Fig. 3.45 are replotted using eq. (3.88) while keeping SiC particle size d constant, equal to 5 μm.

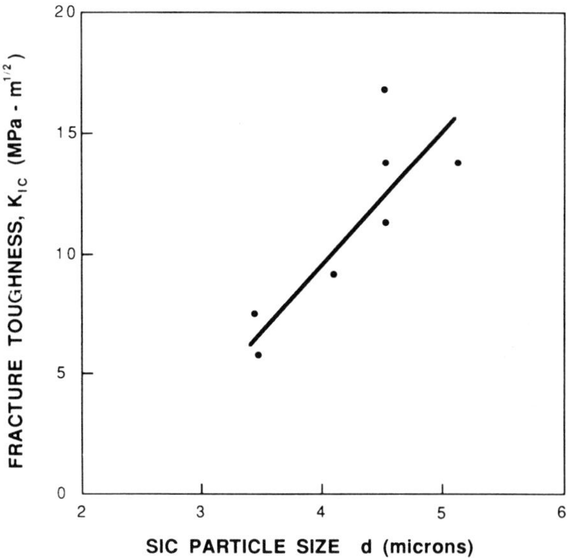

FIG. 3.47 Prediction of fracture toughness as a function of the average SiC particle size d. Volume fraction of SiC is kept constant (20%). Equation (3.88) was employed and values of s were obtained from Fig. 3.46.

Fig. 3.47. We propose that the explanation for the differences between Fig. 3.47 and Fig. 3.42 is as follows. It has been shown that the matrix in these composites has an average tensile residual stress. From the analysis of Eshelby et al. it can be shown that the value of the strain energy in the composite is independent of the particle size provided the volume fraction remains constant. However, if the particle size remains constant and the volume fraction changes, then the value of the stored energy increases as the volume fraction increases.

It has also been shown experimentally that there is no direct evidence of void nucleation and growth at the SiC particle. The fracture seems to appear almost exclusively through the matrix, although there are separations at the particle–matrix interface.

In the case where there is a constant volume fraction but varying particle size, the value of the elastic stored energy is a constant; therefore the fracture toughness (K_{IC}) is independent of particle size. However, in a case where there is an increase in volume fraction but constant particle size, there is a decrease in K_{IC} as particle spacing decreases (the volume fraction increases) due to the fact that magnitude of the elastic stored energy and the dislocation density increases as the volume fraction increases. The stored elastic energy is an addition to the strain energy component due to external load at the crack tip. The presence of stored energy is not taken into consideration in the existing theories.

It can be argued that since the value of the elastic stored energy has to balance out to zero, the compression must balance out the tension, and that is indeed true. However, the reinforcement is in compression, but the reinforcement, *per se*, does not enter into the fracture process; i.e., the fracture occurs through the matrix.

An additional point should be raised in regard to attempting to fit the data from discontinuous SiC/Al composites to existing theories [100, 101–104]. The existing theories assume that the matrix is in the annealed condition, i.e., there is no plastic zone about the second phase. In other words, the dislocation density in the matrix is assumed to be very low. However, in the case of SiC/Al discontinuous composites there is a very high dislocation density within the matrix, which means that the matrix is in a cold worked state prior to testing.

The work of fracture (γ_F) given by eq. (3.84) is also applicable to a short fiber and particulate metal matrix composite. Daimaru et al. [80] measured the γ_F of a SiCp/6061 Al composite with particulate volume $V_f = 20$ and 30%, and the results of γ_F are plotted as filled circles against V_f in Fig. 3.48. In this figure the prediction based on eq. (3.85) is shown by the solid line and the law of mixtures prediction, $\gamma_F = \gamma_{Ff} V_f + \gamma_{Fm} V_m$ is also plotted as a dashed line where γ_{Ff} and γ_{Fm} are the work of fracture of SiC particle and 6061 Al, respectively. It can be concluded from Fig. 3.48 that γ_F increases slightly with V_f for the range investigated, and the prediction by eq. (3.85) is reasonably

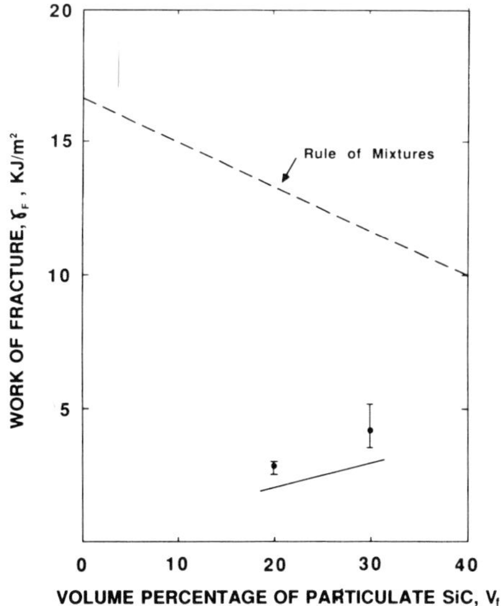

FIG. 3.48 A work of fracture γ_F of a SiC particulate/6061 aluminum composite vs. the volume fraction of a SiC particulate V_f (%): experiment (filled circles), predicted by eq. (3.86) (solid line) and that by law of mixture (dashed line) [80].

accurate, but the law of mixtures prediction significantly overestimates γ_F. This observation that γ_F increases with increasing V_F, i.e., decreasing inter-particle spacing, is in disagreement with the results obtained by Flom and Arsenault [112], which showed that the work of fracture as measured by the tearing modulus increased as the inter-particle spacing increased (Fig. 3.49).

3.5 Thermal Stress

Two types of thermal stresses can be included in metal matrix composites. The first type is due to the temperature gradient within a composite and may be termed 'temperature gradient induced thermal stress.' The second type is induced under uniform temperature within a composite due to the mismatch in the coefficient of thermal expansion (CTE) between the matrix metal and fiber, thus it may be called a 'CTE mismatch-induced thermal stress.' Clearly, CTE mismatch-induced thermal stress is a characteristic with metal matrix composites since the CTE mismatch for most of the metal matrix composites, particularly the ceramic fiber/metal matrix system, is large. Therefore, we will discuss CTE mismatch-induced thermal stress here. An exact analysis of the thermal stress problems in metal matrix composites usually requires a rigorous numerical solution because constituent phases, particularly matrix metal, behave as an elastic/plastic/creep body under thermal loading. However, in this section the main consideration will be the analysis based on an

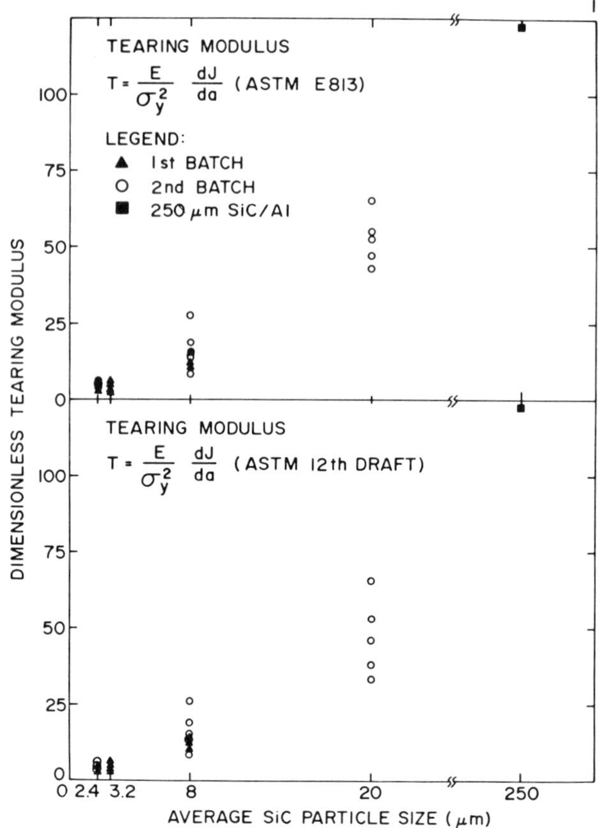

FIG. 3.49 Tearing modulus vs. average particle size of SiC/Al composites.

elastic body model because the thermal stress based on elastic analysis is the upper bound on the actual thermal stress in the composite, and it is relatively easy to trace its computation scheme.

First, let us estimate thermal stresses induced in a continuous fiber metal matrix composite by a simple one-dimensional model (Fig. 3.50). A model for a continuous fiber metal matrix composite is constructed based on the assumption that it is initially stress-free (Fig. 3.50(a)) where the moduli and CTEs of the matrix metal and fiber are denoted by E_m and E_f, and α_m and α_f, respectively. Let this model composite be subjected to a uniform temperature rise (ΔT). Then each phase in the composite would expand freely by $\alpha_i \Delta T$ with $i = \mathrm{m}$ or f phase under no constraint (Fig. 3.50(b)). Due to the requirement of isostrain (continuity of the axial displacement at the interface), the matrix is compressed while the fiber is stretched (Fig. 3.50(c)). Denoting the isostrain measured from the initial stage by e_c, one can establish the equilib-

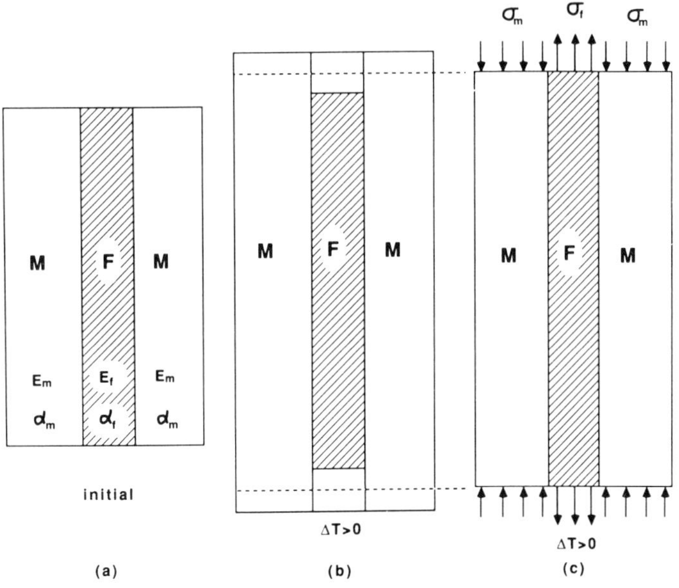

FIG. 3.50 One-dimensional model for the analysis of the thermal stress in a continuous fiber metal matrix composite subjected to temperature change $\Delta T(>0)$ where $\alpha_m > \alpha_f$ was assumed; (a) initial configuration, (b) constraint-free configuration and (c) final equilibrium configuration.

rium in force along the axial direction, i.e., the total force caused by the internal stresses must be balanced by the external force (in this case, zero)

$$-\sigma_m V_m + \sigma_f V_f = 0 \tag{3.89}$$

Hooke's law holds each phase:

$$\alpha_m \Delta T - e_c = \frac{\sigma_m}{E_m} \tag{3.90}$$

$$e_c - \alpha_f \Delta T = \frac{\sigma_f}{E_f} \tag{3.91}$$

From eqs. (3.89)–(3.91) the internal stresses in the matrix (σ_m in compression) and in the fiber (σ_f in tension) are obtained [37, 113] as:

$$\begin{aligned} \sigma_m &= \frac{V_f E_m E_f \Delta\alpha \Delta T}{E_c} \\ \sigma_f &= \frac{V_m E_m E_f \Delta\alpha \Delta T}{E_c} \end{aligned} \tag{3.92}$$

where $\Delta\alpha = \alpha_m - \alpha_f$ and E_c is the modulus of the composite and given by the law of mixtures equation, eq. (3.1). The above internal stresses are indeed 'the thermal stress' in the composite. The solutions given by eq. (3.92) illustrate the key variables that directly influence the magnitude of the thermal stresses, and they are the CTE mismatch ($\Delta\alpha$), temperature change (ΔT) and fiber volume fraction (V_f).

The analysis of the thermal stress field in actual metal matrix composites is not so simple as above, and it requires more accurate models. In the following we will discuss these models for two cases of reinforcement geometry, continuous fiber and short fiber metal matrix composites.

3.5.1 Continuous fiber metal matrix composites

Thermal stress in a continuous fiber metal matrix composite has been well studied because of its relatively simple geometry. The analysis of the thermal stress in continuous fiber metal matrix composites can be grouped into two models, the concentric cylinder model and the plane-stress lamina model. The former model is aimed at the analysis of the thermal stress field in a cylinder or thick plate made of a continuous fiber metal matrix composite, thus the macroscopic properties of such a composite will become transversely isotropic. The latter model is strictly applicable to thin plate made of continuous fiber metal matrix composite (lamina) where the out-of-plate stress components can be neglected (plane-stress condition).

Among the various concentric cylinder models [114–118], the four concentric cylinders model proposed by Mikata and Taya [118] is a generalization of this group and can be used for several purposes. Mikata and Taya applied the four concentric cylinders model to a coated continuous fiber metal matrix composite (Fig. 3.51) and obtained the distribution of various stress components in the composite. Their analytical results of the thermal stress field in a composite, particularly in the coating zone, were used to optimize the choice of material and thickness of the coating on fiber. If the coating zone (shaded area in Fig. 3.51) is interpreted as a reaction zone between the matrix metal and fiber, then the thermal stress in and around the reaction zone can also be calculated by this model. The results will be useful for predicting the formation of circumferential cracks in the reaction zone. The detailed formulation of the four concentric cylinders model is given in Ref. 118. Hecker et al. [114] used a similar concentric cylinder model to study the thermal residual stress induced in a model composite cylinder made of two components (i.e., steel wire core surrounded by OHFC copper). The model of Hecker et al. assumes that components are thermoelastic/plastic material. Dvorak [119] has recently proposed a simple continuum model to predict the thermoelastic/plastic behavior of a continuous fiber metal matrix composite where the assumption of transversely isotropic unit cells is used and the stress and strain are given in incremental form.

Thermal (residual) stress problems in a lamina have been studied exten-

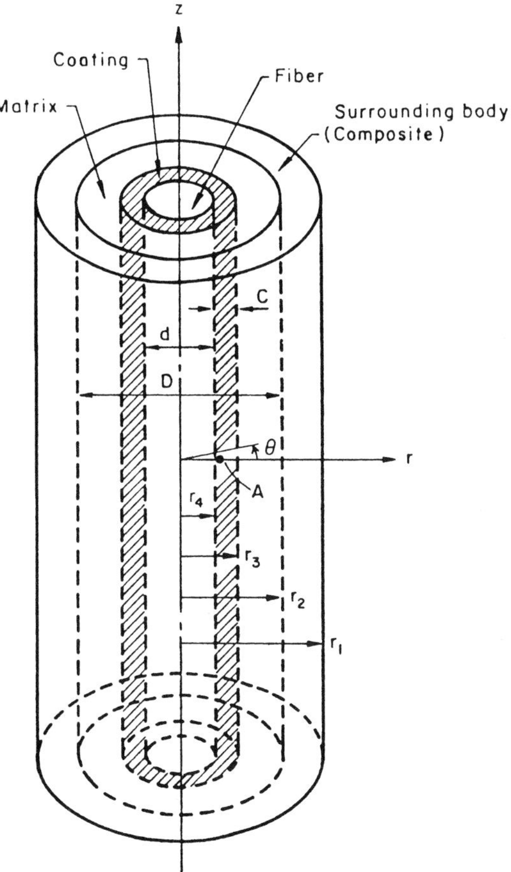

FIG. 3.51 A four concentric cylinders model for the analysis of the thermal stress in a continuous coated fiber metal matrix composite [118].

sively with most of the applications devoted to polymeric composites, and the standard models for a lamina and laminates have been described elsewhere [120]. Min and his co-workers [121, 122] applied the lamina model to study the thermoelastic/plastic behavior of metal matrix composite lamina and laminates. The above lamina model is based on a plane stress assumption, thus it is easy to use and extend to any type of laminate by following a standard lamination theory (for example, see details in Chap. 6 of Ref. 120). However, the lamina model tends to be a crude approximation of the transverse behavior of the lamina.

An attempt has been made by Tsai et al. [123] to measure the thermal residual stresses in a continuous fiber metal matrix (graphite/aluminum) composite by the x-ray diffraction method.

3.5.2 Short fiber metal matrix composites

Unlike continuous fiber metal matrix composites, only a limited number of studies of the thermal (residual) stress problems in a short fiber (including particulate) metal matrix composites have been conducted. This is partly because the entry of a short fiber and particulate metal matrix composite is relatively new and also because its reinforcement geometry requires a three-dimensional stress analysis. The Eshelby model discussed in Section 2.3 is thus suited for the analysis of the thermal stress in a short fiber and particulate metal matrix composites. Short fibers embedded in the matrix are usually misoriented due to processing, but all of them are assumed to be aligned to facilitate the analysis. The analytical model of the Eshelby type for the thermal stress analysis is shown in Fig. 3.52 where all short fibers are converted to prolate spheroids of length l and diameter d. The original Eshelby model discussed in Section 2.3 must be modified to account for the interaction between fibers which will increase with fiber volume fraction V_f. Thus, the thermal stress in a composite, $\boldsymbol{\sigma}$ can be solved from the following equations:

$$\begin{aligned} \boldsymbol{\sigma} &= \mathbf{C}_f \cdot (\bar{\mathbf{e}} + \mathbf{e} - \boldsymbol{\alpha}) \\ &= \mathbf{C}_m \cdot (\bar{\mathbf{e}} + \mathbf{e} - \mathbf{e}^*) \end{aligned} \tag{3.93}$$

$$\mathbf{e} = \mathbf{S} \cdot \mathbf{e}^* \tag{3.94}$$

$$\bar{\mathbf{e}} = -V_f(\mathbf{e} - \mathbf{e}^*) \tag{3.95}$$

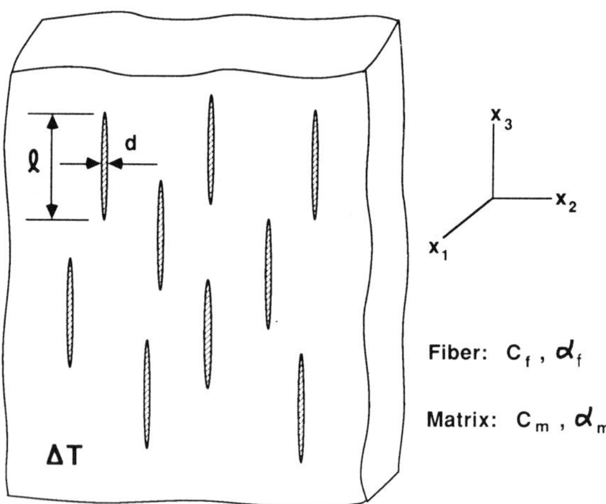

FIG. 3.52 Analytical model for the analysis of the thermal residual stress in a short fiber metal matrix composite subjected to uniform temperature change ΔT.

Equation (3.95) was derived from the fact that the volume integration of $\boldsymbol{\sigma}$ over the entire composite domain vanishes. The details of the Eshelby method modified for finite fiber volume fractions are given in subsection 2.4.4 (or see Refs. 11, 12, 124–126). In eq. (3.93) $\mathbf{C}_f$ and $\mathbf{C}_m$ are the elastic tensors of the fiber and matrix, respectively, $\boldsymbol{\alpha}$ is the CTE mismatch strain tensor and given by $(\boldsymbol{\alpha}_f - \boldsymbol{\alpha}_m)\Delta T$ where $\boldsymbol{\alpha}_f$ and $\boldsymbol{\alpha}_m$ are the CTE tensors of the fiber and matrix, respectively and ΔT is the temperature change. After eliminating $\bar{\mathbf{e}}$ and $\mathbf{e}$ in eq. (3.93) by use of eqs. (3.94) and (3.95), one can solve for eigenstrain $\mathbf{e}^*$ (see details in subsection 2.4.4). Then, the stress field in a composite can be obtained from eq. (3.93). For example, the average stress in the matrix $\langle \sigma_{ij} \rangle_m$ is given in index form,

$$\langle \sigma_{ij} \rangle_m = -V_f \{ (S_{\kappa\kappa mn} e^*_{mn} - e^*_{\kappa\kappa}) \lambda \delta_{ij} + 2\mu (S_{ijmn} e^*_{mn} - e^*_{ij}) \} \tag{3.96}$$

where both fiber and matrix are assumed to be isotropic, λ and μ are Lame constants of the matrix and the repeated sum on subscripts is to be summed over 1, 2 and 3 (Einstein's summation rule). Equation (3.96) was applied to the case of an aligned SiC whisker/Al composite subjected to $\Delta T = -200\,°\mathrm{C}$ and the analytical results of the longitudinal (along the x_3-axis in Fig. 3.52) and transverse (along the x_1- or x_2-axis) components of $\langle \sigma \rangle_m$ are plotted as a function of V_f in Fig. 3.53 [22]. It is clear from Fig. 3.53 that the averaged

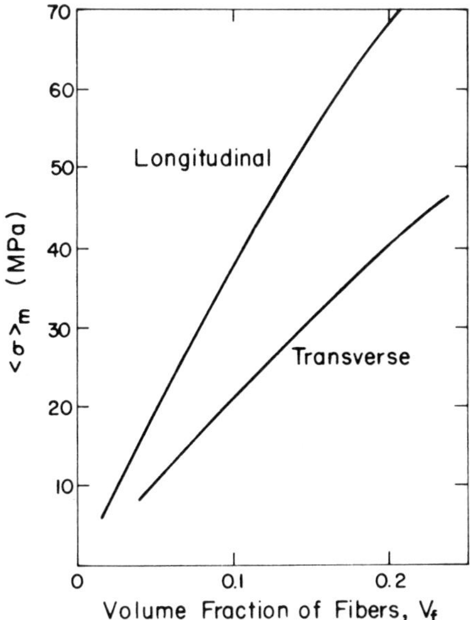

FIG. 3.53 Averaged thermal stresses $\langle \sigma \rangle_m$ in the matrix in a SiC whisker/Al composite predicted by eq. (3.96) as a function of V_f longitudinal and transverse components are along the x_3- and x_1-(or x_2-)axis, in Fig. 3.52, respectively [22].

thermal stresses in the matrix are tensile and they increase with V_f. This tensile thermal residual stress is known to reduce the room temperature tensile yield stress of the composite [22]. The above model is also applied to the case of a 30% V_f SiC particulate/Al composite and the analytical prediction of the average thermal stress in the matrix $\langle\sigma\rangle_m$ and in the fiber $\langle\sigma\rangle_f$ are given in Table 3.2, where the experimental results measured by x-ray diffraction method [127] are also shown for comparison. It is noted in Table 3.2 that the state of the average stresses in the fiber and matrix is isotropic since the shape of SiC particulates is considered nearly equal to spherical.

Table 3.2 *Comparison between experimental and analytical results on the thermal residual stress in 30% V_f SiC particulate/Al composite.*

Thermal stress	Experimental results by x-ray analysis [127]	Predicted values by eq. (3.94)
$\langle\sigma\rangle_m$	206 MPa	129 MPa
$\langle\sigma\rangle_f$	−219 MPa	−302 MPa

Although x-ray diffraction analysis has been the standard method to measure the residual stress in a material, it has a limitation in that only the region near the surface can be detected. This may have contributed to the poor agreement between the experimental and analytical results as evident in Table 3.2 [129]. It should also be noted in Table 3.2 that the experimental data do not satisfy the requirement of internal stress in a composite:

$$V_f\langle\sigma\rangle_f + (1 - V_f)\langle\sigma\rangle_m = 0 \qquad (3.97)$$

where the analytical results do satisfy the above requirement.

On the other hand, the neutron diffraction method appears to be best suited to measure the residual stress well within a thick composite specimen. Allen et al. measured the residual stresses in several different specimens made of monolithic metals by the neutron diffraction method, and confirmed the accuracy of this method by comparing it with the known distribution of the residual stresses [128]. The neutron diffraction method has also been applied to short fiber metal matrix composites [129]. The preceding models for thermal stresses in a short fiber metal matrix composite are based on the assumption of an elastically deforming matrix. When a temperature change ΔT and/or CTE mismatch $\Delta\alpha$ are large, then the plastic deformation of the metal matrix is more likely to occur in order to relax the large elastic residual stress that would otherwise exist in the matrix. This relaxation, in the form of punched-out dislocations, has been described in Section 3.3 [42]. The plastic deformation (or equivalently the punched-out dislocations) are considered to be localized in the matrix near the matrix–fiber interface for long fiber metal matrix composites, or in the case of matrix metal being alloy instead of pure metal. There have been no models to calculate the exact thermal stress field in the matrix which deforms plastically around the matrix–fiber interface and elastically in the region outside this plastic domain, except for the case of a spherical reinforcement embedded in an infinite matrix [130].

3.6 Problems

(1) Why does the law of mixtures fail to predict the transverse modulus E_T of a unidirectional metal matrix composite? What method would you then suggest?

(2) Discuss the effect of fiber misorientation and non-uniform fiber length on the stiffness of a short fiber metal matrix composite.

(3) Discuss the pros and cons of two approaches for the strengthening mechanism in metal matrix composites: the continuum type model and the dislocation type model.

(4) Give a possible reason for the large values of σ_{yc}/σ_{ym} of annealed SiCw/Al composites compared with those of heat-treated composites (see Fig. 3.10a).

(5) In the case of continuous fiber metal matrix composites, what would be the strengthening mechanism based on a dislocation theory?

(6) What would be the further advancements in the Rosen–Zewben type fracture models modified by Fukuda [55], Kimpara and Ozaki [56], and Ochiai [57]?

(7) What is the formula for Y_b (a/w) for a chevron-notched three-point bend specimen? How can one obtain it if it is not available in the literature?

(8) What is the relation between γ_F and K_{IC} (or K_R) for a given unidirectional metal matrix composite?

(9) How do you apply the weak point theory in fibers proposed by Cooper [72] to the transverse strength σ_T of a unidirectional metal matrix composite?

(10) Discuss the difference between Piggott's model [73] and Taya et al. model [79–81] for the prediction of γ_F for a given unidirectional metal matrix composite.

(11) If a temperature drop ($\Delta T < 0$) is large enough to cause the yielding of the matrix metal, what models should be used to estimate the stress field in the matrix? In this case, what would be the tensile yield stress of the metal matrix composite?

(12) Discuss possible methods to measure the residual stresses in metal matrix composites.

(13) Discuss possible approaches to the elimination or reduction of the residual stresses in a metal matrix composite.

(14) What would be the limitation of Eshelby's method for the analysis of the size effect of particulates or whiskers on basic mechanical properties?

References

1. McDanels, D. L., Jech, R. W. and Weeton, J. W., *Trans. Metal. Soc. AIME*, Vol. 233, April 1965, pp. 636–642.
2. Kelly, A. and Lilholt, H., *Phil. Mag.*, Vol. 20, 1969, pp. 311–328.
3. Jackson, P. W., Braddick, D. M. and Walker, P. J., *Fibre Sci. Tech.*, Vol. 5, 1972, pp. 219–236.
4. Prewo, K. M. and Kreider, K. G., *Metal. Trans.*, Vol. 3, August 1972, pp. 2201–2211.
5. Webster, D., *Metal. Trans.*, Vol. 13A, 1982, pp. 1511–1519.
6. Rack, H. J., Baruch, T. R. and Cook, J. L., *Progress in Science and Engineering of Composites*, edited by T. Hayashi et al., ICCM-IV, Tokyo, Japan Soc. Comp. Mater. 1982, pp. 1465–1472.
7. Nieh, T. G. and Chellman, D. J., *Scripta Metal.*, Vol. 18, 1984, pp. 925–928.
8. Nardone, V. C. and Prewo, K. M., *Scripta Metal.*, Vol. 20, 1986, pp. 43–48.
9. Taya, M. and Arsenault, R. J., *Scripta Metal.*, Vol. 21, 1987, pp. 349–354.
10. Eshelby, J. D., *Proc. Roy. Soc. Lond.*, Vol. A241, 1957, pp. 376–396.
11. Mori, T. and Tanaka, K., *Acta Metal.*, Vol. 21, May 1973, pp. 571–574.

12. Taya, M. and Mura, T., *J. Appl. Mech.*, Vol. 48, 1981, pp. 361–367.
13. Cox, H. L., *Br. J. Appl. Phys.*, Vol. 3, 1952, p. 72.
14. Wolfenden, A. and Arsenault, R. J., unpublished work.
15. Patterson, W. G., unpublished work.
16. Takao, Y., Chou, T. W. and Taya, M., *J. Appl. Mech.*, Vol. 49, 1982, pp. 536–540.
17. Arsenault, R. J., *Mater. Sci. and Eng.*, Vol. 64, 1984, pp. 171–181.
18. Takao, Y. and Taya, M., *J. Comp. Mater.*, Vol. 21, 1987, pp. 140–156.
19. Taya, M., *J. Comp. Mater.*, Vol. 15, May 1981, pp. 198–210.
20. Takao, Y., Taya, M. and Chou, T. W., *Int. J. Solids Structs.*, Vol. 18, No. 8, 1982, pp. 723–728.
21. Daimaru, A. and Taya, M., *Progress in Science and Engineering of Composites*, edited by T. Hayashi *et al.*, ICCM-IV, Tokyo, Japan *Soc. Comp. Mater.*, 1982, pp. 1099–1106.
22. Arsenault, R. J. and Taya, M., *Acta Metal.*, Vol. 35, No. 3, 1987, pp. 651–659.
23. Hedyes, S. M. and Mitchell, J. W., *Phil. Mag.*, Vol. 44, 1953, pp. 223–224.
24. Ashby, M. F. and Johnson, L., *Phil. Mag.*, Vol. 20, 1969, pp. 1009–1022.
25. Weatherly, G. C., *Metal Sci. J.*, Vol. 2, 1968, pp. 25–27.
26. Ashby, M. F., Gelles, H. and Tanner, L. E., *Phil. Mag.*, Vol. 19, 1969, pp. 757–771.
27. Flom, Y. and Arsenault, R. J., *Mater. Sci. Eng.*, Vol. 77, 1986, pp. 191–197.
28. Vogelsang, M., Fisher, R. M. and Arsenault, R. J., *Mat. Trans.*, Vol. 17A, 1986, pp. 379–389.
29. Chawla, K. K. and Metzger, M., *J. Mater. Sci.*, Vol. 7, 1972, pp. 34–39.
30. Williams, J. J. and Garmong, G., *Metal. Trans.*, Vol. 6A, 1975, pp. 1699–1709.
31. Arsenault, R. J. and Fisher, R. M., *Scripta Metal.*, Vol. 17, 1983, pp. 67–71.
32. Hoffman, C. A., *J. Eng. Mater. Tech.*, Vol. 95, 1973, pp. 55–62.
33. Garmong, G., *Metal. Trans.* Vol. 5, 1974, pp. 2183–2190.
34. Dvorak, G. J., Rao, M. S. and Tarn, J. Q., *J. Comp. Mater.*, Vol. 7, 1973, pp. 194–216.
35. Dvorak, G. J. and Bahei-El-Din, Y. A., *J. Appl. Mech.*, Vol. 49, 1982, pp. 327–335.
36. Mehan, R. L., *Metal Matrix Composites*, ASTM STP 438, 1968, pp. 29–58.
37. Koss, D. A. and Copley, S. M., *Metal. Trans.*, Vol. 2, 1971, pp. 1557–1560.
38. Arsenault, R. J. and Shi, N. *Mater. Sci. Eng.*, Vol. 81, 1986, pp. 175–187.
39. Shi, N. and Arsenault, R. J., ASME Winter Annual Meeting, Bound Volume edited by S. S. Wang and Y. Rajapakse, in press.
40. Hansen, N., *Acta Metall.*, Vol. 25, 1977, pp. 863–869.
41. Metallurgical Materials Laboratory, University of Maryland, unpublished research, 1985.
42. Taya, M. and Mori, T., *Acta Metal.*, Vol. 35, 1987, pp. 155–162.
43. Arsenault, R. J. and Wu, S. B., *Mater. Sci. Eng.*, Vol. 96, 1987, pp. 77–88.
44. Arsenault, R. J. and Feng, C. R., to be submitted for publication.
45. Flom, Y. and Arsenault, R. J., *J. Metals*, 1986, Vol. 38, pp. 31–34.
46. Kelly, A., *Strong Solids*, 2nd edn. Oxford University Press, Oxford, 1973.
47. Hull, D., *An Introduction to Composite Materials*, Cambridge University Press, Cambridge, 1981.
48. Piggott, M. R., *Load-bearing Composites*, Pergamon Press, Oxford, 1980.
49. Kagawa, Y. and Choi, B. H., *Composites '86: Recent Advances in Japan and the United States*, edited by K. Kawata et al., Japan Society for Composite Materials, 1986, pp. 537–543.
50. Rosen, B. W., *AIAA Journal*, Vol. 2, 1964, pp. 1985–1991.
51. Zweben, C., *AIAA Journal*, Vol. 6, 1968, pp. 2325–2331.
52. Coleman, B. D., *J. Mech. Phys. Solids*, Vol. 7, 1959, pp. 60–70.
53. Daniels, H. E., *Proc. Roy. Soc. Lond.*, Vol. A183, 1945, pp. 405–435.
54. Hedgepeth, J. M., *Stress Concentrations in Filamentary Structures*, TN D-882, May 1961, NASA.
55. Fukuda, H., *Recent Advances in Composites in the United States and Japan*, ASTM STP 864, eidted by J. R. Vinson and M. Taya, American Society for Testing and Materials, Philadelphia, 1985, pp. 5–15.
56. Kimpara, I. and Ozaki, T., *Composites '86: Recent Advances in Japan and the United States*, edited by K. Kawata et al., Japan Society for Composite Materials, 1986, pp. 93–100.
57. Ochiai, S. and Osamura, K., ibid., pp. 751–759.
58. Kyono, T., Hall, I. W. and Taya, M., *J. Mater. Sci.*, Vol. 21, 1986, pp. 4269–4280.
59. Jackson, P. W. and Cratchley, D., *J. Mech. Phys. Solids*, Vol. 14, 1966, pp. 49–64.
60. Azzi, V. D. and Tsai, S. W. *Exp. Mech.* Vol. 5, 1965, pp. 283–288.

61. Friend, C. M., *Proc. ICCM-6/ECCM-2*, edited by F. L. Matthews et al., Elsevier, Vol. 2, 1987, pp. 2.402–2.411.
62. Sakamoto, A., Hasegawa, H. and Minoda, Y., *Proc. ICCM-5*, edited by W. C. Harrigan, Jr., et al., AIME, 1985, pp. 699–707.
63. Kohara, S. and Muto, N., *Composites '86: Recent Advances in Japan and the United States* edited by K. Kawata et al., Japan Society for Composite Materials, 1986, pp. 491–496.
64. Kelly, A., *Proc. Roy. Soc. Lond.*, Vol. A282, 1964, pp. 63–79.
65. Kelly, A. and Tyson, W. R., *J. Mech. Phys. Solids*, Vol. 13, 1965, pp. 329–350.
66. Kelly, A., *Proc. Roy. Soc. Lond.* Vol. A319, 1970, pp. 95–116.
67. Broek, D., *Elementary Engineering Fracture Mechanics*, 3rd edn, Martinus Nijhoff, 1982, Chapter 7.
68. Dadkhah, M. S., unpublished work.
69. Wright, M. A., Welch, D. and Jollay, J., *J. Mater. Sci.*, Vol. 14, 1979, pp. 1218–1228.
70. Kanninen, M. F., Rybicki, E. F. and Brinson, H. F., *Composites*, January 1977, pp. 17–22.
71. Shibata, S., Mori, T. and Mura, T., *Mechanical and Physical Behavior of Metallic and Ceramic Composites*, edited by S. I. Andersen, et al, Risø National Laboratory Denmark, 1988, pp. 469–474.
72. Cooper, G. A., *J. Mater. Sci.*, Vol. 5, 1970, pp. 645–6545.
73. Piggott, M. R., *J. Mater Sci.*, Vol. 5, 1970, pp. 669–675.
74. Ishikawa, H. and Taya, M., *Progress in Science and Engineering of Composites*, edited by T. Hayashi et al., Japan Society for Composite Materials 1982, pp. 675–680.
75. Nakayama, J., *Amer. Ceram. Soc. J.*, Vol. 48, 1965, pp. 583–587.
76. Tattersall, H. G. and Tappin, G., *J. Mater Sci.*, Vol. 1, 1966, pp. 296–301.
77. Gordon, J. E. and Jeronimidis, G., *Phil. Trans. Roy. Soc. Lond.*, Vol. A294, 1980, pp. 545–550.
78. Skinner, A., Koczak, M. J. and Lawley, A., *Metal. Trans.*, Vol. 13A, 1982, pp. 289–297.
79. Taya, M. and Daimaru, A., *J. Mater. Sci.*, Vol. 18, 1983, pp. 3105–3116.
80. Daimaru, A., Hata, T. and Taya, M., *Recent Advances in Composites in the United States and Japan*, ASTM STP 864, edited by J. R. Vinson and M. Taya, American Society for Testing and Materials, Philadelphia, 1985, pp. 505–521.
81. Kyono, T., Hall, I. W. and Taya, M., *J. Mater. Sci.*, Vol. 21, 1986, pp. 1879–1888.
82. Divecha, A. P., Fishman, S. G. and Karmarkar, S. D., *J. Metals*, 1981, Vol. 33, pp. 12–17.
83. Logsdon, W. A. and Liaw, P. K., Tensile, fracture toughness and fatigue crack growth rate properties of SiC whisker and particulate reinforced Al metal matrix composites, Westinghouse Scientific Paper 83-1D3-NODEM-P1, December 1983.
84. Crowe, C. R., Gray, R. A. and Hasson, D. F., *Proc. of ICCM-5*, edited by W. C. Harrigan Jr. et al., AIME, 1985, pp. 843–866.
85. McDanels, D. L., *Met. Trans.*, Vol. 16A, 1985, pp. 1105–1115.
86. Nieh, T. G., Rainen, R. A. and Chellman, D. J., *Proc. ICCM-5*, edited by W. C. Harrigan, Jr. et al., AIME, 1985, pp. 825–842.
87. Flom, Y. and Arsenault, R. J., *Proc. ICCM-6/ECCM-2*, Vol. 2, edited by F. L. Matthews et al., Elsevier, 1987, pp. 189–198.
88. Edelson, B. I. and Baldwin, W. M., *Trans. ASM*, Vol. 55, 1962, pp. 230–250.
89. Gurland, J. and Plateau, J., *Trans. ASM*, Vol. 56, 1963, pp. 442–454.
90. Gangullee, A. and Gurland, J., *Trans. Met. Soc. AIME*, Vol. 239, 1967, pp. 269–272.
91. Gurland, J., *Trans. Met. Soc. AIME*, Vol. 227, 1963, pp. 1146–1150.
92. Liu, C. T. and Gurland, J., *Trans. Met. Soc. AIME*, Vol. 224, 1968, pp. 1535–1542.
93. Liu, C. T. and Gurland, J., *Trans. ASM*, Vol. 61, 1968, pp. 156–167.
94. Gurland, J., *Fracture of Metal-Matrix Particulate Composites, Modern Composite Materials*, edited by L. J. Broutman and R. H. Krock, Addison-Wesley, 1969.
95. Mogford, I. L., *Met. Reviews*, Vol. 12, 1967, pp. 49–65.
96. Van Stone, R. H., Merchant, R. H. and Low, J. R., *ASTM STP* 556, 1974, pp. 93–124.
97. Cox, T. B. and Low, J. R., *Metal. Trans.*, Vol. 5, 1974, pp. 1457–1470.
98. Hahn, G. T. and Rosenfield, A. R., *Metal. Trans. A*, Vol. 6, 1975, pp. 653–668.
99. Garret, G. G. and Knott, J. F., *Metal. Trans. A*, Vol. 9, 1978, pp. 1187–1201.
100. Schwalbe, K. H., *Eng. Frac. Mech.*, Vol. 9, 1977, pp. 795–832.
101. LeRoy, G., Embury, J. D., Edwards, G. and Ashby, M. F. *Acta Metal.*, Vol. 29, 1981, pp. 1509–1522.
102. Bates, R. C., Modeling of Ductile fracture by microvoid coalescence for prediction of

fracture toughness, *Proc. 113th AIME Conf.*, edited by J. M. Wells and J. B. Landes, 1985, pp. 117–155.
103. Gerberich, W. W., *Proc. 113th AIME Conf.*, edited by J. M. Wells and J. B. Landes, 1985, pp. 49–74.
104. Firrao, D. and Roberti, R., *Proc. 113th AIME Conf.*, edited by J. M. Wells and J. B. Landes, 1985, pp. 165–177.
105. Longston, W. A. and Liaw, P. K., Westinghouse Sci. Paper, 83-1D3-NODEM-P1, December 1983.
106. Schoutens, J. E., *Discontinuous Silicon Carbide Reinforced Aluminum Metal Matrix Composites Data Review*, MMCIAC Pub. No. 000461, 1984.
107. Hsieh, R. J., Effect of Precracking Procedure on Fracture Toughness of SiC/Al Composite, M.S. paper, University of Maryland, 1985.
108. Nutt, S. R. and Duva, J. M., *Scripta Metal.*, Vol. 20, 1986, pp. 1055–1058.
109. Arsenault, R. J. and Flom, Y., *Proc. Symp., Phase Boundary Effects on Deformation*, TMS AIME, Toronto, Canada, 1985, pp. 13–17.
110. Flom, Y., Parker, B. H. and Chu, C., Paper in preparation, NASA, Goddard Space Flight Center, 1987.
111. Underwood, E. E., *Quantitative Stereology*, Addison-Wesley, 1970.
112. Flom, Y. and Arsenault, R. J., *Acta Metal*, in press.
113. Laszlo, F., *J. Iron Steel Inst.*, Vol. 148, 1943, pp. 173–199.
114. Hecker, S. S., Hamilton, C. H. and Ebert, L. J., *J. Mater.*, Vol. 5, No. 4, 1970, pp. 868–900.
115. Uemura, M., Iyama, H. and Yamaguchi, *J. Thermal Stress*, Vol. 2, 1979, pp. 393–412.
116. Iesan, D., *J. Thermal Stress*, Vol. 3, 1980, pp. 495–508.
117. Christensen, R. M. and Lo, H., *J. Mech. Phys. Solids*, Vol. 27, 1979, pp. 315–330.
118. Mikata, Y. and Taya, M., *J. Comp. Mater.*, Vol. 19, 1985, pp. 554–578.
119. Dvorak, G. J., *J. Appl. Mech.*, Vol. 53, 1986, pp. 737–743.
120. Tsai, S. W. and Hahn, H. T., *Introduction to Composite Materials*, Technomic Pub. Co., p. 389.
121. Min, B. K., *J. Mech. Phys. Solids*, Vol. 29, No. 4, 1981, pp. 327–352.
122. Min, B. K. and Crossman, F. W., *Composite Materials: Testing and Design*, 6th Cont. ASTM STP 788, edited by I. M. Daniel, 1982, pp. 371–392.
123. Tsai, S. D., Mahulikar, D., Marcus, H. L., Noyan, I. C. and Cohen, J. B., *Mater. Sci. Eng.*, Vol. 47, 1981, pp. 145–149.
124. Weng, G. J., *Int. J. Eng. Sci.*, Vol. 22, 1984, pp. 845–856.
125. Brown, L. M. and Stobbs, W. M., *Phil. Mag.*, Vol. 23, 1971, pp. 1185–1199.
126. Pedersen, O. B., *Acta Metal.*, Vol. 31, 1983, pp. 1795–1808.
127. Ledbetter, H. M. and Austin, M. W., *Mater. Sci. Eng.*, Vol. 89, 1987, pp. 53–61.
128. Allen, A. J., Hutchings, M. T., Windsor, C. G. and Audreani, C., *Adv. Phys.*, Vol. 34, No. 4, 1985, pp. 445–473.
129. Withers, P. J., Stobbs, W. M., Jensen, D. J. and Lilholt, H., *Proc. ICCM-6/ECCM-2*, edited by F. L. Matthews et al., Elsevier, Vol. 2, 1987, pp. 2.255–2.264.
130. Lee, J. K., Earmme, Y. Y., Aaronson, H. I. and Russell, K. C., *Metal. Trans.*, Vol. 11A, 1980, pp. 1837–1847.

CHAPTER 4

Mechanical Behavior in Use Environments

4.1 Introduction

Although the basic mechanical behavior of a metal matrix composite is proving to be promising, application of metal matrix composites to actual structural components depends on the mechanical behavior in the use environment. In fact, most of the new metal matrix composites that possess good to excellent basic mechanical properties have often been rejected due to their poor performance in the use environment. The use environments that will be discussed in this chapter are thermal, impact, fatigue, wear, oxidation corrosion and radiation environments. Among the above use environments, the thermal environment is considered to be the most important and it includes isothermal, creep, and thermal cycling, each of which will be discussed extensively in this chapter. The results of recent advances in the other use environments will be given in a concise manner at the end of this chapter.

4.2 Mechanical Behavior in Thermal Environments

Most of the metal matrix composites are designed to be used in a thermal environment, for example, as components of turbine engines [1]. Many metal matrix composites undergo high- to room-temperature excursions during their processing. Thus, assessment of the mechanical behavior of a metal matrix composite in various thermal environments will be a prerequisite before a component made of the metal matrix composite is successfully designed. The thermal environments that will be discussed in this section are isothermal exposure, constant load (creep), and thermal cycling. An oxidation environment which often involves high temperature will be discussed in Section 4.4, along with corrosion environments.

4.2.1 Isothermal exposure

The mechanical behavior of metal matrix composites subjected to isothermal temperature (high temperature) has been well studied, with the main focus on high-temperature stiffness and strength of continuous fiber metal matrix composites. Most of the previous studies identified that the degradation of metal matrix composites exposed to high temperature is attributed to a change of the matrix–fiber interface or/and fiber itself.

Since the interfacial reactions play the key role in the degradation of metal matrix composites exposed to high temperature, studies of the interfacial reactions have been conducted reasonably well for the popular metal matrix composite systems, such as boron fiber/alpha titanium alloy composites [2] and boron fiber/aluminum composites [3]. It was found in these studies that only one or two main reaction products for a given metal matrix composite were identified by x-ray diffraction analysis and the thickness x of the reaction product follows a usual diffusion type equation [2].

$$x^2 = 4Dt \tag{4.1}$$

where D is the diffusion coefficient and t is the time. The thickness x of the reaction product was found to play a crucial role in reducing the strength of a continuous fiber metal matrix composite.

Metcalfe and Klein [2] proposed a simple analytical model to predict the strain at failure (e_f) of a continuous fiber metal matrix composite exposed to high temperature (T) for some time (t). A boron fiber/α-titanium composite will be considered for the purposes of illustration of the Metcalfe–Klein model. The theoretical strength of a boron fiber σ_{th} is approximated by

$$\sigma_{th} = E_f/10 \tag{4.2}$$

where E_f is the Young's modulus of the boron fiber. A typical boron fiber presumably contains intrinsic defects whose stress concentration factor K_f is given by

$$K_f = \sigma_{th}/\sigma_B \tag{4.3}$$

where σ_B is the strength of a boron fiber with intrinsic defects. On the other hand, a reaction product TiB_2 which is between the boron fibers and the titanium matrix will be cracked easily upon initial loading. Then the stress concentration factor K_r of the reaction zone with such a crack is given by

$$K_r = B(x/r)^{1/2} \tag{4.4}$$

where B depends on the stress distribution around the crack, particularly on the degree of support provided by the matrix, x is the size of a crack and r is its root radius. For a given B and r, K_r increases with crack size x which is approximately equal to the thickness of the reaction zone (TiB_2 in this case). Figure 4.1 illustrates this case where cracks of size x are located in the TiB_2 reaction zone. When the reaction zone is small, the stress concentration factor of the cracks in the reaction zone K_r also remains small, thus $K_r < K_f$. In this case, boron fibers fracture at intrinsic defects. As the reaction zone grows, thus increasing the crack size x, the stress concentration factor K_r

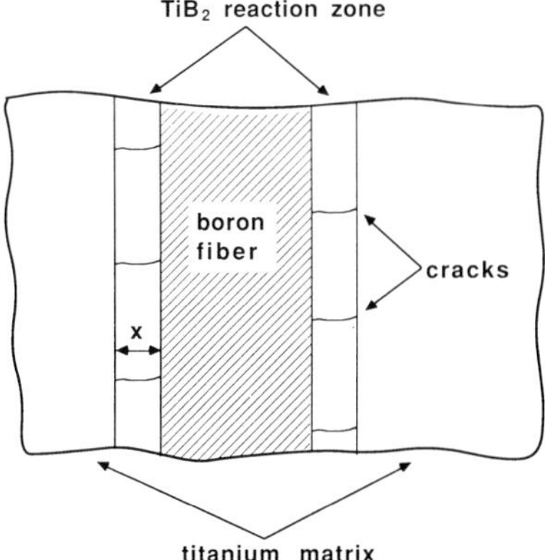

FIG. 4.1 Schematic view of cracks in the reaction zone (TiB_2) in a continuous boron fiber/titanium composite where the width of cracks x is equal to the reaction zone thickness [2].

reaches K_f with the critical crack size x_{c1} given by

$$x_{c1} = \left(\frac{E_f}{10B\sigma_B}\right)^2 r \tag{4.5}$$

For the thickness of the reaction zone $x > x_{c1}$, cracks will be formed first in the reaction zone and then induce the fracture of boron fiber following the Griffith fracture criterion. In this case the fracture strain e_f is given by

$$\frac{\sigma_B}{E_f} = \frac{1}{10B}\left(\frac{r}{x}\right)^{1/2} \tag{4.6}$$

In the case of boron fiber, the fracture strain e_f is proportional to $1/\sqrt{x}$, as seen from eq. (4.6). As the reaction zone grows further ($x = x_{c2}$), the fracture strain of the boron fiber e_f reduces to that of the reaction zone, e_r given by

$$e_r = \frac{\sigma_r}{E_r} \tag{4.7}$$

where σ_r and E_r are the strength and modulus of the reaction zone. The critical thickness (x_{c2}) of the reaction zone in this case can be obtained by equating eq. (4.6) to eq. (4.7), and it is given by

$$x_{c2} = \left(\frac{E_r}{10B\sigma_r}\right)^2 r \tag{4.8}$$

The relationship between the reaction zone thickness (x) and the fracture strain (e_f) of a boron fiber with a TiB_2 reaction zone is shown in Fig. 4.2 where $B = 1$ and $r = 3$ Å were assumed. Figure 4.2 clearly illustrates three stages for fracture strain of a continuous fiber/titanium composite, e_F as a

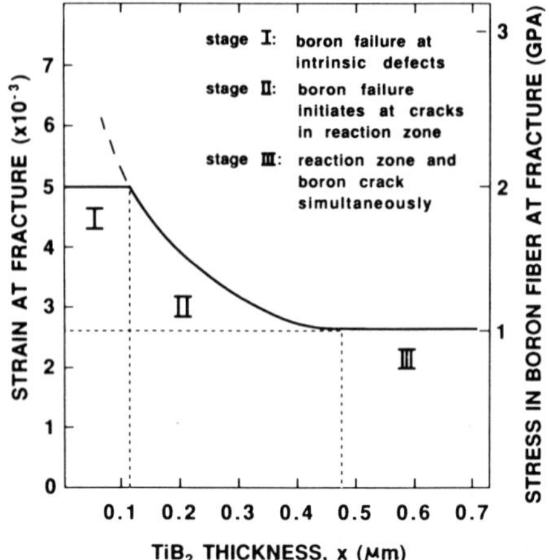

FIG. 4.2 Strain at fracture vs. reaction zone thickness x for a boron fiber/titanium composite. The relationship is characterized by three stages: for Stage I, boron failure at intrinsic defects; for Stage II, boron failure initiates at cracks in the reaction zone; and for Stage III, the boron and the reaction zone crack simultaneously [2].

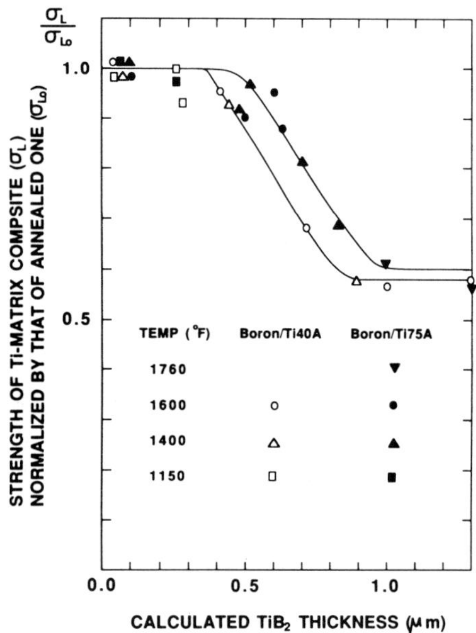

FIG. 4.3 Effect of reaction zone thickness (TiB_2) on the strength of continuous boron fiber/titanium composites, Ti(40A) and Ti(75A). The strength σ_L is normalized by that of the annealed composite (σ_{L0}).

function of a reaction zone thickness, x, i.e., stage I for $0 \leqslant x \leqslant x_{c1}$; stage II for $x_{c1} \leqslant x \leqslant x_{c2}$ and stage III for $x_{c2} \leqslant x$. The experimental results from continuous boron fiber/Ti40A and Ti75A composites seem to support the existence of such three stages (Fig. 4.3) where the TiB_2 thickness x was calculated based on eq. (4.1) and the strength of the composite σ_L is normalized by that of the annealed composite σ_{L0}. The model proposed by Metcalfe and Klein assumed that the thickness of the reaction zone is uniform, which appears to be the case in the boron fiber/titanium composite. For other metal matrix composites, non-uniform reaction zones have been reported; for example, in boron fiber/6061 aluminum [2] and a boron/1100 aluminum matrix composite [3, 4]. Figure 4.4 shows the surfaces of boron fibers extracted from boron fiber/1100 aluminum composite where (a) and (b) denote the cases of as-received and as-exposed composites at 500 °C for 24 hours, respectively [4]. It is obvious from Fig. 4.4(b) that the reaction zone (made of AlB_2 in this case) is not uniform. Despite the non-uniform reaction zone for the boron fiber/aluminum composite, DiCarlo [5] attempted to measure the reaction zone thickness x of the born fiber/6061 aluminum composite exposed at 504 °C for various times t. The measured x are plotted against $\sqrt{t}$ in Fig. 4.5. The results of Fig. 4.5 support the growth kinetic model of parabolic type, eq. (4.1), which can be expressed in terms of time and temperature (T).

$$x = \alpha\sqrt{t}\,\exp\{-Q/(2kT)\} \tag{4.9}$$

where α is a normalizing constant that can be determined from the experiment, Q is the activation energy controlling reaction zone growth, k is the Boltzmann's constant and T is the exposure temperature. The growth kinetic model similar to eq. (4.9) was also found to be applicable to the reaction product (Al_4C_3) for carbon fiber/aluminum composite [6]. DiCarlo [5] derived the formula to predict the average fiber strength σ_{fb}, for the stage II fracture model (reaction zone control, Fig. 4.1) by substituting eq. (4.9) into eq. (4.6).

$$\bar{\sigma}_{fb}' = ct^{-1/4}\exp\{Q/(4kT)\} \tag{4.10}$$

where c is an empirically determined constant. In the case of continuous fiber metal matrix composites, the fracture of fibers will lead directly to the fracture of the composite. Assuming that the fiber strength in the composite at failure σ_L is given by eq. (3.37) and following the Rosen's cumulative weakening model [7], DiCarlo [5] derived the composite fracture strain e_c as

$$e_c = \sigma_L^*/E_f = \bar{\sigma}_{fb}(l)\,\lambda/E_f \tag{4.11}$$

where

$$\lambda = \left(\frac{l}{l_c}\right)^{1/\beta}\{(\beta e)^{-1/\beta}/\Gamma(1+1/\beta)\} \tag{4.12}$$

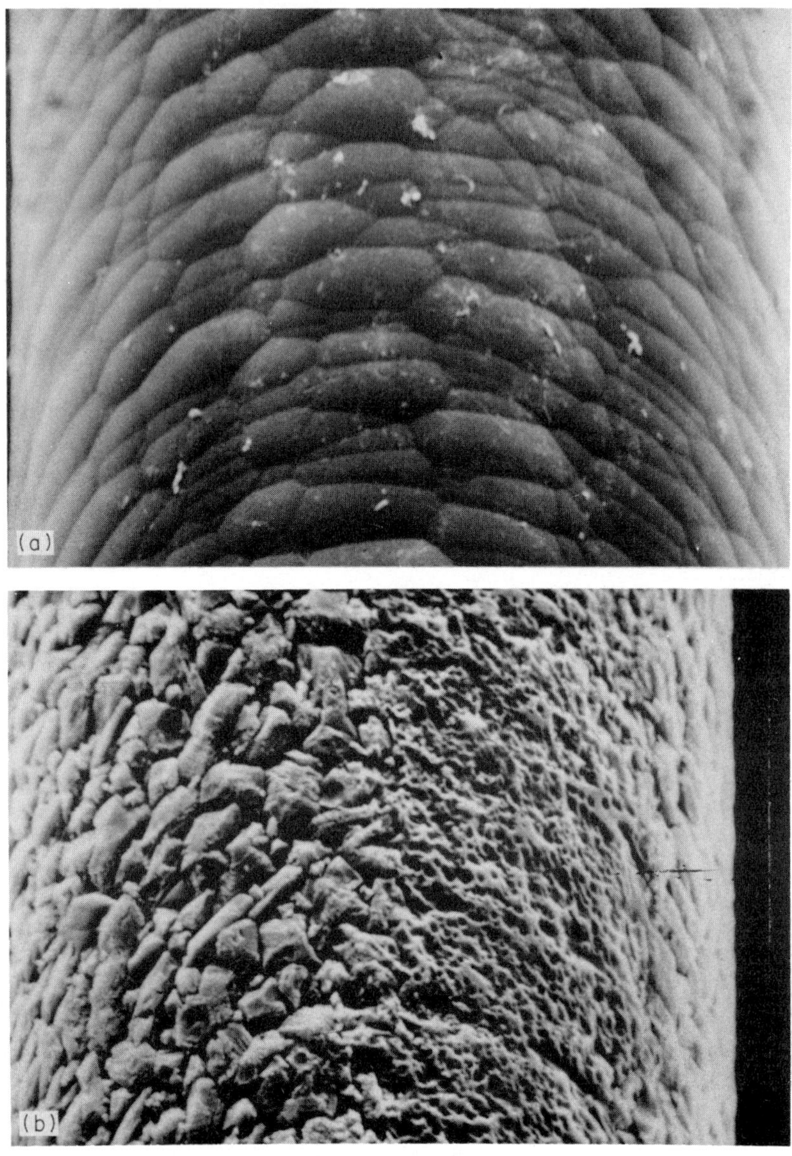

FIG. 4.4 SEM photos of the surfaces of boron fibers extracted from a boron fiber/1100 aluminum composite, (a) as-received, and (b) exposed at 500 °C for 24 hours [4].

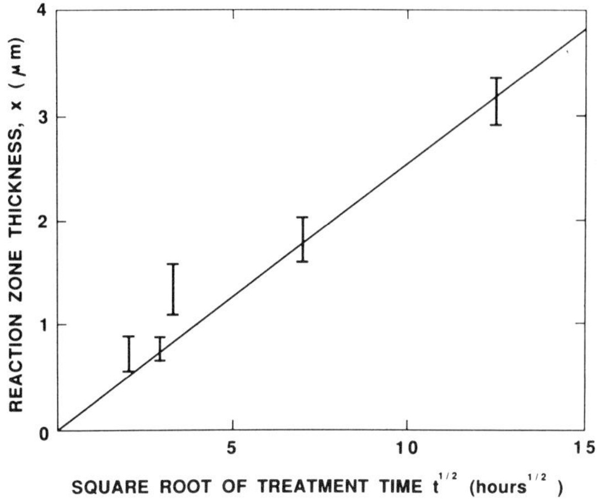

FIG. 4.5 Reaction zone thickness x (in μm) vs. $\sqrt{t}$ (in $\sqrt{\text{hour}}$) for a boron fiber/6061 aluminum composite exposed at 504 °C for duration t (in hours) [5].

and where $\bar{\sigma}_{fb}(l)$ is the average fiber strength where a number of nonfilaments of length l are tested individually; this is given by eq. (3.28), which is a two-point parameter Weibull equation, and l_c is the ineffective length (which is equal to the critical fiber length defined by eq. (3.39)). It should be noted in eq. (4.11) that the fiber bundle strength σ_L^* is related to the average fiber strength σ_{fb} as

$$\sigma_L^* = \lambda \bar{\sigma}_{fb} \tag{4.13}$$

To verify the above equations, i.e., eqs. (4.10) and (4.11), DiCarlo plotted the experimental data of fiber strength as a function of the reciprocal absolute

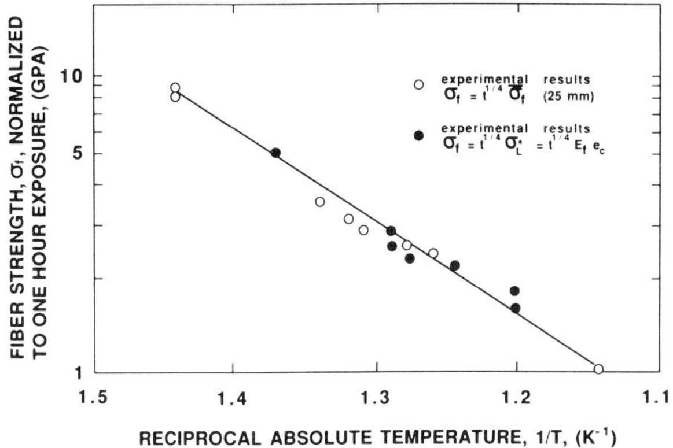

FIG. 4.6 Fiber strength vs. $1/T$ (in K^{-1}) for various boron fiber/aluminum composites [5].

temperature in Fig. 4.6. In Fig. 4.6, open circles denote the experimental results of aluminum-coated fibers [8] and as-extracted fibers [9, 34] calculated from $\sigma_f = t^{1/4}\ \sigma_{fb}$, and filled circles for the fiber bundle strength calculated from $\sigma_L = E_f e_c$ where e_c is the axial fracture strain data of multifilament boron fiber/aluminum composites [2, 10] and E_f is the fiber modulus taken as 400 GPa. To account for differences in exposure time t (hours), all strengths are normalized to a 1-hour exposure by multiplying the experimental data by $t^{1/4}$. It can be concluded from Fig. 4.6 that the above model, which is a combination of the interaction (or reaction) zone theory proposed by Metcalfe and Klein [2], the Rosen's cumulative weakening model [7] and the DiCarlo's reaction zone kinetic model, appears to be reasonably accurate in predicting the strength of a continuous fiber metal matrix composite exposed at high temperature. The fact that open and filled circles follow the same line in Fig. 4.6 indicates $\lambda = 1$ in eq. (4.13), i.e., the fiber bundle strength is equal to the average fiber strength.

The average fiber strength $\bar{\sigma}_{fb}$ predicted by eq. (4.10) is considered to be important basic data, and it should be used for estimating the mechanical properties of as-thermally exposed metal matrix composites. Kyono et al. [4] measured the room temperature work of fracture γ_F (defined by eq. (3.84)) of continuous boron fiber/1100 aluminum composites that were exposed to isothermal environment. The experimental data of γ_F for 30% V_F boron/1100

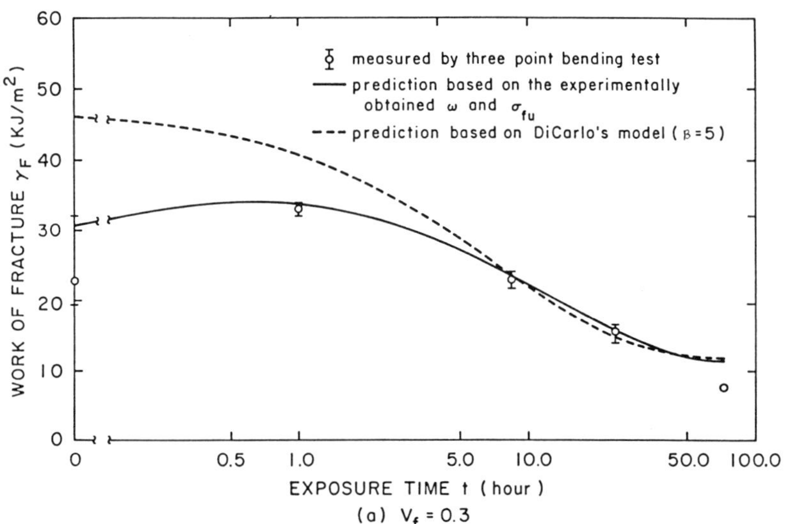

FIG. 4.7 Room temperature work of fracture γ_F of a continuous boron fiber/1100 aluminum composite that was exposed at 500 °C vs. exposure time; experimental results (open circles), prediction by eqs. (3.66) and (3.67) based on DiCarlo's results (dashed curve) [15] and the average strength of the fibers extracted from the composite [4].

aluminum are plotted against exposure time t (hours) as open circles in Fig. 4.7, where two predictions by eqs. (3.85) and (3.86) are shown, one (dashed curve) based on the average fiber stress $\bar{\sigma}_{fb}$ predicted by DiCarlo's model (i.e., Fig. 4.6) where a Weibull modulus $\beta = 5$ was used, and the other (solid curve) based on the $\bar{\sigma}_{fb}$ obtained experimentally which gave rise to $\beta > 5$. In the above predictions, σ_{fb} and σ^*_{fb} in eq. (3.86) are assumed to be the average fiber strength $\bar{\sigma}_{fb}$ and fiber bundle strength σ^*_L, respectively. In the experiment by Kyono et al., $\sigma^*_L < \sigma_{fb}$, due mainly to a relatively large Weibull modulus. Despite some discrepancy between the solid and dashed curves, the DiCarlo's model for predicting the average fiber strength weakened by isothermal exposure has proven to be quite useful in prediction of the mechanical properties of continuous fiber metal matrix composites that are degraded under isothermal exposure.

As discussed above, isothermal exposure at higher temperature and for longer periods of time tends to induce a thicker reaction zone at the matrix–fiber interface. Thus, once a crack within a continuous fiber metal matrix composite has started propagating along the crack plane perpendicular to the fiber axis under a critical applied stress σ_L, it tends to keep propagating along the same crack plane and under the same applied stress if thicker reaction zones were nucleated at the majority of the interfaces. This is particularly true for higher fiber volume fractions. Therefore, the fracture surfaces of continuous fiber metal matrix composites that have undergone isothermal exposure appear to be flat. Figure 4.8(a) and (b) show the fracture surfaces of a boron fiber/1100 aluminum composite for (a) as-fabricated and (b) exposed at 500 °C for 72 hours [11]. The isothermal exposure is expected to have a strong effect on the transverse strength of a continuous fiber metal matrix composite since the transverse strength is controlled directly by the matrix–fiber interface, which itself consists of the reaction bond strength. The formation and growth of the reaction zone is known to increase the interfacial bonding, hence the longer exposure time appears to increase the transverse strength (σ_T) of the composite. This trend is indeed illustrated in Fig. 4.9 where the σ_T of a thermally exposed boron fiber/1100 aluminum composite with two different fiber volume fractions $V_f = 0.3$ (marked by open circles) and 0.5 (marked by open triangles) are plotted as a function of exposure time t at 500°C [11]. The interfacial bonding enhanced by isothermal exposure also influences the fracture mode in the transversely tensile tested composite. This is clearly seen in Fig. 4.10, where (a) and (b) denote the fracture surfaces of the transversely tensile tested as-fabricated and exposed boron fiber/1100 aluminum composite at 500°C for 72 hours [11], respectively. Namely, the as-fabricated composite fractured at the weak matrix–fiber interfaces (interfacial debonding), while the as-thermally exposed composite fractured by a combination of fiber splitting and interfacial debonding.

The mechanical behavior of short fiber metal matrix composites exposed to high temperature have not been studied to the same degree as continuous

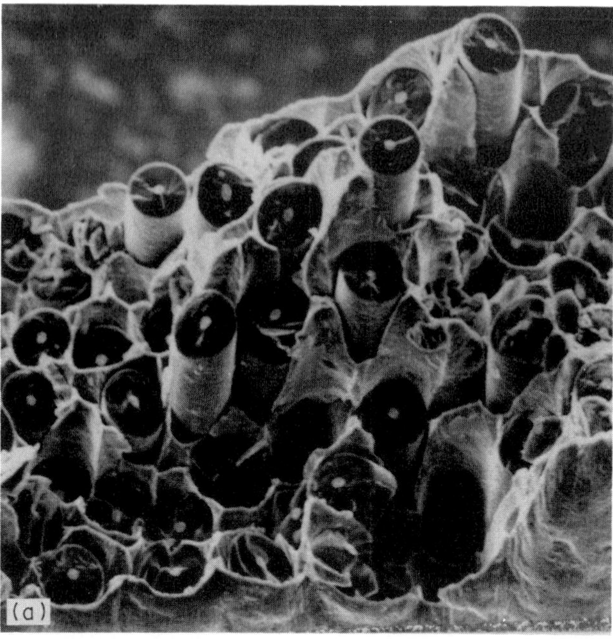

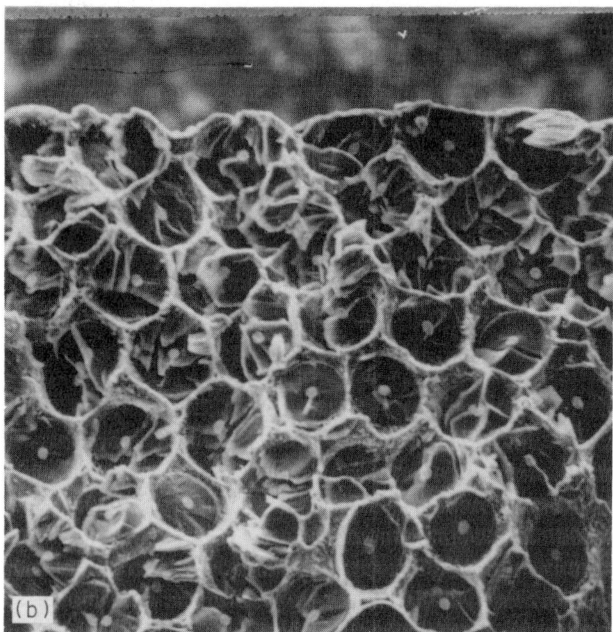

FIG. 4.8 SEM photos of the fracture surfaces of the longitudinally tensile tested boron fiber/1100 aluminum composite for (a) as-fabricated and (b) exposed at 500 °C for 72 hours [11].

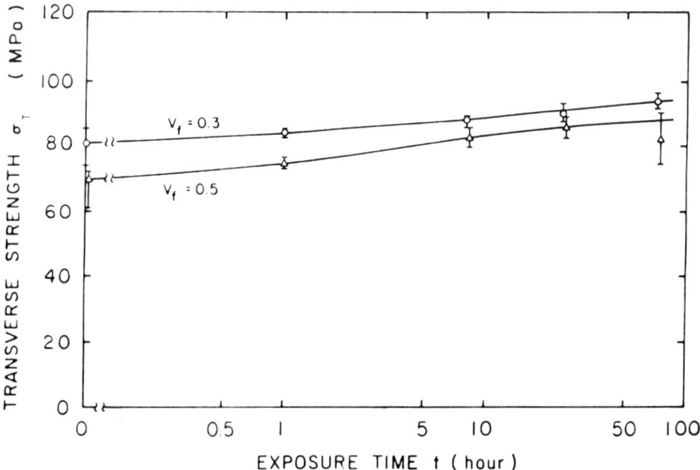

FIG. 4.9 Transverse strength (σ_T) of a thermally exposed boron fiber/1100 aluminum composite vs. exposure time t at 500 °C; (○) $V_f = 0.3$ and (△) $V_f = 0.5$.

fiber metal matrix composites. The lack of mechanical data for thermally exposed short fiber metal matrix composites is attributed to the fact that short fiber metal matrix composites are considered primarily for use in the room to intermediate temperature range, where the degradation at the fiber–matrix interfaces is not so serious. Experimental findings based on a limited number of investigations [12, 13] revealed that the mechanical behavior of short fiber and particulate metal matrix composites exposed to intermediate temperatures is proportional to that of unreinforced metal exposed to the same isothermal condition. For example, as the temperature increases the flow stress of the unreinforced matrix metal decreases, leading to a reduction in the flow stress of a short fiber or particulate metal matrix composite, as easily predicted by the law of mixture equation, eq. (3.3).

4.2.2 Creep behavior

Assessment of the creep behavior of a candidate metal matrix composite for high temperature application is a prerequisite for a component design engineer who wants to use such a candidate metal matrix composite. The creep data of metal matrix composites have often been summarized as the relationship between stress to rupture and time to rupture for a given high temperature, which is considered useful in practice, but it does not provide the vital information by which the mechanism of creep resistance of a metal matrix composite can be understood. To this end it is necessary to obtain creep data in terms of the creep strain–time curve. A typical creep strain–time curve of a 25% V_f SiC whisker/2124 aluminum composite tested at tempera-

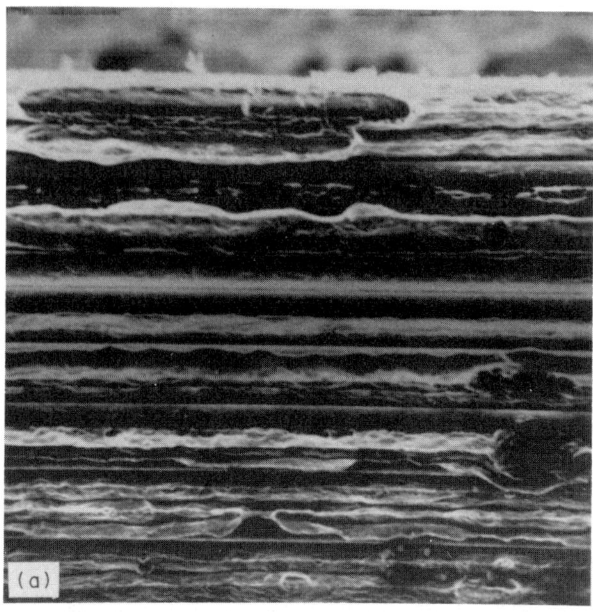

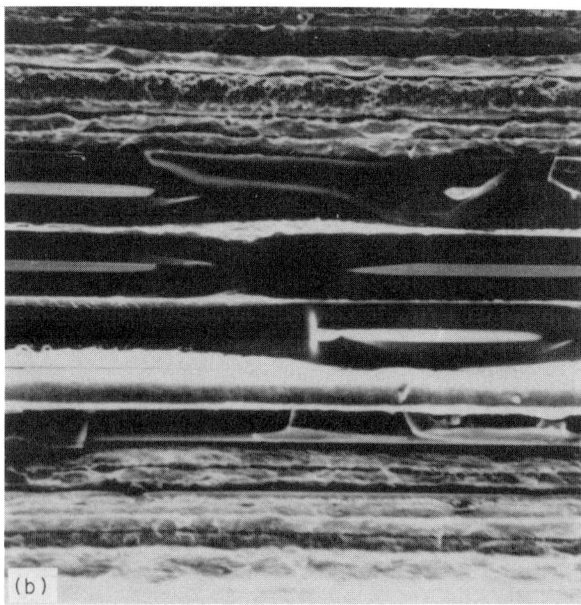

FIG. 4.10 SEM photographs of the fracture surfaces of the transversely tensile tested B/Al for (a) as fabricated and (b) exposed at 500 °C for 72 hours.

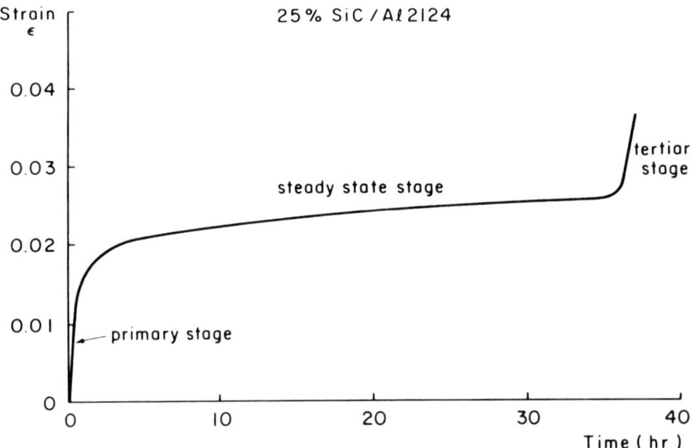

FIG. 4.11 A typical creep strain–time curve for a 25% V_f SiC whisker/2124 aluminum composite where three stages are clearly distinguishable [14].

ture (T) 300°C and applied stress (σ_c) 50 MPa is shown in Fig. 4.11 [14], which clearly exhibits three stages: the first (primary), second (steady state) and third (tertiary) stages. The existence of these three distinct stages depends on T, σ_c and the metal matrix composite system. For example, most of the creep strain—time curves of eutectic (or *in-situ*) composites do not possess these three distinct stages, but consist of an elongated 'S'-shape curve, the details of which have been discussed by Mclean and his co-workers [15–17], and thus will not be discussed in this section. However, the majority of metal matrix composites appear to possess the three stages, partly due to the fact that at these test temperatures the matrix metal creeps, but the reinforcing fiber does not, or creeps very little. This is particularly true for ceramic fiber/metal matrix composites. Among the above three stages, the second stage is considered most important since its duration is the longest. The slope of the second stage is usually constant and it is termed the 'steady state creep rate,' or simply 'creep rate.' The main reason for using non-creeping fiber or less-creeping fiber (more likely refractory fiber) as a reinforcement is to reduce the overall creep rate of the metal matrix composite, thus increasing the creep resistance. Figure 4.12 demonstrates the effect of the reinforcing fiber, i.e., the creep rate of a tungsten fiber/Ag matrix composite being much less than that of the unreinforced Ag matrix metal for a given temperature and applied stress [18]. Therefore, consideration of the creep behavior of metal matrix composites has been centered on a better description of their steady-state creep rates. The main objective of this section is to study the mechanism of the steady-state creep rate of metal matrix composites, which can be described by appropriate simple models.

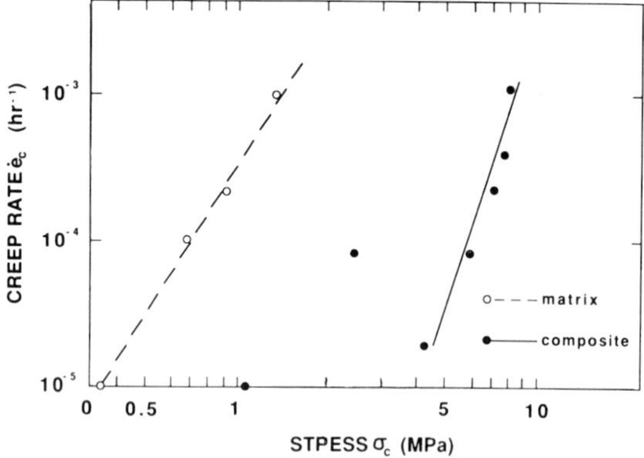

FIG. 4.12 Creep rate (steady-state creep rate) vs. applied stress for a tungsten fiber/Ag matrix composite tested at temperature 600 °C. [18].

As discussed briefly in Section 2.1, the steady-state creep rate ($\dot{e}_m$) of metals has been described mainly by two creep laws, power and exponential laws. The former creep law is well suited for the region of low to intermediate stress level which is typically dominant at higher temperatures, and the latter for the region of higher stress level; it is thus important at relatively lower temperatures. One of the important tasks of the analysis of the steady-state creep of a metal matrix composite is to determine which creep law the experimental data of $\dot{e}_m$ would follow. Then, the important issue is can we predict $\dot{e}_c$ of the composite from given creep-rate data of the unreinforced metal ($\dot{e}_m$) and reinforcing fiber? If we can, by using an appropriate model, then the model can be used to tailor a better metal matrix composite component for high-temperature application which has higher creep resistance.

Models to predict composite creep rate

Analytical models to predict the (steady-state) creep rate of metal matrix composites have centered on the aligned short fiber system [19–25], for the creep rate of aligned short fiber metal matrix composite is much larger than that of continuous fiber metal matrix composites; yet it is much less than that of the unreinforced matrix, thus occupying the practically important creep rate range, and the creep-rate of continuous fiber metal matrix composites can be predicted in a straightforward manner (such as by the law of mixtures type model), and therefore does not require any rigorous analysis. However, the assumption of aligned short fibers does not represent actual cases of a short fiber metal matrix composite where short fibers are usually misoriented. Nevertheless, the model of an aligned short fiber metal matrix composite seems to be the easiest one to use, and provides the upper bound on the creep rate of the composite. Mileiko [20] and Mclean [21] considered a simple shear lag model where rigid short fibers are embedded in the creeping matrix in shear. Kelly and Street [22] extended these models to include tensile creep of the matrix and also considered the case of creeping fibers. In these models the matrix creep rate is assumed to obey a power law. Lilholt [24] modified the Kelly and Street model to include the contribution of friction and mean stresses, and also used several matrix creep laws including the exponential type law. Taya and Lilholt [25] extended the above models to account for debonded interfaces which appear to take place under higher applied stress. For perfectly bonded interefaces the Taya and Lilholt model is basically reduced to the Kelly and Street type model, although in the Taya and Lilholt model the matrix creep law is assumed to follow the exponential law given by

$$\dot{e}_{\mathrm{m}}/\dot{e}_{\mathrm{m0}} = \exp(\sigma_{\mathrm{m}}/\sigma_{\mathrm{m0}}) \tag{4.14}$$

where $\dot{e}_{\mathrm{m}}$ and σ_{m} are the creep rate and stress in the matrix, $\dot{e}_{\mathrm{m0}}$ and σ_{m0} are some reference values for the purpose of normalization and they can be determined by the creep test of the unreinforced matrix metal. Consider aligned short fibers embedded in the matrix metal with its creep law governed by eq. (4.14) where fibers are of the same size and assumed to behave elastically (i.e., non-creeping fiber). Assuming that the distributions of aligned short fibers are random in a three-dimensional space (Fig. 4.13(a)), one can take out a representation unit cell from the composite. The dimensions of the unit cell are shown in Fig. 4.13(b) where the local coordinate z is taken along the fiber axis with its origin at the mid-point. Following the Kelly and Street

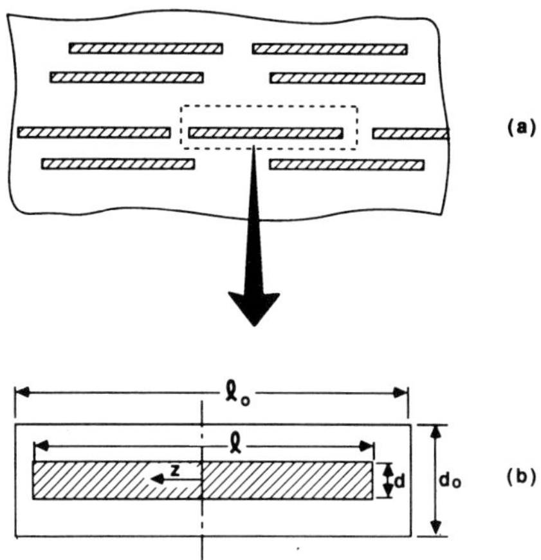

FIG. 4.13 Non-creeping aligned short fibers of the same size are embedded in a creeping matrix (a), from which a unit cell is taken out for the analysis (b) [25].

shear lag model, the average shear strain-rate $\dot{\gamma}$ in the matrix at a point z is given by

$$\dot{\gamma} = (\dot{e}_m z)/h \tag{4.15}$$

where $\dot{e}_m$ is the uniform tensile creep rate of the matrix defined in eq. (4.14) and h given by

$$h = (d_0 - d)/2 \tag{4.16}$$

and where d_0 and d are the diameters of the unit cell cylinder and fiber, respectively and they are related to fiber volume fraction V_f as

$$V_f = d^2 l/(d_0^2 l_0) \tag{4.17}$$

where l_0 is the length of the unit cell.

From eqs. (4.16) and (4.17), h is reduced to

$$h = \frac{d}{2}\left\{\sqrt{(l/l_0 V_f)} - 1\right\} \tag{4.18}$$

By using eqs. (4.14)–(4.18) and assuming that the creep rate of the matrix satisfies the incompressibility requirement, $\dot{e}_m = (2/3)\dot{\gamma}$, and the stress of the matrix follows the Tresca type requirement, $\sigma_m = 2\tau$, one can arrive at

$$\tau = \frac{1}{2}\sigma_{m0} \ln\{\dot{e}_m/\dot{e}_{m0})\eta\tfrac{z}{d}\} \tag{4.19}$$

where

$$\eta = \frac{4}{3}\frac{1}{\{\sqrt{(l/l_c V_f)} - 1\}} \tag{4.20}$$

Equation (4.19) describes the shear stress (τ) in the matrix adjacent to the interface as a function of z and it is considered to be valid for $z/d \geqslant 1/(\eta\dot{e}_m/\dot{e}_{m0})$, hence τ is taken as zero for much smaller values of z/d.

Next let us compute the average fiber stress $\bar{\sigma}_f$. Two cases are considered separately below; the case of perfectly bonded interfaces and that of debonded interfaces.

Case of perfectly bonded interfaces

The stress in the fiber along the fiber axis (z-axis) σ_f is obtained by considering the force equilibrium along the z-axis and the continuity of the normal stress across the fiber-ends.

$$\sigma_f = -\int_{l/2}^{z} \frac{4\tau}{d} dz + \sigma_m \tag{4.21}$$

where σ_m is the matrix tensile creep stress. By substituting eq. (4.19) into (4.21) and performing elementary integration, we arrive at

$$\sigma_f = 2\sigma_{m0}\left[\frac{1}{2}(l/d)\{\ln A + \ln(l/(2d)) - 1\} + \{1 - \ln A\}x - x\ln x + \sigma_m/(2\sigma_{m0})\right] \tag{4.22}$$

where $x = z/d$, and A is defined by

$$A = \eta\, \dot{e}_m/\dot{e}_{m0} \tag{4.23}$$

The average fiber stress $\bar{\sigma}_f$ then can be obtained as

$$\begin{aligned}\bar{\sigma}_f &= 2/l \int_0^{l/2} \sigma_f\, dz \\ &= \sigma_{m0}\cdot\frac{1}{2}\left(\frac{l}{d}\right)\left\{\ln A + \ln\left[\frac{1}{2}\left(\frac{l}{d}\right)\right] - \frac{1}{2}\right\} + \sigma_m\end{aligned} \tag{4.24}$$

Then, the composite stress σ_c can be obtained by using a law of mixtures,

$$\sigma_c = V_f\sigma_f + (1 - V_f)\sigma_m \tag{4.25}$$

From eqs. (4.14), (4.23)–(4.25) we can arrive at the relationship between the applied stress of the composite σ_c and creep rate of the matrix $\dot{e}_m$ as

$$\dot{e}_m = \dot{e}_{m0}\exp\left\{\frac{\sigma_c/\sigma_{m0} - c}{1 + 0.5V_f(l/d)}\right\} \tag{4.26}$$

$$c = 0.5\, V_f(l/d)\{\ln(l/d) + \ln\eta - 1.193\} \tag{4.27}$$

By assuming a non-creeping fiber, the composite creep rate $\dot{e}_c$ is related to the matrix creep rate $\dot{e}_m$ as

$$\dot{e}_c = (1 - V_f)\dot{e}_m \tag{4.28}$$

Hence, the relationship between σ_c and $\dot{e}_c$ is now given by

$$\dot{e}_c = (1 - V_f)\dot{e}_{mo}\exp\left\{\frac{\sigma_c/\sigma_{mo} - c}{1 + 0.5V_f(l/d)}\right\} \tag{4.29}$$

In order to obtain explicitly the relation between $\dot{e}_c$ and σ_c for given material data, eqs. (4.27) and (4.29) are not complete. Namely, there must exist some relation between d/d_0 and l/l_0 (see Fig. 4.13). To this end, we assume a constant parameter k such that

$$d/d_0 = k(l/l_0) \tag{4.30}$$

The value of k can be found experimentally by examining SEM photos of the short fiber metal matrix composite, and its value is expected to vary from 0.3

to 1.0 [25]. With eq. (4.30), η defined by eq. (4.20) is now given explicitly by

$$\eta = \frac{4}{3} \frac{1}{\{1/(kV_f)^{1/3} - 1\}} \tag{4.31}$$

Case of debonded interfaces

Here we consider the case where the matrix–fiber interface is debonded over the region $l\xi/2$ from the fiber end (see Fig. 4.14). Then, the non-debonded fiber length $\bar{l}$ is given by

$$\bar{l} = l(1 - \xi) \tag{4.32}$$

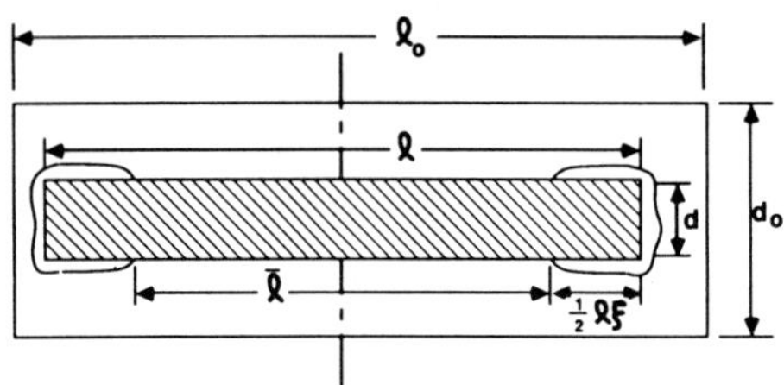

FIG. 4.14 A unit cell model with debonded interface with debonding length $l\xi/2$ from the fiber ends where $0 \leqslant \xi \leqslant 1$ [25].

where ξ is a parameter describing the extent of the debonding and its value ranges from 0 (no debonding) to 1 (complete debonding). The modeling of the case with debonded interfaces assumes that the matrix continues to creep in (its own) second-stage creep, and that debonding only affects the load transfer to the fibers with the effect that the matrix creeps at *higher stress* (but still in the second stage). The formulation to compute the average fiber stress in this case is basically the same as that for the case of perfectly bonded interfaces except that here no load transfer across the fiber ends takes place and l is now replaced by $l(1 - \xi)$. Then, the formula to compute $\dot{e}_c$ is obtained as

$$\dot{e}_c = (1 - V_f)\dot{e}_{mo} \exp\left\{\frac{\sigma_c/\sigma_{mo} - c}{1 - V_f + V_f/2 \cdot (l/d)(1 - \xi)}\right\} \tag{4.33}$$

$$c = V_f/2 \cdot (l/d)(1 - \xi)\{\ln[(l/d) \cdot (1 - \xi)] + \ln \eta - 1.193\} \tag{4.34}$$

where η is defined by eq. (4.31). It is noted here that eq. (4.33) with $\xi = 0$ does not lead to the formula for the case of perfectly bonded interfaces, i.e., eq. (4.29), since it is the result for $\xi = 0$ with no load transfer at fiber ends. The creep rate of a composite with completely debonded interfaces is obtained from eqs. (4.33) and (4.34) with $\xi = 1$,

$$\dot{e}_c = (1 - V_f)\dot{e}_{mo}\exp\left\{\frac{\sigma_c/\sigma_{mo} - c}{1 - V_f}\right\} \tag{4.35}$$

$$c = 0 \tag{4.36}$$

where $\lim_{\xi \to 1} (1 - \xi) \ln(1 - \xi) = 0$ was used.

Let us now examine the effect of several key material and geometric parameters on the creep rate of an aligned short fiber metal matrix composite. The key parameters to be examined are applied stress, σ_c, fiber volume fraction V_f and fiber aspect ratio l/d. The range of the values of these parameters and their typical values are given in Table 4.1.

Table 4.1 *Range and typical values of the key material and geometric parameters*

Parameters	σ_c/σ_{mo}	V_f	l/d	ξ	k
Range	1–20	0–0.6	1–100	0–1	
Typical values	10	0.3	5	0	0.7

When the effect of one of the key parameters is examined, the others are set equal to their typical values.

By using eq. (4.29), the creep rate of a composite with $V_f = 0.3$ and $l/d = 5$, $\dot{e}_c$, normalized by $\dot{e}_{mo}$ is plotted as a solid line against applied stress σ_c, normalized by σ_{mo} in Fig. 4.15 where the creep rate of the unreinforced matrix metal was also given. Figure 4.15 illustrates clearly the effect of the reinforcement, namely that the creep rate $\dot{e}_c$ can be reduced significantly in comparison to the unreinforced matrix for a given applied stress level. The reduction in the creep rate (or for increase in the creep resistance) increases with σ_c.

We now examine in more detail the role of each key parameter on the creep resistance. Figures 4.16 and 4.17 show the results of $\log_{10}(\dot{e}_c/\dot{e}_m)$ as functions of fiber volume fraction V_f, and fiber aspect ratio l/d, respectively where $\dot{e}_m$ is the creep rate of the unreinforced matrix. It follows from these figures that the larger the values of V_f and l/d are, the larger the reduction in the creep of the composite, thus leading to better creep resistance of the composite. The strong dependence of creep resistance on l/d has also been observed experimentally [26, 27].

Next, the effect of the debonding parameter ξ (defined by eq. (4.32)) on the creep rate, $\dot{e}_c$, is studied by using eqs. (4.33) and (4.34). The material data used for this computation are again typical ones given by Table 4.1, except that ξ becomes primary variable here, $0 \leqslant \xi \leqslant 1$. The predicted results of

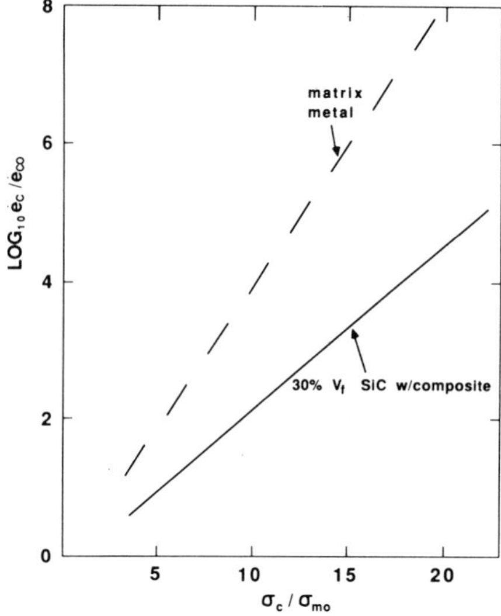

FIG. 4.15 Creep rate ($\dot{e}_c$) of a 30% V_f SiC whisker/aluminum composite as a function of applied stress σ_c (solid line) where the relation between creep rate and applied stress for the unreinforced matrix (aluminum) is shown by dashed line. Creep rate and applied stress are normalized by $\dot{e}_{mo}$ and σ_{mo}, respectively [25].

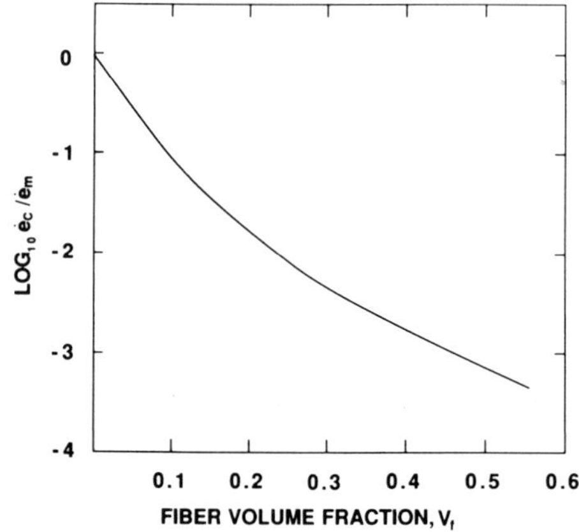

FIG. 4.16 The effect of fiber volume fraction V_f on the composite creep rate $\dot{e}_c$ normalized by $\dot{e}_m$ (the creep rate of the unreinforced matrix) [25].

$\log_{10}(\dot{e}_c/\dot{e}_{co})$ are plotted as a function of debonding parameter ξ in Fig. 4.18, where $\dot{e}_{co}$ is the creep rate of a composite with perfectly bonded interfaces (eq. (4.29)). It is noted in Fig. 4.18 that the creep rate at $\xi = 0$ refers to a composite with fiber ends being debonded, and its value is three times larger than that of a composite with perfectly bonded interfaces $\dot{e}_{co}$. It also follows

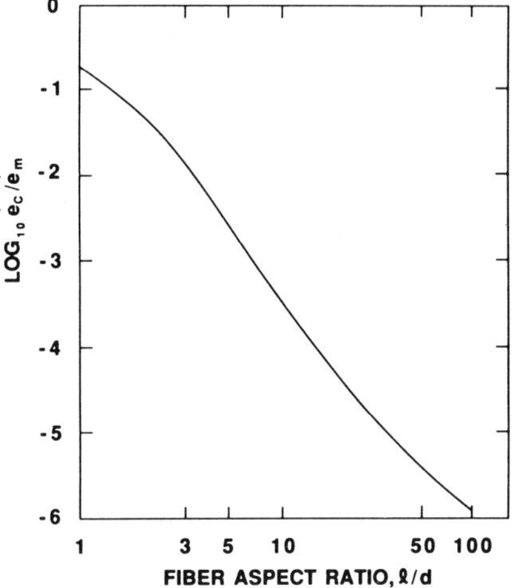

FIG. 4.17 The effect of fiber aspect ratio l/d on the composite creep rate $\dot{e}_c$ normalized by $\dot{e}_m$ (the creep rate of the unreinforced matrix) [25].

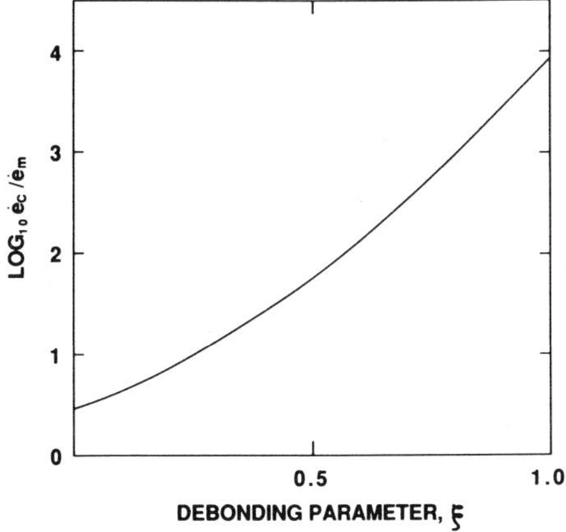

FIG. 4.18 The effect of a debonded interface with debonding parameter ξ on the composite creep rate $\dot{e}_c$ normalized by $\dot{e}_{co}$ (the creep rate of the composite with perfectly bonded interface) [25].

from Fig. 4.18 that the debonding of interfaces, even partial debonding, can result in a large increase in the creep rate, thus weakening the creep resistance.

The creep rate of a continuous fiber metal matrix composite can be obtained by the law of mixtures equation:

$$\dot{e}_c = V_f \dot{e}_f + (1 - V_f)\dot{e}_m \tag{4.37}$$

where $\dot{e}_f$ is the creep rate of the continuous fibers. If continuous fibers do not creep, $\dot{e}_f = 0$, then eq. (4.37) is reduced to eq. (4.33). However, the creep rate of a continuous fiber metal matrix composite must be considered to be zero if non-creeping continuous fibers reach both ends of the composite specimen. Hence, the law of mixtures equation to predict the composite creep rate is not valid except when both fiber and matrix are creeping. Independently of the above shear lag type models, Min [28] developed a plane stress creep model for continuous fiber metal matrix composites. The equations in Min's model are expressed in incremental form, thus the model can accommodate basically any type of matrix creep law. Min's model was successfully applied to the transverse creep of continuous graphite fiber/6061 aluminum composite [29].

When an aligned (short or continuous) fiber metal matrix composite is creep-loaded in an off-axis direction, i.e., the loading axis makes angle θ with respect to the fiber axis (see Fig. 4.19), Johnson [30] showed that the off-axis creep behavior can be estimated by a combination of three different modes of creep behavior, i.e., longitudinal creep, transverse creep and longitudinal shear creep of an aligned fiber metal matrix composite. These three modes of creep loading are basically the same as those illustrated in Fig. 3.17. In order to use Johnson's model, one must know *a priori* the creep rate-stress relations for these three basic modes which are expressed in terms of power law type creep equations [30]:

For longitudinal load (Fig. 3.17(a)):

$$\dot{e}_{33} = \lambda^{m}\sigma_{L}^{2m-1} \tag{4.38}$$

For transverse load (Fig. 3.17(b)):

$$\dot{e}_{11} = \mu^{m}\sigma_{T}^{2m-1} \tag{4.39}$$

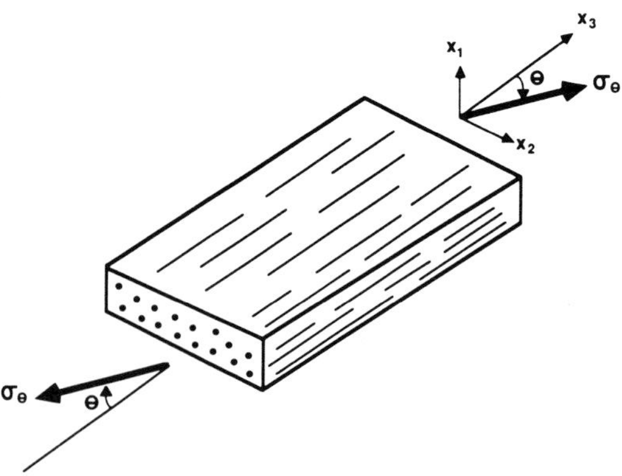

FIG. 4.19 An aligned fiber (unidirectional) metal matrix composite subjected to off-axis creep loading σ_θ.

For longitudinal shear load (Fig. 3.17(c)):

$$\dot{e}_{13} = \frac{1}{2}\nu^{m}\tau^{2m-1} \tag{4.40}$$

where $\dot{e}_{33}$, $\dot{e}_{11}$ and $\dot{e}_{13}$ are creep rates along the longitudinal (fiber axis, or x_3-axis), transverse (x_1-axis) and the $x_1 - x_3$ plane, respectively, and parameters λ, μ, ν and m will be determined by three different creep tests. Then the creep rate-stress relation of an aligned fiber (unidirectional) metal matrix composite loaded along the θ-direction is given by

$$\begin{aligned} \dot{e}_c &= F^{1/2(n+1)}\sigma_\theta^n \\ F(\theta) &= \lambda\cos_4\theta + (\nu - \lambda)\sin^2\theta\cos^2\theta + \mu\sin^4\theta \end{aligned} \tag{4.41}$$

where $n = 2m - 1$. For an isotropic composite, $\mu = \lambda$ and $\nu = 3\lambda$, then $F(\theta) = \lambda$, which is independent of the loading direction θ. The creep rates predicted by eq. (4.41) agreed well with the experimental results of the $(Co,Cr)_7C_3$ fiber/(Co,Cr) matrix composite which was tested at different θ's. A comparison of the experimental data (filled circles) with predicted values of $F(\theta)$ (solid line) is shown in Fig. 4.20.

4.2.3 Thermal cycling

Among various thermal environments, thermal cycling can be considered one of the most severe environments. The damage due to thermal cycling has

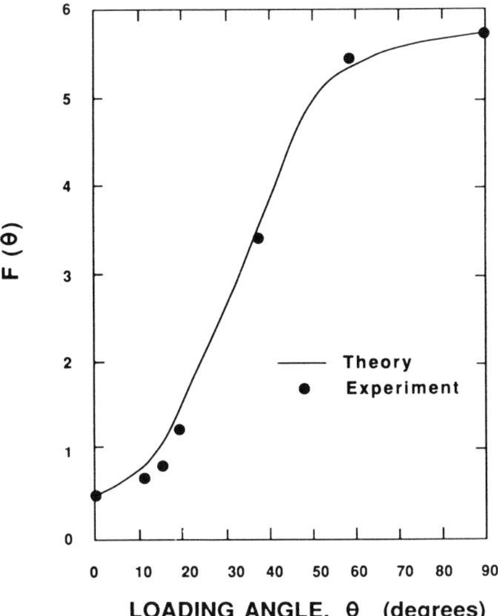

FIG. 4.20 The comparison of experimental data with predicted values of $F(\theta)$ by eq. (4.41) [30].

been found in a number of high-temperature components—for example, brake drum and turbine blades which are made of metals or super-alloys. If these components are made of metal matrix composites, the thermal cycling damage is expected to be worse than that of unreinforced metals or super-alloys due to the difference in coefficients of thermal expansion (CTEs) between the matrix metal and reinforcement. CTE values of fibers and metals for popular metal matrix composite systems are given in Table 4.2. Appendices B4 and B5 provide more complete data of fibers and matrix metals, respectively.

Table 4.2 *Coefficients of thermal expansion (CTE) of fibers and matrix metals at room temperature*

Constituent Materials	CTE $\alpha_L(\times 10^{-6}/°C)$	CTE $\alpha_T(\times 10^{-6}/°C)$
Matrix metals		
Copper	16.5	16.5
Aluminum	23.6	23.6
Fibers		
Boron	5	5
Tungsten wire	4.5	4.5
SiC whisker	2.3 (4.3)*	2.3 (4.3)*
Carbon	−1.1	20

* at 200 °C

It is clear from Table 4.2 that the CTE mismatch $\Delta\alpha$ for popular metal matrix composites are from 4 (W/Ti system) to 24.6 (carbon/Al system) $\times 10^{-6}/°C$. When $\Delta\alpha$ is multiplied by temperature differential ΔT (in typical value ranges from 200 to 800 °C), then the thermal strain due to CTE mismatch ($=\Delta\alpha\Delta T$) can reach 10^{-3} or more, which is approximately the order of yield strain of a typical matrix metal, resulting in the high internal (thermal) stress field. Although some of the high stress as induced can relax by plastic and/or creep deformation of the matrix metal, or the diffusion at the matrix–fiber interfaces, the level of the stress and strain that take place during thermal cycles is most often severe enough to cause damage.

Thermal cycling damage has been reported in several metal matrix composite systems: tungsten fiber/copper [31, 32], tungsten fiber/super-alloys (the results summarized in Ref. 33), boron fiber/aluminum [34–36], Al_2O_3-α (FP) fiber/aluminum [37], FP fiber/magnesium [38], SiC fiber/titanium [39], SiC whisker/aluminum [40], and graphite fiber/aluminum [41] composites.

The thermal cycling damage reported in these investigations is micro-damage associated with the matrix–fiber interfaces, namely microvoids or cracks at the interfaces, often leading to the debonding of the interfaces. An example of the voids formed at the interface in a tungsten fiber/copper

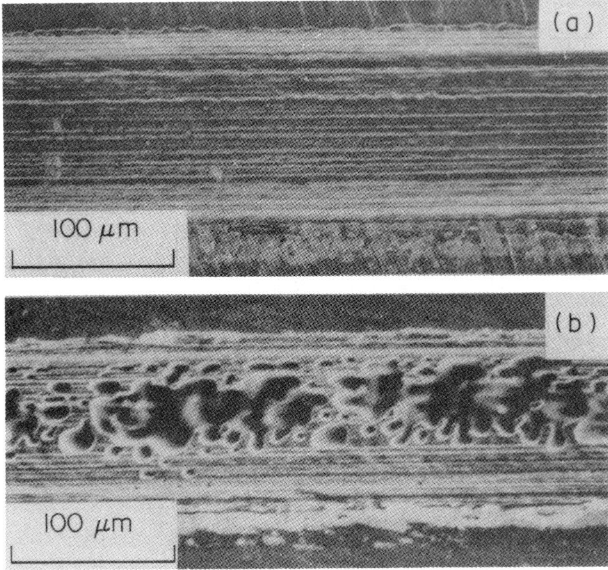

FIG. 4.21 SEM photos of the interface between tungsten fiber and copper matrix, (a) uncycled condition and (b) thermally cycled 200 times between 200 °C and 850 °C [32].

composite is shown in Fig. 4.21 where (a) and (b) denote the SEM photos of uncycled and cycled 200 times between 473 and 1073 K, respectively [32]. This micro-damage results in the degradation of the mechanical properties. Figure 4.22 shows the ultimate tensile strength σ_L of continuous boron fiber/6061 Al composites that were cycled between room temperature and maximum temperature (T_{max}) for various cycles [34] where σ_L of the uncycled composite also is shown as an open square box. It is clearly seen from Fig. 4.22 that the residual strength of the as-cycled composite is reduced compared with the data of the uncycled composite, and the reduction in σ_L is enhanced for larger T_{max} (shown as dashed line, $T_{max} = 420\,°C$) and at larger number of cycles.

The micro-damage often results in dimensional changes, which have been observed in several continuous fiber metal matrix composites [31, 36, 38, 41] and short fiber metal matrix composite [40]. Although the majority of the above investigations reported that the major dimensional changes occurred along the fiber axis, Kyono et al. [41] observed the dimensional changes of a continuous graphite fiber/5056 Al composite along the transverse direction (perpendicular to the fiber axis). This transverse dimensional change appears to have been caused by the complete debonding of the matrix–fiber interfaces at the early stage of cycling and then followed by the transverse swelling. Figure 4.23 indicates the transverse section view of a graphite fiber/5056 Al

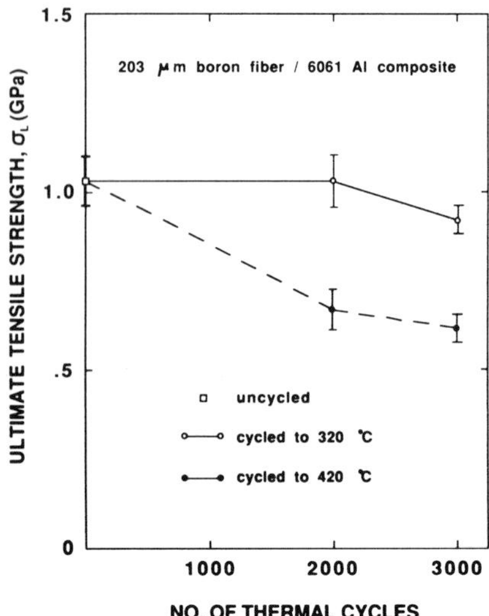

FIG. 4.22 The room temperature ultimate tensile strength (circles) of 203 μm boron fiber/6061 Al composites that were thermally cycled between room temperature and T_{max} (320 °C and 420 °C) as a function of a number of cycles where the data of the uncycled composite are shown as a open square [34].

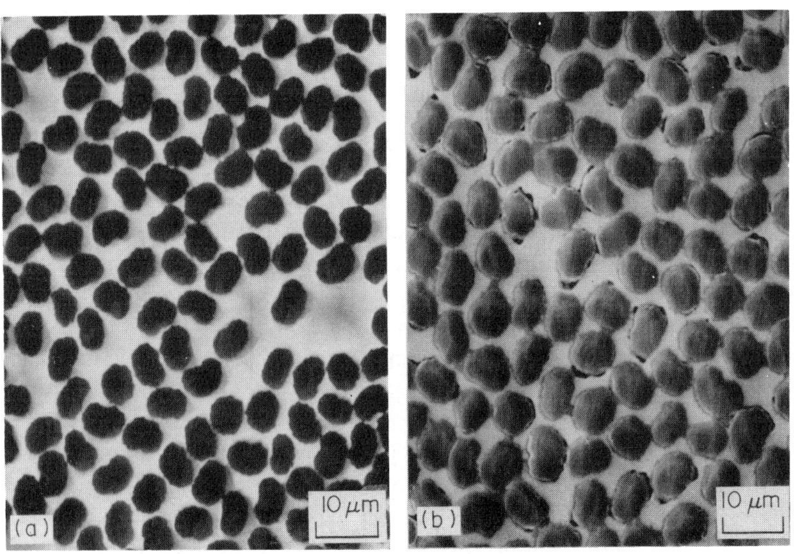

FIG. 4.23 Transverse section views of a graphite fiber/5056 Al composite, (a) uncycled, and (b) cycled 1000 times between room temperature and T_{max} = 350 °C.

composite, uncycled (a) and cycled 1000 times between room temperature and $T_{max}= 350\,°C$ (b). In the case of short fiber metal matrix composites, only limited experimental data have been reported to date, one under pure thermal cycling [40] and the other under a combination of thermal cycling and constant stress [42].

Models to predict dimensional changes

Some attempts have been made to model the dimensional changes observed in the thermally cycled metal matrix composites [31, 43–47]. Among these models the following are worth noting:

(1) Garmong's model [43, 44]
(2) Yoda et al.'s model [31]
(3) Taya-Mori model [46]
(4) Min's model [47]

Garmong assumed in his model [43] that the state of the stress and strain in the matrix and fiber phases is one-dimensional, and the matrix deforms as an elastic/plastic/creep material while the fiber deforms only elastically. The matrix creep is assumed to follow the Dorn type law [48] given by

$$\dot{e}_{mc} = \pm A\left\{\frac{|\sigma_m|}{G}\right\}^n \frac{Gb}{kT} D_0 \exp\{-Q/(RT)\} \tag{4.42}$$

where $\dot{e}_{mc}$ and σ_m are the creep rate and stress of the matrix metal, A and n are contants that can be determined experimentally, G is the matrix shear modulus, b is the Burger's vector, k is Boltzmann's constant, and D_0 and Q are the pre-exponential constant and the activation energy for self-diffusion. The sign in front of A is positive if $\sigma_m > 0$ and negative if $\sigma_m < 0$. The time-independent matrix flow stress σ_m is assumed to follow the Ludwik relation,

$$\sigma_m = \sigma_0 + K^*(e_{mp})^m \tag{4.43}$$

where σ_0 is the yield stress of the matrix, K^* and m are the material constants. The total strain in the matrix is the sum of elastic (e_{me}), plastic (e_{mp}) and creep (e_{mc}) strains, i.e.,

$$e_m = e_{me} + e_{mp} + e_{mc} \tag{4.44}$$

where e_{me} is defined by

$$e_{me} = \frac{\sigma_m}{E_m} \tag{4.45}$$

The fiber strain e_f is related to fiber stress σ_f as

$$e_f = \frac{\sigma_f}{E_f} \tag{4.46}$$

The composite stress σ_c is given by the law of mixtures equation (eq. (3.3)) and the requirement of strain continuity along the fiber axis renders

$$\int_{T_{min}}^{T_{max}} \alpha_m \, dT + e_m = \int_{T_{min}}^{T_{max}} \alpha_f \, dT + e_f \tag{4.47}$$

where α_m and α_f are CTEs of the matrix and fiber, respectively. From eqs. (4.42)–(4.47), one can arrive at the governing equation:

$$\int_{T_{min}}^{T_{max}} (\alpha_m - \alpha_f) \, dT = \frac{\{\sigma_c - \sigma_m(1 - V_f)\}}{V_f E_f} - \left\{ \frac{\sigma_m}{E_m} + \left[\frac{(\sigma_m - \sigma_0)}{K*} \right]^{1/m} \right.$$

$$\left. \pm \int_{T_{min}}^{T_{max}} A \left(\frac{|\sigma_m|^n}{G} \right) \frac{Gb}{kT} D_0 \exp\{-Q/(RT)\} \left(\frac{dt}{dT} \right) dt \right\} \tag{4.48}$$

Garmong calculated the plastic strain per cycle (Δe_p) of a hypothetical eutectic composite by using the above governing equation and arrived at the following findings [44]:

(1) the larger T_{max}, V_f and the difference of the CTE and the smaller the heating/cooling rate dT/dt, the larger Δe_p will become, leading to larger dimensional changes along the fiber axis;
(2) Δe_p will become constant after a few cycles. This will enable us to use the low cycle fatigue damage model of the Manson–Coffin law [49] which is usually described as

$$\Delta e_p = C N_F^z \tag{4.49}$$

where C and exponent z are material constants and N_F is the number of cycles at failure (of the matrix–fiber interface in this case). In practice the determination of such material constants of interfacial substances such as reaction products would be difficult because the low cycle tests on an interfacial substance whose size is usually extremely small are difficult, if not impossible, to conduct.

The Yoda et al. model [31], on the other hand, allows a jump in the axial strain (e_s) at the matrix–fiber interface. Hence, the continuity condition eq. (4.47) must be replaced by

$$\int_{T_{min}}^{T_{max}} \alpha_m \, dT + e_m = \int_{T_{min}}^{T_{max}} \alpha_f \, dT + e_f + e_s \tag{4.50}$$

Instead of assuming the matrix creep behavior at higher temperature (given by eq. (4.42)), Yoda et al. assumed that the relaxation of high stress at higher temperature takes place by the sliding of the matrix–fiber interface which has a thin layer. Thus, the Yoda et al. model consists of three phases: fiber, thin interface layer and matrix where the fiber deforms only elastically, the matrix metal deforms only as an elastic/plastic body, and the interface slides as a Newtonian viscous flow.

By applying the principle of virtual work to three different works, elastic and plastic work, and also the work due to the interfacial sliding, they obtained the permanent strain (dimensional change) per cycle, Δe_p as

$$\Delta e_p = \left(\frac{V_m}{q}\right)\sigma_0[1 - \exp(-t_{max}/\beta_{max})] \tag{4.51}$$

where V_m and σ_0 are the volume fraction and yield stress of the matrix metal, t_{max} is the duration of T_{max} (see the idealized temperature–time curve of the Yoda et al.'s model in Fig. 4.24), and β_{max} is given by

$$\beta_{max} = \frac{1}{3}\left(\frac{V_f}{q}\right)\left(\frac{l^2}{d}\right)\left(\frac{\mu_0}{h}\right)\exp\left(\frac{Q}{RT_{max}}\right) \tag{4.52}$$

$$q = \frac{V_f(1 - V_f)E_m E_f}{E_c}$$

and where μ_0 and h are the viscosity and thickness of the interfacial layer, l and d are the length and diameter of the fiber and E_c is the composite modulus defined by eq. (3.1). It can be concluded from eqs. (4.51) and (4.52) that the main findings obtained by Garmong's model are also applicable to the model of Yoda et al., except for the effect of V_f. Namely, the model of Yoda et al. predicts that the larger V_f induces the smaller Δe_p, leading to the smaller dimensional change in the thermally cycled continuous fiber metal matrix composite. This is just opposite to the conclusion obtained by Garmong's model.

Yoda et al. compared the prediction by eq. (4.51) with their experimental results of a continuous tungsten fiber/copper (W/Cu) composite [31] and confirmed that the predicted dependence of Δe_p on fiber length (l) and diameter (d) and also volume fraction (V_f) agreed with the experiment. This good agreement supports the validity of the Yoda et al. model, i.e., that of interfacial sliding at T_{max} which may have actually taken place in the W/Cu composite, for this composite system is known to have good wettability at the matrix–fiber interface without forming brittle metallic compounds. The validity of the Yoda et al. model to other metal matrix composites remains to be seen, since the majority of the other metal matrix composites do not possess such an interfacial layer which would behave as a Newtonian viscous material at high temperatures.

Unlike cases involving continuous fiber metal matrix composites, the modeling for the dimensional change of short fiber metal matrix composites requires a three-dimensional (3D) analysis. Taya and Mori [46] proposed such a 3D model to predict the dimensional change of short fiber and particulate metal matrix composites. The Taya–Mori model assumes that the temperature–time relation is of the step function type as in Fig. 4.24, where a metal matrix composite initially (at T_{max}) is stress-free and the stress relaxation at T_{min} occurs by the plastic deformation of the matrix and that at T_{max} by the matrix creep and the diffusion at the matrix–fiber interfaces. Thus, the Taya–Mori model is similar to the Garmong model except that it is a 3D model as opposed to the one-dimensional Garmong model, and it allows the relaxation at T_{max} by the interfacial diffusion where Garmong's model does not. The Taya–Mori model uses the Eshelby type model (see subsection 2.4.4 for details of the Eshelby model) to analyze the 3D state of stress and strain in an aligned short fiber metal matrix composite (Fig. 3.52) where the matrix and fiber are assumed to be isotropic both in stiffness (E_i, ν_i) and in CTE (α_i) with $i = \text{m}$ or f. During the cooling process, the temperature drop ($\Delta T = T_{max} - T_{min}$) is assumed to be large enough to cause the yielding of the matrix metal first at temperature T_y (see Fig. 4.24) and followed by the increase in the plastic strain, reaching the final averaged plastic strain in the matrix at temperature T_{min}, $\mathbf{e}^p$. The corresponding average stress in the matrix at T_{min} is the sum of the stress due to temperature drop Δ_T, $\boldsymbol{\sigma}^T$ and that induced by the plastic strain $\mathbf{e}^p$ in the matrix, $\boldsymbol{\sigma}^p$. These stresses are easily obtained by using the Eshelby's method modified for finite volume fraction V_f (see subsection 2.4.4). For example, the average stress $\boldsymbol{\sigma}^p$ in the matrix, $\langle\boldsymbol{\sigma}^p\rangle_m$ and that in the fiber, $\langle\boldsymbol{\sigma}^p\rangle_f$ are solved as

$$
\begin{aligned}
\langle\boldsymbol{\sigma}^p\rangle_m &= -V_f\mathbf{C}_m\cdot(\mathbf{S}\cdot\mathbf{e}^* - \mathbf{e}^*) \\
\langle\boldsymbol{\sigma}^p\rangle_f &= (1 - V_f)\mathbf{C}_m\cdot(\mathbf{S}\cdot\mathbf{e}^* - \mathbf{e}^*)
\end{aligned}
\tag{4.53}
$$

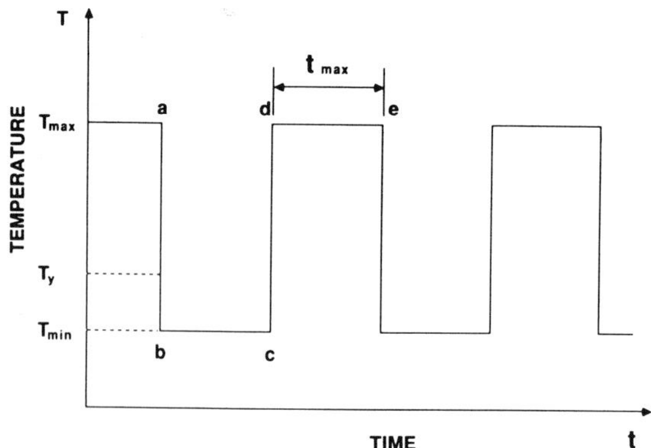

FIG. 4.24 The idealized temperature-time curve for thermal cycling used in the models of Yoda et al. [31] and of Taya and Mori [46].

where e^* will be solved from

$$\mathbf{C}_f \cdot (\bar{\mathbf{e}} + \mathbf{e} + \mathbf{e}^p) = \mathbf{C}_m \cdot (\bar{\mathbf{e}} + \mathbf{e} - \mathbf{e}^*)$$
$$\mathbf{e} = \mathbf{S} \cdot \mathbf{e}^* \tag{4.54}$$
$$\bar{\mathbf{e}} + V_f(\mathbf{e} - \mathbf{e}^*) = 0$$

In the above equations, $\langle \quad \rangle_i$ and C_i denote the quantity averaged over the ith phase domain and stiffness tensor of the ith phase with i = m (matrix) or f (fiber), **S** is the Eshelby's tensor (see Appendix C), **e** and $\bar{\mathbf{e}}$ are the strain disturbance due to a single fiber and the average strain disturbance in the matrix due to all the other fibers, respectively. The total stress ($\boldsymbol{\sigma} = \boldsymbol{\sigma}^T + \boldsymbol{\sigma}^p$) must satisfy the yield criterion (in this case the Von Mises type),

$$\langle \sigma_{33} \rangle_m - \langle \sigma_{11} \rangle_m = Y \tag{4.55}$$

where $\langle \sigma_{ij} \rangle_m$ is a component of the total average stress in the matrix and Y is the matrix yield stress which is assumed to be constant.

The averaged plastic strain in the matrix at T_{min}, $\mathbf{e}^p$ will be computed from eq. (4.55). This plastic strain will contribute to the macroscopic strains (dimensional change) along the fiber axis (L or x_3-axis), $\langle \varepsilon^p_L \rangle$ and along the transverse direction to the x_3-axis (T or x_1, x_2-axis), $\langle \varepsilon^p_T \rangle$. To the first order approximation, they are obtained as

$$\langle \varepsilon^p_L \rangle = (1 - V_f)e_p$$
$$\langle \varepsilon^p_T \rangle = -(1 - V_f)e_p/2 \tag{4.56}$$

where e_p is the x_3-component of $\mathbf{e}^p$.

Next consider the heating process (c → d in Fig. 4.24). The thermal stress induced by temperature change ΔT is just negative of $\boldsymbol{\sigma}^T$ that was induced during the cooling stage (a → b in Fig. 4.24). Thus at stage d, the only stress remaining in the composite is that due to plastic strain $\mathbf{e}^p$, i.e., $\boldsymbol{\sigma}^p$ given by eq. (4.53). The equivalent stress of $\boldsymbol{\sigma}^p$ in the matrix, σ_m is defined by

$$\sigma_m = \langle \sigma^p_{33} \rangle_m - \langle \sigma^p_{11} \rangle_m \tag{4.57}$$

where the sign of σ_m is negative for $\alpha_f < \alpha_m$. This stress σ_m in the matrix at stage d (the beginning of the high-temperature plateaux with T_{max}) will be relaxed by the matrix creep governed by a power law; eq. (4.42). During the T_{max} plateaux with duration t_{max}, the matrix creep progresses, resulting in the creep strain averaged over the matrix domain, $\mathbf{e}^c$. This creep strain will contribute to the macroscopic strain of the composite, $\langle \varepsilon^c_I \rangle$ with I = L (along the x_3-axis) and T (along the x_1-axis), and they are given by

$$\langle \varepsilon^c_L \rangle = (1 - V_f)e_c$$
$$\langle \varepsilon^c_T \rangle = -(1 - V_f)e_c/2 \tag{4.58}$$

where e_c is the x_3-component of creep strain $\mathbf{e}_c$ which will be computed from eq. (4.42) by iterative method. The above creep strain will decrease σ_m, but σ_m may remain as a non-zero flow stress in the matrix at the end of the high-temperature plateaux. This non-zero σ_m, however, can be reduced to zero if the diffusional relaxation at the interfaces occurs at the same time [50]. This diffusional relaxation can be modeled by the Eshelby's model if eigenstrain $\mathbf{e}^D$ is introduced in the fiber domain. The diffusional relaxation requires that the total stress in the fiber becomes hydrostatic, leading to the solutions to $\mathbf{e}^D$, i.e.,

$$\mathbf{e}^D = \mathbf{e}^p + \mathbf{e}^c \tag{4.59}$$

The contribution of $\mathbf{e}^D$ to the macroscopic strain of the composite is obtained as

$$\begin{aligned} \langle \varepsilon_L^D \rangle &= V_f(e_p + e_c) \\ \langle \varepsilon_T^D \rangle &= -\ V_f(e_p + e_c)/2 \end{aligned} \tag{4.60}$$

Thus, the sum of the contributions, plastic strain, creep strain and diffusional strain results in the dimensional change per cycle, $\Delta\varepsilon_L$ (along the fiber axis) and $\Delta\varepsilon_T$ (along its transverse direction).

$$\begin{aligned} \Delta\varepsilon_L &= e_p + e_c \\ \Delta\varepsilon_T &= -(e_p + e_c)/2 \end{aligned} \tag{4.61}$$

It should be noted that $\Delta\varepsilon_L$ is equivalent to Δe_p obtained by the preceding two models and the transverse dimensional change is just half of the axial dimensional change with opposite sine, thus satisfying the incompressibility requirement for the matrix strain.

Taya and Mori [46] applied the above model to SiC whisker/2124 Al composite whose experimental data are available [40]. The material data for SiC whisker/2124 Al composite are given in Table 4.3 and the creep data of the matrix metal in Table 4.4.

By using the above model, Taya and Mori examined the effect of fiber volume fraction V_f and fiber aspect ratio l/d on the longitudinal dimensional change per cycle $\Delta\varepsilon_L$, which are shown in Figs. 4.25 and 4.26, respectively. Figs. 4.25 and 4.26 illustrate the strong dependence of $\Delta\varepsilon_L$ on V_f and l/d. Namely, the larger the values of V_f and l/d are, the larger the longitudinal

Table 4.3 *Material data of SiCw/2124 Al composite*

Parameters	Symbols	Units	Matrix (2124Al)	Fiber (SiCw)
Shear modulus	G	GPa	25.4 at 300 K	182.5
Poisson's ratio	ν	1	0.33	0.17
CTE	α	/K	24.7×10^{-6}	4.3×10^{-6}
Yield stress	Y	MPa	50.8 (0.2% str.)	—
Fiber aspect ratio	l/d	1	—	$2 \sim 100$
Fiber volume fraction	V_f	1	—	$0.1 \sim 0.55$

Table 4.4 *Creep data of 2124 Al matrix*

Parameters	Symbols	Units	Data	References
Dorn parameter	A	1	3.4×10^{-6}	51
Shear modulus at T_{max}	G	GPa	20.2	52
Pre-exponent	D_0	m^2/s	1.7×10^{-4}	51
Activation energy	Q_v	kJ/mole	142	51
Burger's vector	b	m	2.86×10^{-10}	51
Gas constant	R	J/mole/K	8.314	51
Boltzmann's constant	k	J/K	1.381	51
Exponent	n	1	$n = 5 + .025\sigma_m$	25

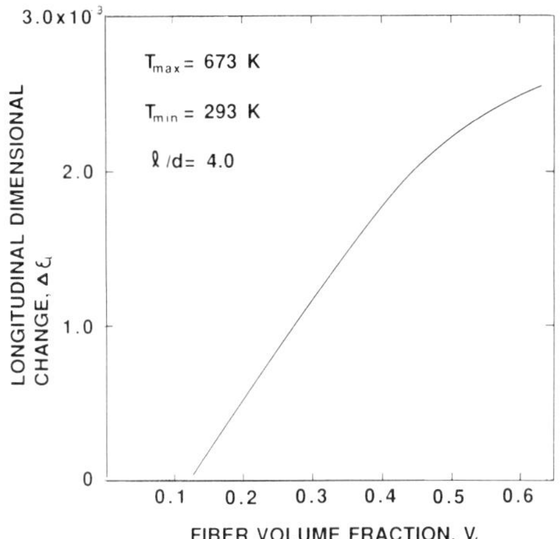

FIG. 4.25 The longitudinal dimensional change per cycle $\Delta\varepsilon_L$ of the thermally cycled aligned short fiber metal matrix composite as a function fiber volume fraction V_f [46].

dimensional change that will occur, which is basically in agreement with the prediction by Garmong's model. If the Taya and Mori model is compared with the analytical results given by eq. (4.61) with the experimental ones for 15% V_f SiCw/2124 Al composite [40]. The results of the comparison are shown in Table 4.5, where the dimensional changes at 1000 cycles with $T_{max} = 400\,°C$ and $T_{min} = 20\,°C$, along three directions (longitudinal (L), width (W) and thickness (TH) of the composite plate) are compared. It is noted here that SiC whiskers in the actual SiCw/2124 Al composite plate are misoriented in the plane parallel to the L–W plane although the majority are aligned along the extrusion direction, i.e., L-direction (or x_3-direction in Fig. 3.52) while all the SiC whiskers in the model are assumed aligned along the L-direction, which results in $\varepsilon_w = \varepsilon_{TH}$ in the prediction. If we account for the difference in the actual and the model short fiber metal matrix composites, the Taya–Mori

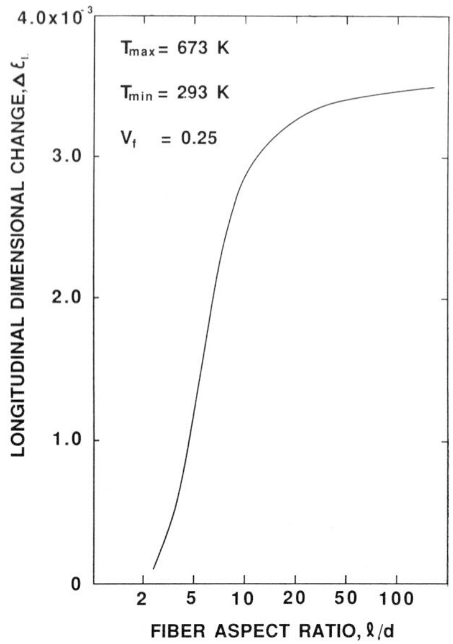

FIG. 4.26 The longitudinal dimensional change per cycle $\Delta\varepsilon_L$ of the thermally cycled aligned short fiber metal matrix composite as a function of fiber aspect ratio l/d [46].

Table 4.5 *Comparison between the analytical and experimental results of 15% SiCw/2124 Al composite with $T = 673$ K*

ε at 1000 cycles	Experiment [40]	Analytical results [46]	
		$n = 5.08$	$n = 4.9$
ε_L	0.055	0.158	0.129
ε_W	0.02	− 0.079	− 0.0645
ε_{TH}	− 0.079	−0.079	− 0.0645

model proved to be reasonably accurate for the prediction of the dimensional change of the thermally cycled short fiber metal matrix composite.

The above models have been developed for unidirectional (short and continuous fiber) metal matrix composites, and are thus not applicable to laminated composites, though in principle they can be modified to laminated composites through the lamination theory. Min [47] developed a simple plane stress model to predict the dimensional change in the thermally cycled laminated composite plates. A lamina, which is the same as a unidirectional (continuous fiber) metal matrix composite, consists of thermoelastic fiber and thermoelastic/plastic matrix metal. It is further assumed in Min's model that the stress and strain in the domain of each phase are uniform and both phases maintain equal longitudinal strain (isostrain), equal transverse stress (isostress) and equal shear stress, and otherwise follow the law of mixtures type relationships, i.e.,

$$\begin{aligned} \sigma_1 &= V_f \sigma_1^f + V_m \sigma_1^m \\ \sigma_2 &= \sigma_2^f = \sigma_2^m \\ \tau_{12} &= \tau_{12}^t = \tau_{12}^m \end{aligned} \tag{4.62}$$

and

$$\begin{aligned} e_1 &= e_1^f = e_1^m \\ e_2 &= V_f e_2^f + V_m e_2^m \\ \gamma_{12} &= V_f \gamma_{12}^f + V_m \gamma_{12}^m \end{aligned} \tag{4.63}$$

where subscripts 1 and 2 denote the direction along and perpendicular to the fiber axis, respectively, and serve as a local coordinate system for a lamina. Min used an incremental method in computation. Thus the total strain increments in both phases are expressed as

$$\begin{aligned} \mathrm{d}\mathbf{e}^{\mathrm{f}} &= \mathrm{d}\mathbf{e}^{\mathrm{fe}} + \boldsymbol{\alpha}^{\mathrm{f}}\,\mathrm{d}T \\ \mathrm{d}\mathbf{e}^{\mathrm{m}} &= \mathrm{d}\mathbf{e}^{\mathrm{me}} + \mathrm{d}\mathbf{e}^{\mathrm{mp}} + \boldsymbol{\alpha}^{\mathrm{m}}\,\mathrm{d}T \end{aligned} \tag{4.64}$$

In the above equations the bold symbols refer to a vectorial quantity having three components, superscripts f and m for the fiber and matrix phases, respectively, e and p for elastic and plastic components, and $\boldsymbol{\alpha}^{\mathrm{f}}$ and $\boldsymbol{\alpha}^{\mathrm{m}}$ are the CTE vectors of the fiber and matrix, respectively. $\boldsymbol{\alpha}^{\mathrm{r}}$ has the following components

$$\boldsymbol{\alpha}^{\mathrm{r}} = [\alpha_1^{\mathrm{r}}, \alpha_2^{\mathrm{r}}, 0]^{\mathrm{T}}; \quad \mathrm{r} = \mathrm{f}, \mathrm{m} \tag{4.65}$$

The plane strain–stress relation for r-phase is given by

$$\mathrm{d}\mathbf{e}^{\mathrm{re}} = \mathbf{C}^{\mathrm{r}}\,\mathrm{d}\boldsymbol{\sigma}^{\mathrm{r}}; \quad \mathrm{r} = \mathrm{f}, \mathrm{m}$$

$$\mathbf{C}^{\mathrm{r}} = \begin{bmatrix} C_{11}^{\mathrm{r}} & C_{12}^{\mathrm{r}} & 0 \\ C_{12}^{\mathrm{r}} & C_{22}^{\mathrm{r}} & 0 \\ 0 & 0 & C_{33}^{\mathrm{r}} \end{bmatrix} = \begin{bmatrix} 1/E_1^{\mathrm{r}}, & -\nu_{12}^{\mathrm{r}}/E_1^{\mathrm{r}}, & 0 \\ -\nu_{12}^{\mathrm{r}}/E_1^{\mathrm{r}}, & 1/E_2^{\mathrm{r}}, & 0 \\ 0, & 0, & 1/G_{12}^{\mathrm{r}} \end{bmatrix} \tag{4.66}$$

where $\mathbf{C}^{\mathrm{r}}$ is the plane strain compliance matrix of the r-phase material which is related to the plane stress engineering elastic constants. Although it is not necessary to specify the plastic constitutive equations due to the incremental formulation used in Min's model, Min used the plastic law of White–Basseling type, which seems to be consistent with cyclic plasticity where two cases, the hardening and non-hardening types, are examined. Once the governing equations are set up for each lamina, then those of lamina (or laminates) can easily be obtained following the classical lamination theory since the governing equations are expressed in incremental form.

By using the above model, Min studied analytically the dimensional change in graphite fiber/6061 Al composite laminates which are subjected to thermal cycling between $-250°F$ and $250°F$. Two different laminates, $[\pm 26°]$s and $[0°/90°]$s, are used to study the effect of layups on the dimensional change.

The numerical results of the thermal strain for the first ten cycles in $[\pm 26°]$s and $[0°/90°]$s laminates are shown in Fig. 4.27(a) and (b), respectively. It follows from Fig. 4.27 that the convergence to a stable cyclic loop in

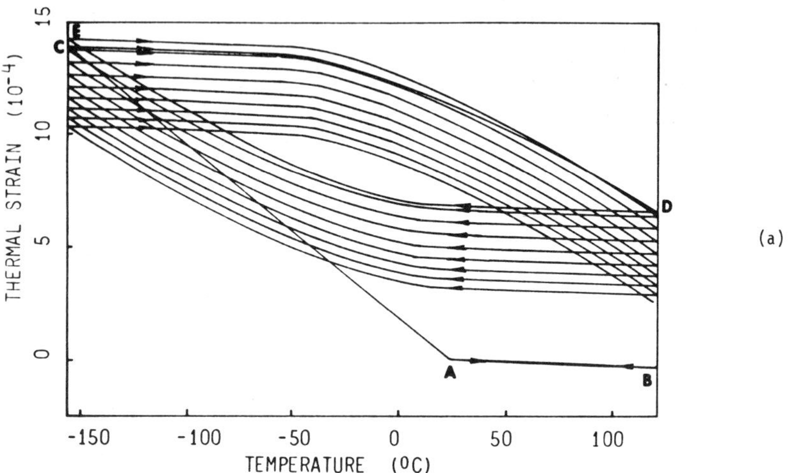

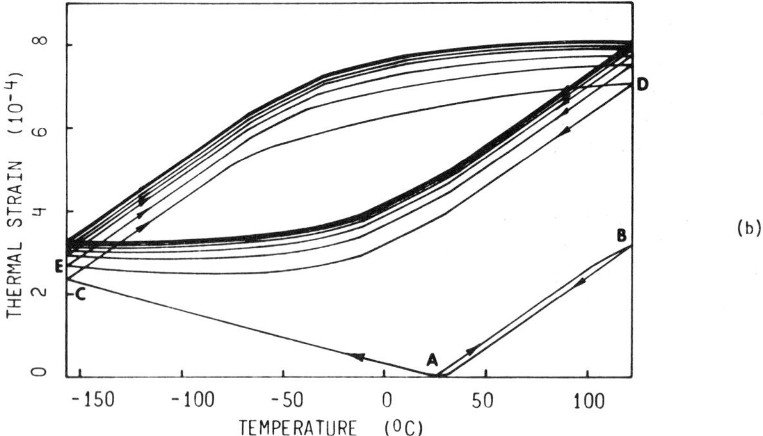

FIG. 4.27 The numerical results of the thermal strain for the first 10 cycles in (a) $[\pm 26°]$s and (b) $[0°/90°]$s laminates [47].

the [0°/90°]s laminate was much faster than the [±26°]s laminate. Another important finding in his work was that a non-hardening type plasticity model gave rise to faster convergence to a stable cyclic loop than a hardening type. It should be noted that the stress relaxation by matrix creep and the diffusion at the matrix–fiber interfaces are not considered in Min's model. Hence the validity of Min's model appears to be for a low to intermediate temperature excursion where the stress relaxation by the plastic deformation in the matrix is considered dominant.

4.3 Mechanical Behavior under Dynamic Loadings

Dynamic loading is considered as important as thermal loading for a structural component made of a metal matrix composite, but the mechanical behavior of the metal matrix composite under dynamic loadings has not been studied as thoroughly as that of thermal loadings. The dynamic loadings that are often encountered in use environments are impact loading, fatigue (or cyclic) loading and erosion and wear. As to the first, two cases of impact loading can be considered: low-velocity and high-velocity impact loadings. The effect of low-velocity impact loadings is considered less severe than high-velocity impact loadings, thus only the case of high-velocity impact loadings, which are often termed as 'high strain rate impact loadings,' will be considered here.

4.3.1 High strain rate impact loadings

Knowledge of the mechanical behavior of a metal matrix composite under high strain rate impact loadings is a prerequisite if a component made of the metal matrix composite is subjected to possible high-velocity impact loadings, for example, the impact of a bird on a turbine engine of a flying airplane, or a space station impacted by various flying objects. However, the mechanical behavior of metal matrix composites under high strain rate impact loadings has not been investigated to any greater extent.

Earlier investigations were focused on the continuous boron fiber/aluminum (B/Al) composite system [53, 54]. Schuster and Reed [53] reported on the experimental results of plate impact testing of a B/Al composite where spall fractures occurred at and above a certain threshold impact velocity, V_{th}. The V_{th} of a B/Al composite was found to be three times larger than that of unreinforced aluminum, resulting in the enhancement of the spall resistance. Gray [54] also reported on the impact damage of a B/Al composite, particularly on the residual fatigue resistance of the impacted composite, which was found to be much less than that of the impacted Ti–6Al–4V alloy. The above investigations indicate two important aspects to be considered in developing an impact-resistant metal matrix composite, namely the threshold impact velocity V_{th} and the assessment of the residual mechanical performance of the impacted metal matrix composite.

As for short fiber metal matrix composites, the SiC whisker/aluminum (SiCw/Al) composite has been relatively well studied in terms of its mechanical behavior at intermediate to high strain rates [55, 56]. Marchand et al. [55] have recently studied experimentally the high strain rate mechanical

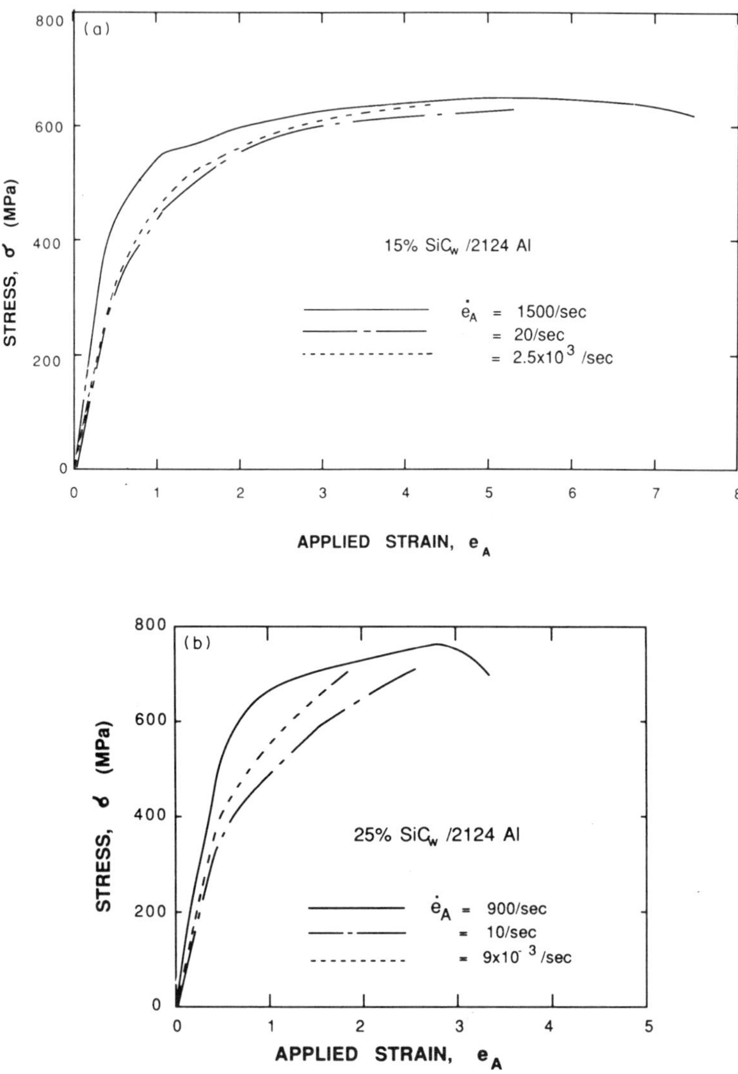

FIG. 4.28 The uniaxial stress–strain curves of a SiCw/2124 Al composite tested at various strain rates: (a) $V_f = 15\%$ and (b) $V_f = 25\%$ [56].

behavior of SiCw/Al by split Hopkinson bar, and found that the dynamic loading raises the fracture toughness K_{IC} of the composites while no such change was observed in the unreinforced aluminum. They also found that the strain rate has little influence on the stress–strain behavior in shear. A similar investigation was conducted by Harding et al. [56], who examined the uniaxial tensile mode of a SiCw/Al composite at various strain rates. The stress–strain curves of the SiCw/Al composite tested by split Hopkinson bar at three different strain rates are shown in Fig. 4.28, where (a) and (b) denote the case of $V_F = 15$ and 25%, respectively. Figure 4.28 indicates that the SiCw/Al composite exhibits little strain rate sensitivity at smaller strain rates, but it appears to show some degree of strain rate sensitivity for the ranges of higher strain rates. Namely, Young's modulus, yield stress (0.2% offset) and failure strain (e_f) were found to increase with applied strain rate $\dot{e}_A$ beyond a certain high strain rate as shown in Fig. 4.29(a), (b) and (c), respectively. Fracture surfaces of both the quasi-statically and impact-loaded composite exhibited cup and cone type dimples, implying a ductile fracture mode, and thus they provide no clues to the fact that e_f of the impact-loaded composite is larger than that of the quasi-statically loaded one.

4.3.2 Fatigue loadings

The fatigue behavior of a metal matrix composite is usually superior to that of the unreinforced matrix. This is particularly true for a unidirectional

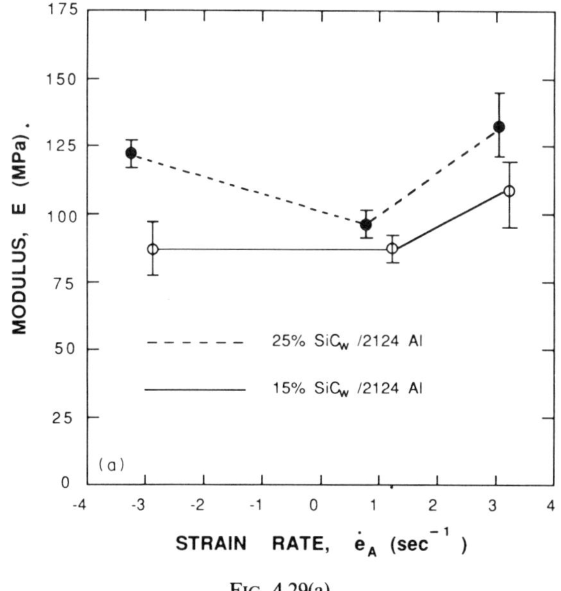

FIG. 4.29(a)

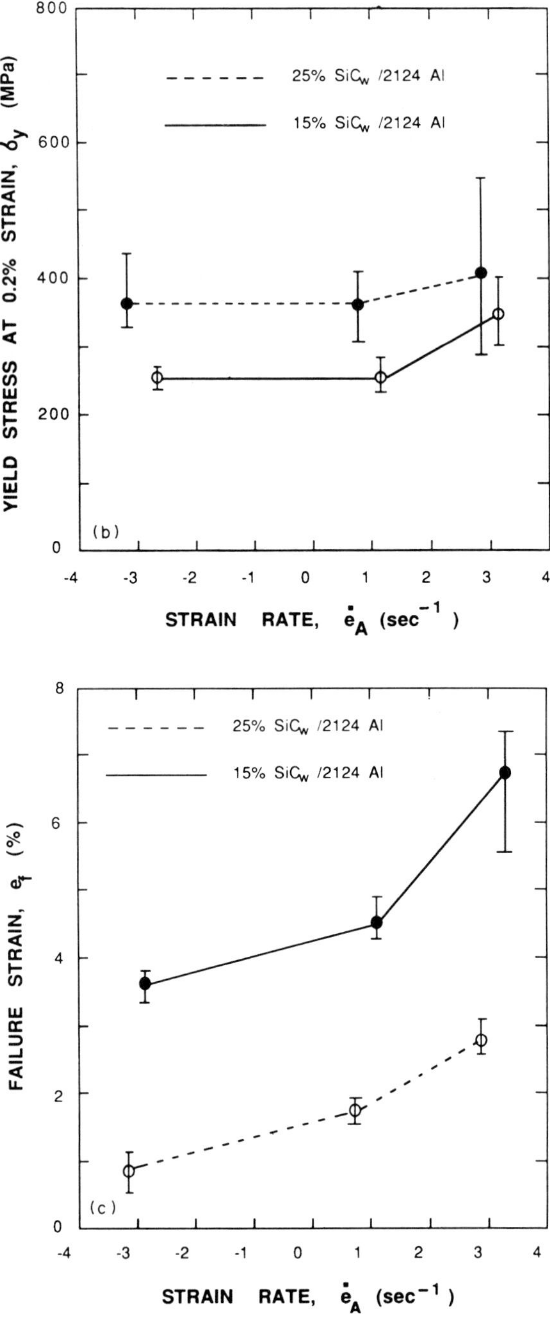

FIG. 4.29 The effect of strain rate on Young's modulus (a) 0.2% offset yield stress (b) and failure strain e_f (c) [56].

metal matrix composite system [57, 58]. The degradation of fatigue behavior of a metal matrix composite was studied earlier for boron fiber/aluminum (B/Al) composites [59–62]. The main findings in these works are that fibers break at defects at early cycles and thus nucleate cracking in the matrix and also the interfacial debonding. The matrix cracking and interfacial debonding will grow under fatigue load, leading to the final fracture of the composites.

Dvorak and Johnson [61] found that the endurance limit of a unidirectional B/Al composite is strongly dependent on the stress amplitude, but not the mean stress. Gouda et al. [62] have also found that the fatigue life is independent of the mean stress and have proposed a simple model to predict a S–N curve of a unidirectional metal matrix composite. This model accounts for two micromechanical damage modes during the fatigue, growth of matrix cracks in the direction perpendicular to the fiber axis (type A) and that of interfacial debonding (type B). Fatigue crack growth of type A is similar to the usual Mode I of growth observed in alloys in which the rate of crack growth as a function of ΔK, the range of stress intensity factor, is sigmoidal between the limits of ΔK_{th} (threshold level for crack growth) and K_{IC} (the fracture toughness of Mode I). An S–N curve computed from such crack growth rate data will lead to the usual form of the S–N curve for alloys. For crack growth of type B, however, the shape of the S–N curve may be quite different. If the interfacial debonding (or crack) progresses along a fiber with no weak points (called perfect-fiber), then the growth of this interfacial debonding will not contribute to final fracture. If all fibers are of the perfect fiber type, then the S–N curve will be flat, i.e., the endurance limit being the static strength. However, in actual unidirectional metal matrix composites some fibers may be perfect, but the majority of fibers usually have weak points. Then the S–N curve will be a mixture of those of types A and B. Three types of S–N curves—types A and B, and actual type (a mixture of types A and B)—are shown schematically in Fig. 4.30, where σ_{max} is normalized by the static strength of the composite σ_L. The prediction based on this model was compared with a limited amount of experimental data on a B/6061 Al composite, resulting in a reasonably good agreement. The shapes of the S–N curves proposed by Gouda et al. have also been confirmed by Nunes et al. [58], who studied the fatigue behavior of a unidirectional FP fiber/EZ41A magnesium (FP/Mg) composite. Namely, when the matrix–fiber interface is very strong and the fiber/matrix strength ratio is relatively low, the transverse crack growth (type B in Fig. 4.30) appears to occur leading to a sigmoidal S–N curve, as seen from the S–N curve at 24 °C in Fig. 4.31. On the other hand, when the matrix–fiber interface is weak and the fiber/matrix strength ratio is relatively high, a longitudinal crack growth along the matrix–fiber interface occurs, and under ideal conditions failure occurs at the bundle strength of fibers resulting in a horizontal S–N curve (type A in Fig. 4.30). However, most fibers fracture have weak points which initiate microcracks, leading to an interconnected network of the microcracks at the weak fiber

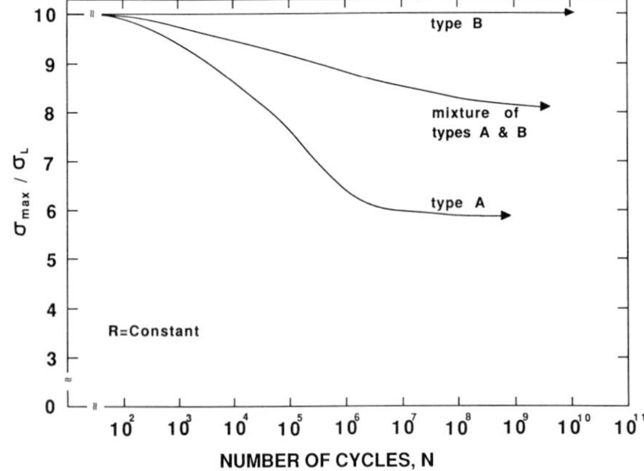

FIG. 4.30 Three types of S–N curve of metal matrix composites, sigmoidal (type A), flat (type B), and mixture (type A and type B) [62].

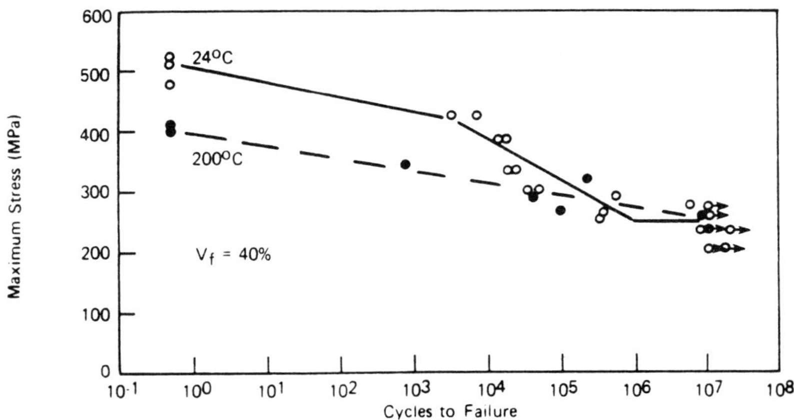

FIG. 4.31 S–N curves of a unidirectional FP/Mg composite tested at 24 and 200°C [58].

sites and matrix cracking. This will lead to a straight S–N curve (mixture type in Fig. 4.30) as seen from the S–N curve at 200 °C in Fig. 4.31).

The extent of fatigue damage can be assessed by measuring the stiffness of the as-fatigued metal matrix composite specimen, since the micro-damage is induced in the metal matrix composite specimen during the fatigue loading. Particularly, matrix cracking is known to reduce the stiffness of the composite. The reductions of the elastic moduli (E) of as-fatigued B/Al specimens are clearly demonstrated in Fig. 4.32 [61] where the experimental results of the percentage of the initial Young's modulus are plotted as a function of cycles N for various σ_{max}. The reduction in E becomes enhanced at larger

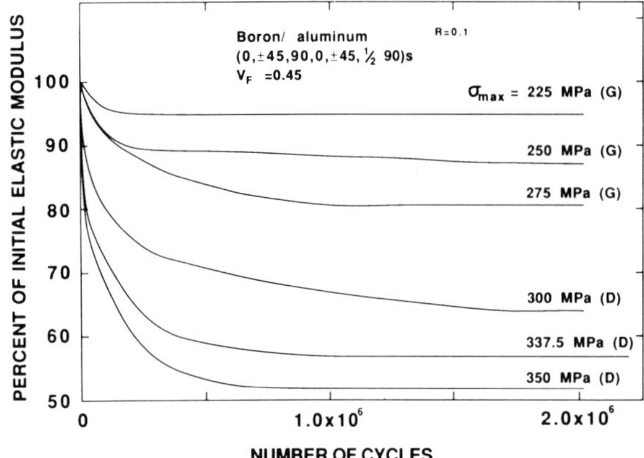

FIG. 4.32 The percentage of the initial elastic modulus vs. number of cycles of 45% V_f B/Al composite for various σ_{max} [61].

σ_{max}. Fatigue loading does not necessarily lead to the extensive matrix cracking. This is the case where fibers of smaller diameter with weak points are strongly bonded to the matrix metal, as in the case of FP fiber/Al composite. In this case extensive fiber fracture tends to occur during fatigue loading, and a fatigue crack propagates along a relatively straight plane. Figure 4.33 shows such an example of a multiple fiber fracture in the as-fatigued FP/Al composite [57].

The fatigue behavior of a short fiber metal matrix composite has hardly been studied, except for the work by Williams and Fine [63], who reported detailed fatigue data of SiC whisker/2124 Al (SiCw/Al) composites. The fatigue loading used for SiCw/Al composites is $R = -1$ (full reverse mode) in contrast to $R = 0.1 \sim 0.2$ for the continuous fiber (unidirectional) metal matrix composite discussed above. At different number of cycles (N), Williams and Fine took replicas from the as-fatigued composite specimens. These replicas were then used to examine the initiation and growth of cracks at different cycles. Since numerous microcracks of different lengths appeared in the as-fatigued specimens, the crack lengths (L) are assumed to obey the Weibull model with a probability density function $f(L)$. The results of $f(L)$ vs. L at different cycles (N) for 20% V_f SiCw/2124 Al-T6 are shown in Fig. 4.34, where $N = 1000$ (dots), 10,000 (dashes) and 16,000 cycles (solid lines). It is clearly seen from Fig. 4.34 that crack length increases with N. The growth of microcracks in a SiCw/Al composite during fatigue loading was then compared with that in the unreinforced matrix (Al 2124–T6) and the results are given in Fig. 4.35 where dots, dashes and solid lines denote the top 1%, and 10% of the Weibull cumulative distribution function, and the average crack length, respectively. Figure 4.35 clearly indicates the advantage in the fatigue crack resistance of the composite compared to the unreinforced

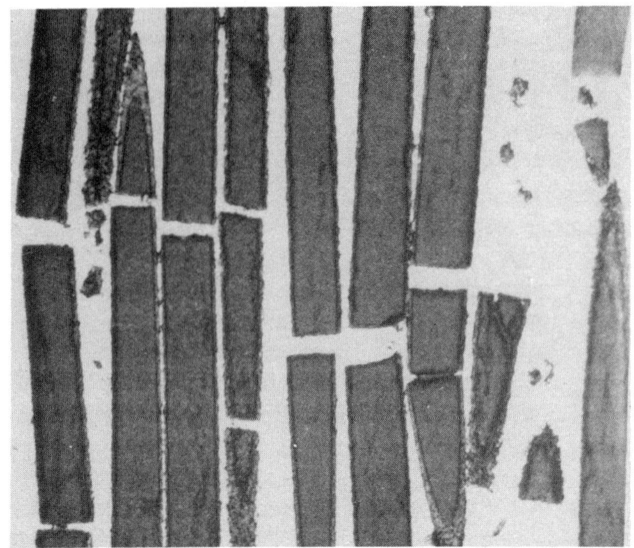

FIG. 4.33 A SEM photo of an as-fatigued runout FP/Al composite specimen, revealing an extensive multiple fiber fracture [57].

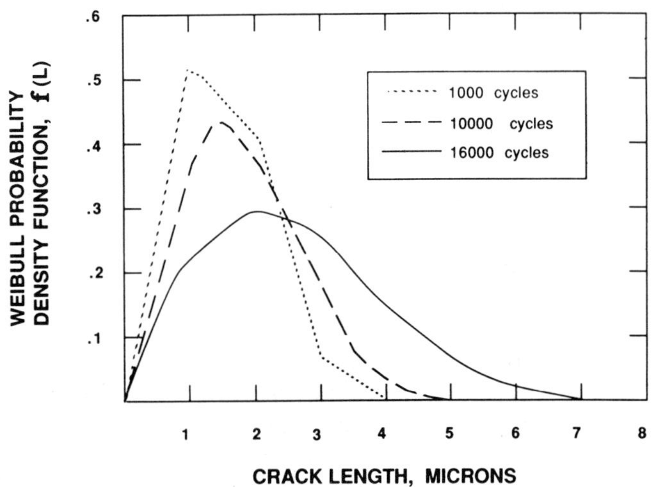

FIG. 4.34 The Weibull probability density function $f(L)$ as a function of crack length L for a 20% V_f SiCW/2124-T6 composite at different cycles [63].

matrix. Another important finding in their study was that microcracks of small length link to each other, leading to final fracture. This lack of damage tolerance may place severe restrictions on the use of this material in fatigued structures since cracks of this small size cannot be observed by *in situ* non-destructive methods.

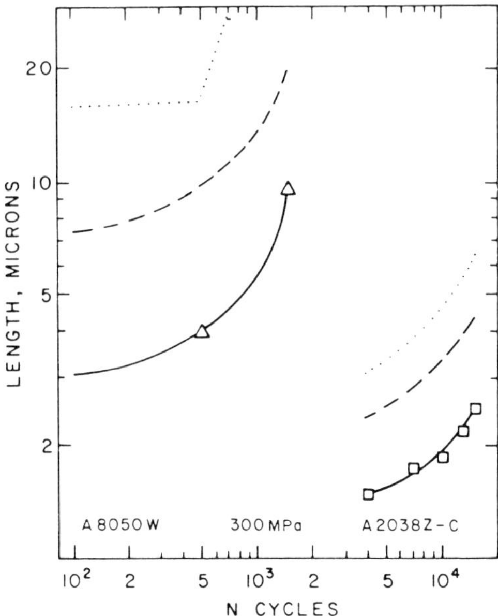

FIG. 4.35 Crack lengths corresponding to the top 1% (dot) and 10% (dash) values of the Weibull cumulative distribution function, and the average crack length (solid line) [63].

4.3.3 Wear

When a machine component is in contact with another component, and they move against each other, progressive loss of substance from the surface of the machine component can occur. This is called 'wear' of the component. If the component is made of a metal matrix composite then its wear resistance must be carefully assessed before the metal matrix composite can be justified as a structural material for the component.

There are a variety of methods to measure the wear resistance of a target material. The most popular method is the pin and disc (or cylinder) type. The apparatus of the pin and disc type is schematically shown in Fig. 4.36, where the case of the stationary pin and rotating disc is illustrated. In Fig. 4.36 the pin can be designed to rotate while the disc remains stationary, and the target component can be either the pin or the disc. Moreover, the pin and disc type can operate with or without lubricating oil. Based on one of these wear-resistance tests, one can measure the loss of the weight (or volume) of the target metal matrix composite, which depends on a number of parameters. The total wear volume V for most of the metals is known to be given by

$$V = kNs/H \tag{4.67}$$

Standard Wear Test

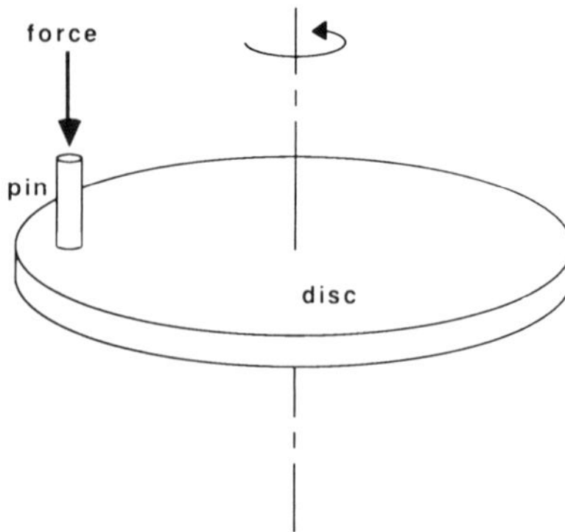

FIG. 4.36 A schematic view of a pin and disc type apparatus for measurement of wear resistance where a target metal matrix composite can be either pin or disc plate.

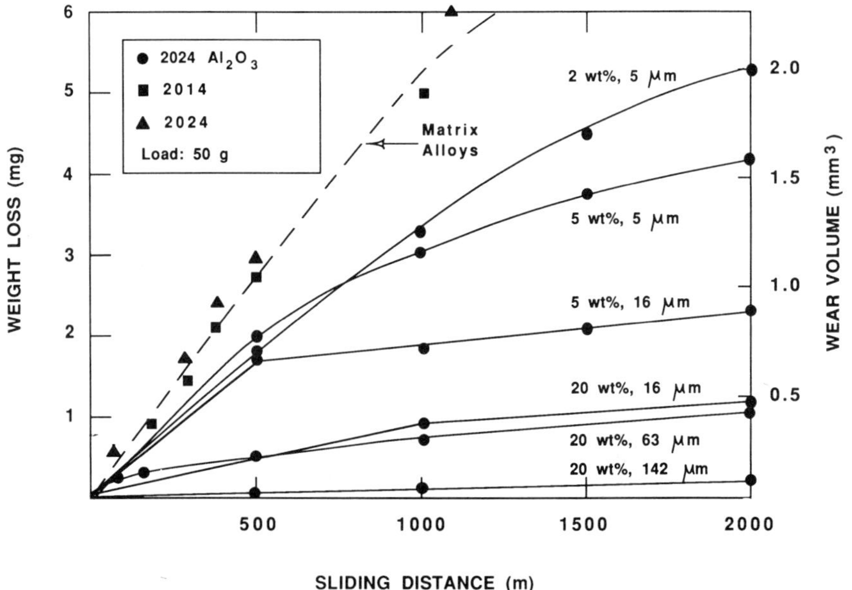

FIG. 4.37 The weight loss and wear volume of aluminum alloys with and without Al_2O_3 particles as a function of sliding distance [65].

where k is the wear coefficient, N is the normal load applied to the surface, s is the sliding distance and H is the hardness of a target material [64].

The metal matrix composites whose wear behavior has been studied so far are all particulate types [65–70]. Hosking et al. [65] studied the wear behavior of an Al_2O_3 particle/aluminum alloy (Al_2O_3/Al) composite and a

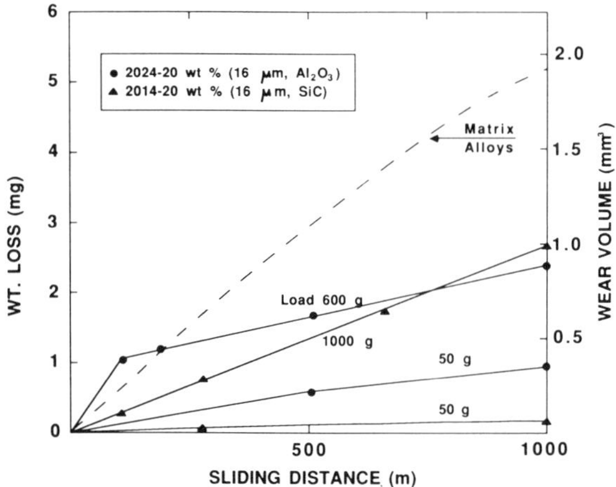

FIG. 4.38 The weight loss and wear volume of Al_2O_3/2024 Al and SiC/2014 Al composites as a function of sliding distance [65].

SiC particle/aluminum alloy (SiC/Al) composite. The weight percent of these particles ranges from 2 to 30%, and the size of the particles ranges from 1 to 142 μm. The wear test was conducted by the pin and disc type method where the discs are made of the target metal matrix composites and the pin tip is a E-5200 ballbearing (AISI type) of 6.4 mm in diameter. The discs slide at a speed of 0.1 m/s under the constant normal load with the range of 0.5 to 1.0 N. The results of the weight loss (or wear volume) of the unreinforced aluminum alloy and Al_2O_3/Al alloy composites are plotted as a function of sliding distance for various weight percents and sizes of particles (Al_2O_3 and SiC) in Fig. 4.37 where a constant normal load 50 g was used. It follows from Fig. 4.37 that the weight loss of the unreinforced aluminum alloy is larger than those of the Al_2O_3/Al composites, and among the composites those with greater weight percents and sizes of Al_2O_3 particles exhibit better wear resistance. The weight loss for all cases in Fig. 4.37 increases with sliding distances. The larger weight percents and sizes of Al_2O_3 particles will increase the hardness (H) of the composite. They also conducted additional wear tests on 20% SiC particle/2024 (SiC/Al) composites and the results of the weight loss (or wear volume) are also plotted as filled triangles against sliding distances in Fig. 4.38, where those of the unreinforced matrix metal (dashed line) and Al_2O_3/Al composites are also plotted for a comparison. It can be concluded from Figs. 4.37 and 4.38 that SiC particles (Vickers hardness 2600) are a more effective reinforcement than Al_2O_3 particles (Vickers hardness 1800) in improving wear resistance. The results of the work by Hosking et al. appear to support the wear resistance model of eq. (4.67). Namely, the longer sliding distances and the smaller the hardness of the composite become, the greater the weight (or volume) loss of the composite. Similar results have been reported by Anand and Kishore [68]. Using pin and disc with grinding paper attached, Surappa et al. [66] and Banerji et al. [67] have studied the abrasive wear resistance of two types of aluminum matrix composites, one with zircon and the other with alumina as reinforcing particles. It was found in their study that alumina particles are more effective reinforcement in increasing the abrasive wear resistance than zircon.

Volume (or weight) loss is controlled by applied normal force N, sliding distance s and hardness H, but it depends also on sliding (or friction) coefficient k (eq. 4.67). Thus, the wear resistance can be improved by reducing friction coefficient k. Akutagawa et al. [70] have recently developed a new particulate metal matrix composite which has low friction coefficient and weight loss under dry conductions. The friction and wear test used is the pin and disc type where the pin made of a target metal matrix composite is stationary while the steel disc plate of bearing quality rotates at speed of 2 m/s. All the tests were conducted for 1 hour under the normal pressure 0.1 MPa. They measured the friction coefficients and wear losses of SiC whisker/aluminum alloy (SC 114A) composites with several different additives which are self-lubricating types of material. First they demonstrated the effect of fiber volume fraction (V_f) on the friction coefficient (k) and weight loss (w). The results are shown in Fig. 4.39(a), which indicates that the addition of SiC whiskers to the matrix aluminum (SC 114A) reduces weight loss whereas it increases friction coefficient, where MMC 1, 2 and 3 denote the SiC whisker/Al (SC 114A) composite with $V_f = 20$, 30 and 40%, respectively. This increase in friction coefficient would require additional power, thus it is not economical from the viewpoint of fuel or energy saving. Thus, they added self-lubricating powders to the SiCw/Al composite. The results of the friction and wear tests are given in Fig. 4.39(b), where MMC 2 is the 30% V_f SiCw/Al without additives and MMC 4, 5 and 6 are the 30% V_f SiCw/Al

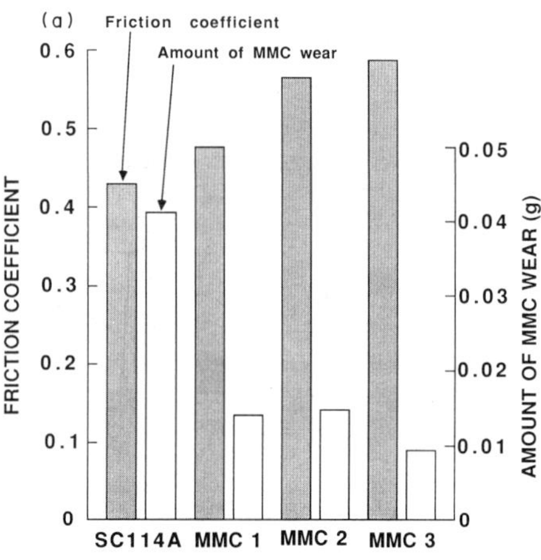

FIG. 4.39(a)

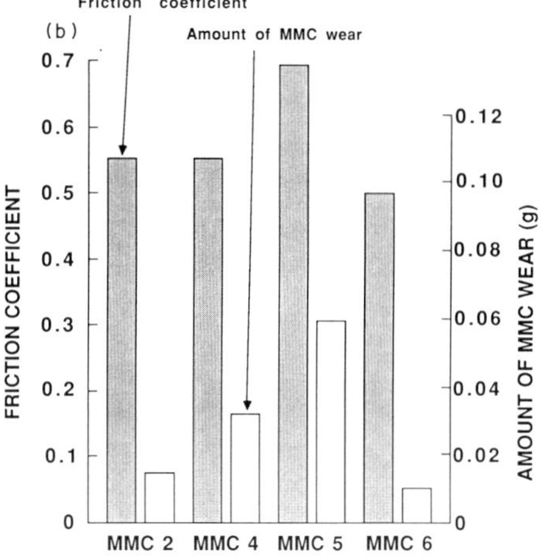

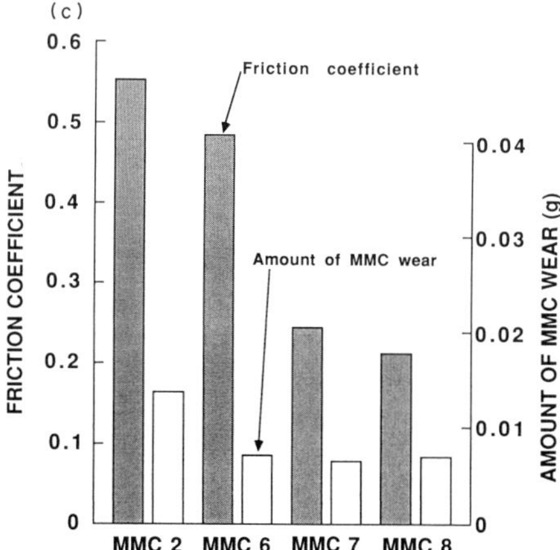

FIG. 4.39 The friction coefficient (k) and weight loss (W) of SiCW/Al composites; (a) unreinforced aluminum alloy and SiCW/Al composites $V_f = 20$, 30, and 30%, (b) 30% V_f SiCW/Al composites with and without self-lubricating additives, BN, graphite and SiC powders, and (c) 30% V_f/Al composites with and without self-lubricating materials, SiC powder, WS_2 precipitates [70].

composite with additives, BN, graphite and SiC powders, respectively. It follows from Fig. 4.39(b) that among the composites with self-lubricating additives, SiC powder (MMC 6) appears to be the best in reducing the weight loss as well as friction coefficient, although the reduction in the friction coefficient is small. These self-lubricating powder additives were pre-mixed with SiC whisker and thus the preforms were made of a mixture of SiC whisker and these powder additives. Finally, they attempted to use another self-lubricating additive, which was mixed with SiCw/Al composite through the precipitation route. The results of the friction and wear tests on the 30% V_f SiCw/Al composites with WS_2 precipitation (MMC 7 and 8) are shown in Fig. 4.39(c) where those of a SiCw/Al composite without any self-lubricating additives (MMC 2) and that with SiC powder (MMC 6) are also shown for a comparison. It can be concluded from Fig. 4.39(c) that the SiCw/Al composite with 1.5% V_f WS_2 precipitates provides the best performance in reducing both weight loss and friction coefficient under dry conditions.

Donomoto et al. [69] have also developed a new aluminum reinforced with alumina fibers (Al_2O_3/Al composite) with its application to the top corner portion of pistons for high-performance diesel engines. The wear test used consists of stationary block (target composite) and a rotating ring of cast iron. They also conducted the seizure tests to assess the seizure stress by using a stationary cylinder (target composite) and a rotating disc made of graphite iron, increasing the applied stress until the seizure starts occurring. The results of the wear and seizure tests indicated that the Al_2O_3/Al composite exhibits better wear and seizure resistance if the volume fraction of alumina fiber exceeds 4%. Use of alumina fibers as a reinforcement for the top corner of the diesel engine pistons can be justified also from the viewpoint of the thermal behavior of the composite.

4.4 Mechanical Behavior in Other Severe Environments

In addition to thermal and dynamic loadings, other severe environments to which the metal matrix composites may be subjected are oxidation, corrosion, and irradiation.

If a metal matrix composite is to be used as a component material for gas turbine engines or marine structures and vessels, then the assessment of its oxidation and corrosion behavior will become a prerequisite before the component made of this metal matrix composite can be justified as a reliable and stable material system for these applications. Both oxidation and hot corrosion occur at high temperatures, while corrosion can also be active at room temperature if a metal matrix composite is immersed in aqueous environments. When a metal matrix composite is designed to be used as a structural component in fusion reactors or in space environments when neutrons (and others) would bombard the surface of the metal matrix composite, the composite is expected to suffer from the irradiation. This is called 'irradiation damage.' We will discuss the behavior of metal matrix composites in these environments separately, below.

4.4.1 Oxidation

Oxidation will become active in a hot-air environment, and the oxidation of a metal matrix composite depends on the range of use temperature and matrix and fiber materials. The matrix metal that has been chosen for a given use temperature range is at least oxidation resistant. Fibers as a reinforcement for a given matrix metal can be categorized into two types: oxidation susceptible type and non-oxidation susceptible types.

Ceramic fibers are usually not susceptible to oxidation at the use temperatures of metal matrix composites which are considered lower than the threshold temperature at and above which the oxidation will become active, while most metal (refractory metal) fibers are susceptible to oxidation because their use temperatures are relatively high and also their reactivity with oxygen is strong. Table 4.6 shows the oxidation resistance of some of the refractory metals, alloys and silicon at 2200 °F in dried flow air [71] where the comparative factor denotes the degree of oxidation resistance, i.e., the larger the factor, the better the resistance to oxidation. It is clear from Table 4.6 that popular refractory fibers such as tungsten exhibit much weaker oxidation resistance in comparison with stainless steel and nickel alloys.

Table 4.6 *Oxidation of various materials at 2200 °F in dried, flowing air [71]*

Material (wt-%)	Exposure. (hr)	Metal loss (in./side)	Comparative factor
Molybdenum	0.5	0.018	1
Tantalum	1.25	0.032	1.4
Columbium	1.25	0.020	2.3
95Cb-5Cr	18.25	0.063	10
Hafnium	18.25	0.047	14
Tungsten	4.0	0.010	14
93Cb-7Mo	1.5	0.003	18
Zirconium	18.25	0.024	27
Titanium	66.25	0.044	54
Chromium	66.25	0.0075	320
67Cr-33Ni	66.25	0.0028	850
310 Stainless Steel	66.25	0.0015	1600
81Ni-12.5Cr-6.5Al	66.25	0.0006	4000
Silicon	66.25	<0.0001	large

Veltri and Galasso [72] reported on the room temperature strength of several popular fibers which were exposed to various high temperatures in air and argon, and the results are shown in Fig. 4.40 where (a) and (b) denote the results for argon and air immersion, respectively. It is noted in Fig. 4.40 that the room temperature strength of most of the fibers exposed to air is lower than those exposed to argon. The degradation in the room temperature strength of the as-air exposed tungsten fibers becomes severe above 700 °C, while that of boron fibers above 500 °C.

The mechanical behavior of metal matrix composites in an oxidation environment has been centered on the tungsten fiber (W) superalloy composite system [73]. The results have been summarized by Petrasec and Signorelli

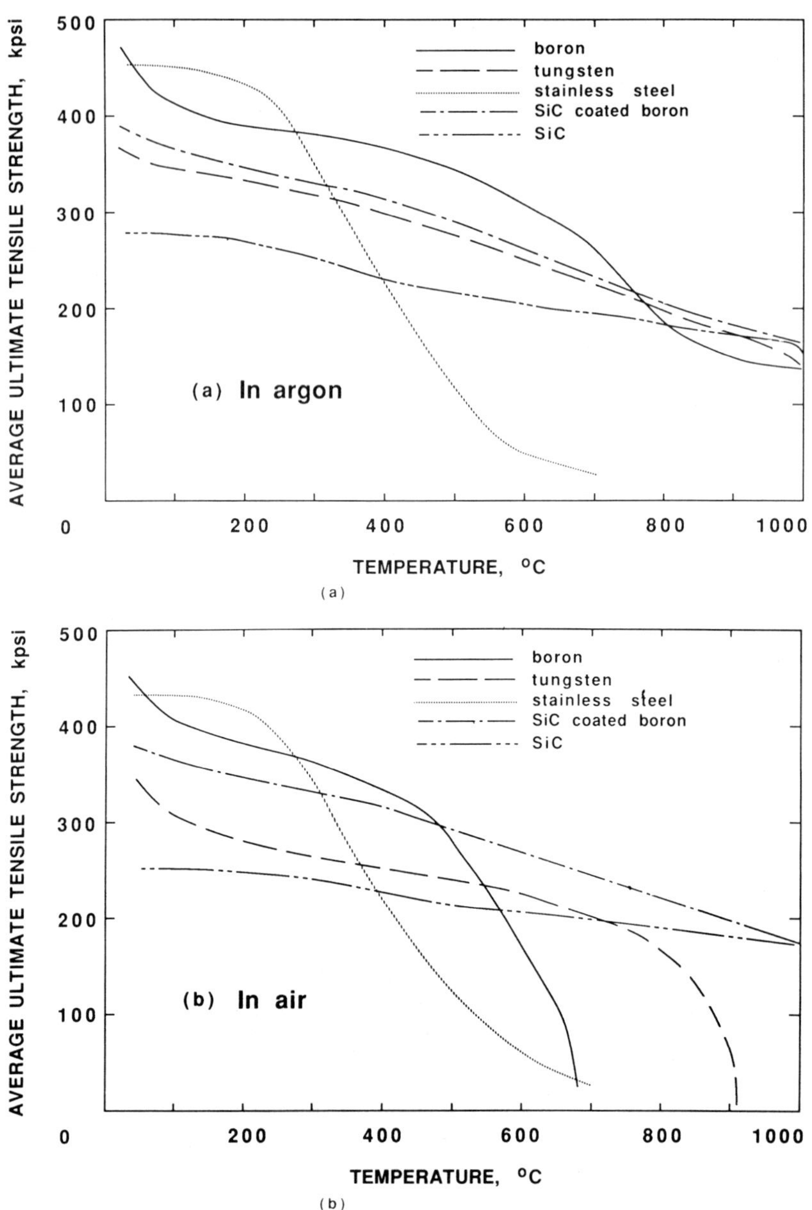

FIG. 4.40 The room temperature strength of several fibers that were exposed to various hight temperatures in argon (a) and in air (b) [72].

[33], who arrived at two conclusions: the matrix (super-alloy) itself must be sufficiently oxidation-resistant, and the tungsten fiber should be surrounded by such an oxidation-resistant matrix. Despite the requirement that all fibers are to be surrounded by a matrix metal, there exists a strong possibility that

some fibers are exposed directly to the oxidation environment. This can happen during the fabrication process or in service. Petrasec and Signorelli discussed two types of oxidation paths, one along the longitudinal (fiber axis) direction and the other along the transverse direction (perpendicular to the fiber axis). In the former type, oxidation is initiated at fiber ends particularly at the matrix–fiber interfaces, and it will penetrate progressively along the matrix–fiber interfaces. The depth (d) of the oxidation penetration along the fiber axis is usually not large, for example, $d = 2.5$ mm for a W-1% ThO_2 fiber/FeCrAlY composite exposed at 1200 °C for 10 hours, and $d = 1.3$ mm (in air) and 2.5 mm (in a simulated engine exhaust gas stream environment) for a W fiber/Nimocast 258 composite exposed at 1100 °C for 100 hours. The degradation mechanism for the longitudinal oxidation path suggested by El-Dahshan et al. [74] is as follows; during the exposure to oxidation environment the interdiffusion between matrix and fibers occurs, resulting in the formation of intermetallic compounds at the matrix–fiber interfaces. The formation of the intermetallic compounds involves a relatively large volume change, thus inducing a high stress field locally, which will, in turn, cause the nucleation of cracks in and around the intermetallic compounds. The cracks so nucleated will propagate into the matrix adjacent to the intermetallic compounds and fibers and then further into the major part of the matrix, thus leading to the shattering of the composite at the ends, as seen from Fig. 4.41 where the tungsten/Ni–20% Cr alloy composite exposed to

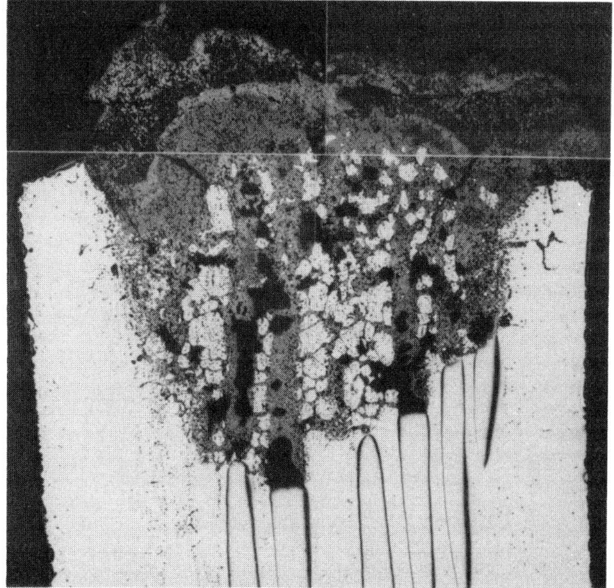

FIG. 4.41 A longitudinal cross-section of the damage induced in a W/Ni–20% Cr composite that was exposed to 1200 °C in air for 44 hours [74].

1200 °C in air for 44 hours was sectioned along the fiber axis [74]. There is the possibility that the volume change due to the formation of the intermetallic compounds will induce the relatively large shear stress along the interfaces between the intermetallic compounds and matrix, leading to the nucleation of the shear cracks at the interfaces [75]. The above mechanisms may contribute to the further penetration of the longitudinal oxidation path. On the other hand, the oxidation along the transverse direction may not be considered as serious as in the oxidation along the longitudinal path because a few thousandths of an inch of an oxidation-resistant matrix material is considered sufficient to protect the fibers close to the surface from oxidation under the oxidation environment of 2000 °F and 300 hours, and most of the fibers are fully embedded in the matrix and thus will not be degraded.

4.4.2 Corrosion

Corrosion can take place at both room and high temperatures, which can be called 'room temperature corrosion,' and 'hot corrosion,' respectively. Two types of corrosion can occur at room temperature—galvanic corrosion between the matrix metal and fibers, and pitting corrosion of the matrix metal itself. Whichever the type, room temperature corrosion often occurs when a metal matrix composite is immersed in aqueous solutions such as salt water. Galvanic corrosion is characteristic of metal matrix composites since the mismatch of the electrochemical potential between the matrix and fiber provides a local cell (short circuit) if they are immersed in aqueous solution. Galvanic corrosion damage has been reported on a number of metal matrix composite systems [76–79]. Trzaskoma studied the galvanic corrosion behavior of tungsten-fiber/depleted uranium (W/DU) composites [76] and

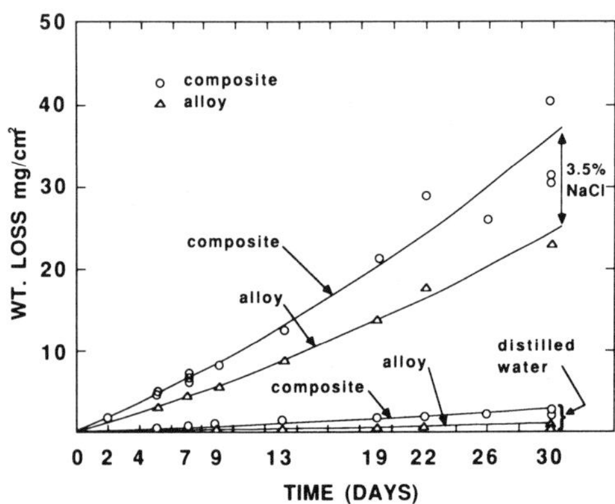

FIG. 4.42 The weight loss of the matrix (DU-0.75 Ti), fiber (W), and 50% V_f W/DU composite due to loose corrosion products in distilled water and 3.5% NaCl solutions [76].

graphite fiber/magnesium alloy (C/Mg) composites [77]. Figure 4.42 shows the weight loss of a DU alloy and 50% V_f W/DU composite subjected to immersion in a solution of distilled water and 3.5% NaCl [76]. Figure 4.42 illustrates the strong effect of immersion time and the immersion in 3.5% NaCl on the weight loss of the composite, which is identified as loose corrosion products. It is also noted from Fig. 4.42 that the weight loss was more significant for the composite than for the unreinforced matrix. This supports the occurrence of galvanic corrosion, which can be further verified by Fig. 4.43 where a relatively large mismatch of open circuit potential between the matrix (DU-4.75Ti) and fiber (W) exists [76]. Trzaskoma found also that the galvanic current between the matrix (DU-0.75Ti) and fiber (W) increases with the ratio of the exposed surface area to the total surface of the composite plate.

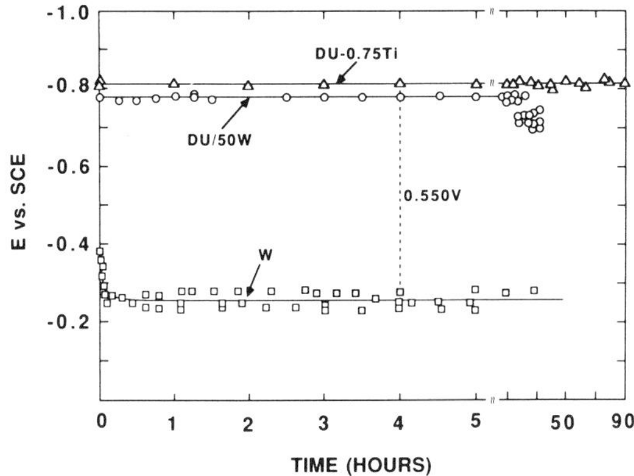

FIG. 4.43 Open circuit potentials of a DU-0.75 T_t (matrix), tungsten fiber and 50% V_f W/DU composite as a function of time [76].

Aylor and Kain [78] investigated the corrosion behavior of graphite fiber/aluminum (Gr/Al) composites and SiC/Al composites where SiC takes various forms, particulate (SiCp), whisker (SiCw) and continuous fiber (SiC_F). All of the composite specimens were immersed in various seawater conditions. SiC/Al composites were found to be more corrosion resistant than Gr/Al composites, which is attributed to the absence of any substantial galvanic driving force between the SiC and Al matrix. It was also found in this study that pitting from the surfaces of the composite plates is the dominant mode initially, followed by penetration of the pitting through the surface aluminum foil and then into the composite area (a mixture of graphite fibers and aluminum). A similar observation was also reported by Aylor et al. [80] and Vassilaros et al. [81]. Once the surface aluminum foils were broken by pitting, then the galvanic corrosion will be active, leading to the depletion of

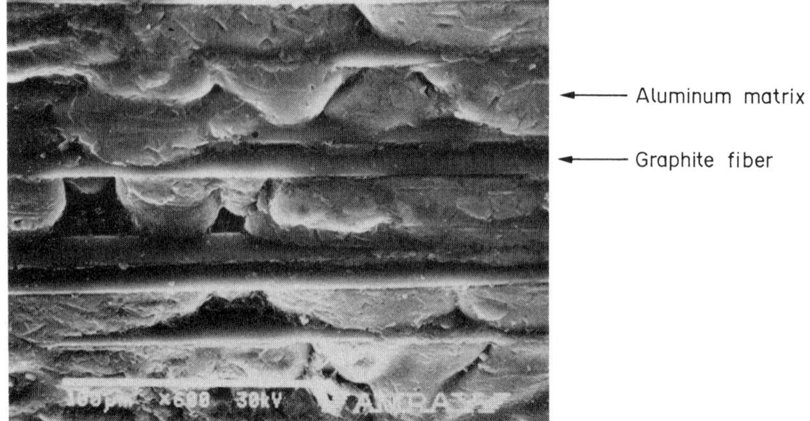

FIG. 4.44 Depleted aluminum in a Gr/Al composite that was subjected to seawater splash/spray test for 30 days [78].

aluminum, as evident in Fig. 4.44. Despite the fact that corrosion resistance is better for SiC/Al composites than Gr/Al composites, the SiC/Al composites exhibited a corrosion process similar to that in Gr/Al. Namely, the surface 1100 Al foils (the thickness is smaller than that of Gr/Al composite) were broken by pitting after 230 days tidal seawater immersion. After penetration of the surface foils, corrosion of the aluminum matrix continued and was concentrated around SiC particulates. Figure 4.45 illustrates a typical cross-section view through a SiCp/Al composite plate subjected to tidal seawater

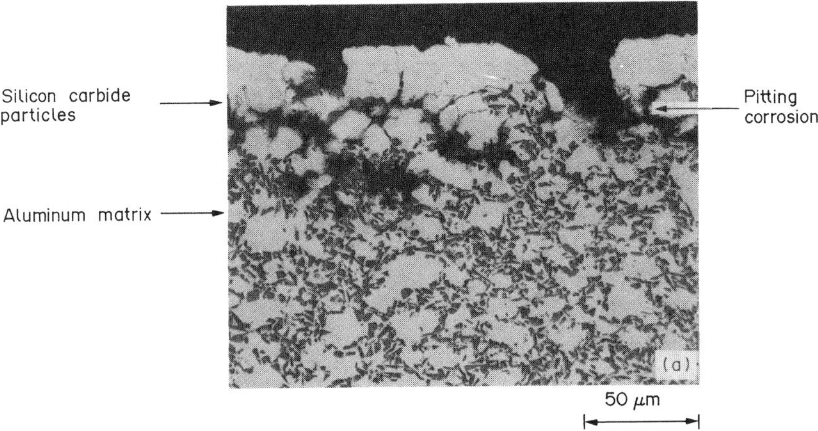

FIG. 4.45 A cross-section through a SiCp/Al composite plate subjected to tidal seawater immersion for 230 days [78].

immersion for 230 days, which reveals both the pitting site and the attack oriented around the SiC particulates (dark Blands in Fig. 4.45). This suggests that the crevices formed at the aluminum matrix–SiCp interfaces may contribute to corrosion.

Oxidation and hot corrosion can occur at high temperatures in a gaseous environment in the gas turbine. The gaseous environment contains not only oxygen, but also non-oxygen agents. The degradation of a material subjected to non-oxygen agents at high temperatures is called 'hot corrosion.' El-Dahshan et al. [74] have reported on the hot corrosion of a W/Ni–20% Cr composite. Figure 4.46 shows the hot corrosion damage in the W/Ni–20% Cr composite that was immersed in a H_2–10% H_2S gas environment at 900 ° C for 1 hour. The damage takes the form of sulfide scale in the matrix (Ni–20% Cr) but not in tungsten fibers. This is because the sulfide rate of tungsten is more than 100 times as slow as that of the Inconel corrosion resistant alloy, while the Ni–20% Cr matrix metal reacted with sulfur, resulting in the reaction products, nickel and chromium sulfides.

The degradation due to oxidation and corrosion can shorten the use life of the component. One of the effective methods of minimizing oxidation and corrosion damage is to put a thin coating zone on the surfaces of a metal matrix composite, or directly on the fibers. The coating of the former type will be used to retard the formation of pitting, whereas the coating of the second

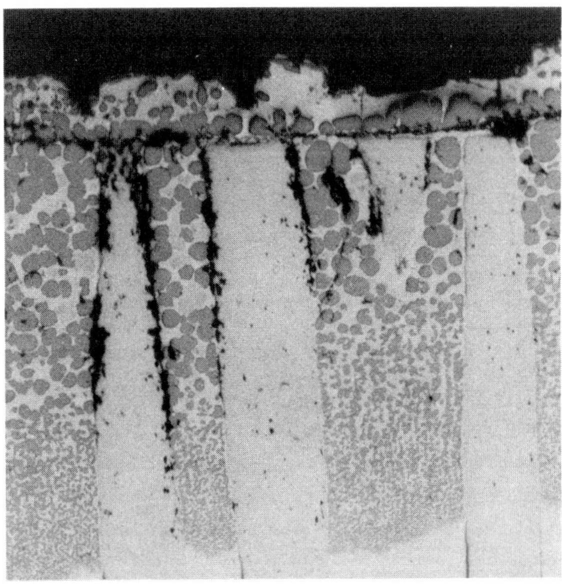

FIG. 4.46 A longitudinal cross-section of the W/N_i–20% Cr composite that was exposed to H_2–10% H_2S gas at 900 °C for 1 hour where sulfide scale was formed in the matrix, but not in the tungsten fiber [74].

type will be effective in protecting fibers from oxidation as shown in Fig. 4.40, where the strength of the SiC coated boron fiber was found to be larger than that of the boron fiber [72].

4.4.3 Irradiation

The effect of irradiation of neutrons, electrons, protons, and cosmic rays on the mechanical properties of metal matrix composites has hardly been studied despite the fact that metal matrix composites are considered for use in irradiation environments. Irradiation environments are unavoidable if metal matrix composites are designed to be used as a first-wall material of a fusion reactor [82] or as structural components of space structures [83]. As to the former application, Japanese researchers have made extensive studies of the effects of neutron irradiation on SiC fibers [84, 85] and SiC fiber/aluminum (SiC/Al) composites [86, 87]. A summary of the above studies will be reviewed below.

Okamura et al. [85] investigated the effect of neutron influence on the Young's modulus (E) and tensile strength (σ_{fb}) of SiC (Nicalon) fibers. As-suppled Nicalon SiC fibers were subjected to neutron irradiation of up to 2×10^{24} n/m^2 below 300 °C in an He gas environment and the results of E and σ_{fb} are summarized in Fig. 4.47. It follows from Fig. 4.47 that E and σ_{fb} increase slightly with neutron fluence. They also subjected SiC fiber/

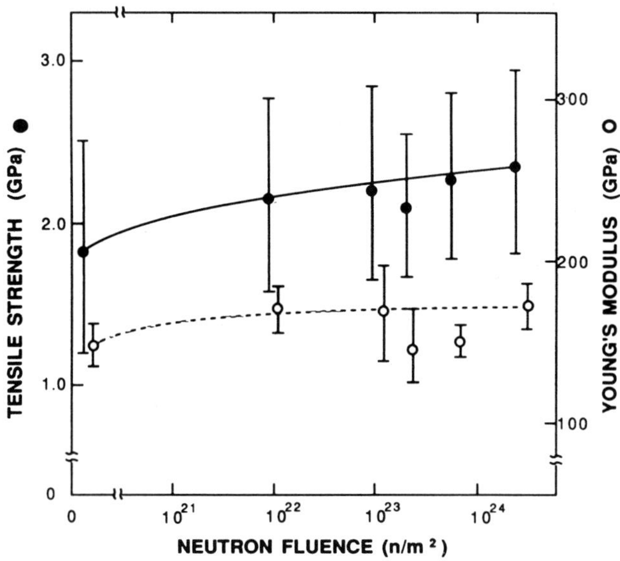

FIG. 4.47 Effect of neutron influence on the tensile strength (filled circles) and Young's modulus (open circles) of SiC (Nicalon) fiber [85].

aluminum composite wires to neutron fluence of up to $1 \times 10^{25}\,n/m^2$ at 450 °C. SiC fibers were then extracted from the SiC/Al composite wires and were subjected to tensile tests to measure E and σ_{fb}. A similar conclusion was obtained for as-extracted SiC fibers, i.e., E and σ_{fb} increased slightly with neutron fluence.

Kohyama et al. [86, 87] further elaborated on the work of Okamura et al. [85] by subjecting SiC fibers and SiC/Al composite wires to neutron irradiation at higher temperatures (420 and 720 K). Exposure to neutron irradiation at a higher temperature (720 K) appears to give rise to smaller flexural strength than at a lower temperature (420 K), but neutron fluence was again found to increase the flexural strength. Such evidence is shown in Fig. 4.48 where the average flexural strength of as-neutron fluenced SiC/Al composite wires (open square boxes) and the tensile strength of as-extracted SiC fibers (open circles) are shown as a function of neutron fluence at 720 K. (JOYO is the fast breeder reactor used for the study.) In the same figure the law of mixtures prediction is indicated by a solid curve, and it appears to agree with

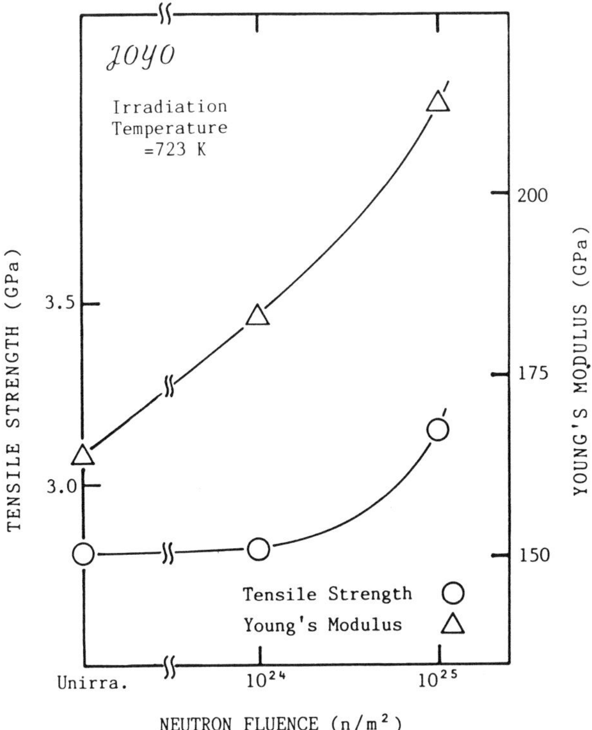

FIG. 4.48 Influence dependence of tensile strength and Young's modulus of SiC (Nicalon) monofilaments irradiated at 72 K by JOYO reactor [87].

the experimental values (open boxes). The strength of fiber (σ_{fb}) used for the law of mixtures prediction (eq. (3.3)) obtained by using the results of as-extracted SiC fibers (open circles in the figure) agree well with the experiment. Kohyama et al. further studied the effect of thermal exposure and neutron

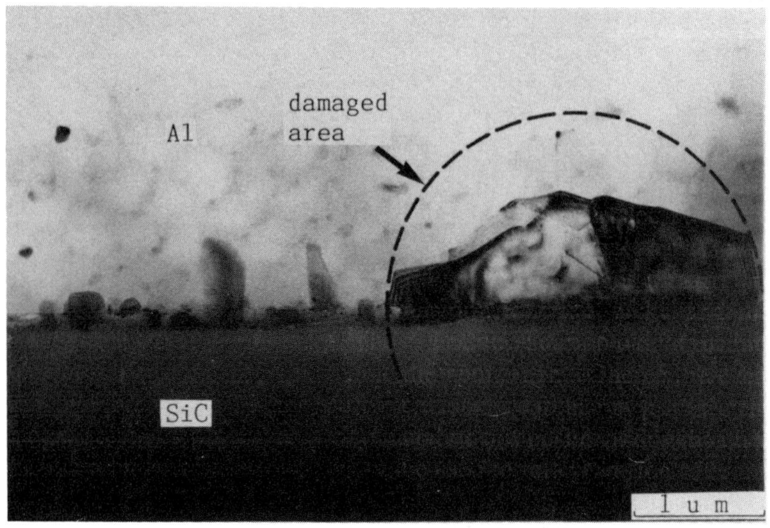

FIG. 4.49 Difference of microstructure in SiC/Al composite under HVEM with (inside the dotted circle) and without (outside the dotted circle) 1 meV electron bombardment for 93 minutes at 760 K [87].

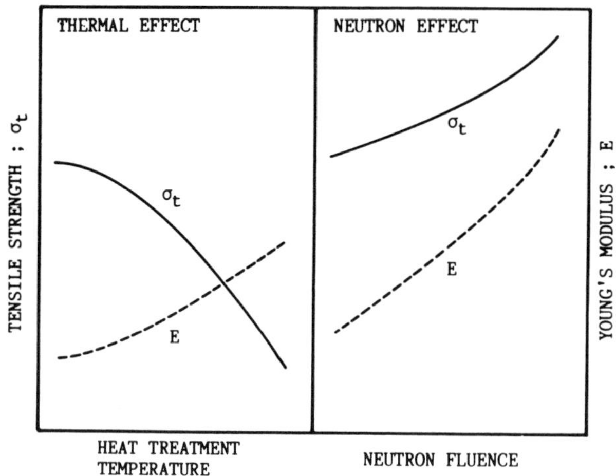

FIG. 4.50 Schematic diagrams of thermal and neutron effects on the tensile strength and Young's modulus of preform wires [87].

irradiation combined with thermal effect on the flexural strength and stiffness of SiC fiber/aluminum composite sheets, and also that of electron bombardment on the composite [87]. Based on these experiments they found that electron bombardment tends to suppress the formation of SiC crystals at the matrix–fiber interfaces, which is the case with thermal exposure, and also to reduce the void formation in the aluminum matrix. Figure 4.49 shows a TEM photo of the SiC fiber–aluminum interface which consists of two regions—a thermally affected zone (dotted circle) and the other damaged by electron beam (relatively smooth interface). They concluded that the thermal exposure increases the stiffness (E), but decreases the strength of the composite (σ_t) while the neutron irradiation increases both E and σ_t as illustrated in Fig. 4.50. Therefore, a SiC fiber/aluminum composite appears to be an ideal candidate for a structural material for the first wall of a fusion reactor.

4.5 Problems

(1) Why does the transverse strength σ_T of a continuous fiber metal matrix composite exposed to high temperature increase with the exposure time t, while the longitudinal strength σ_L of the same composite exposed to the same high temperature decreases with t?

(2) Describe the micro-mechanism for the creep behavior of an aligned short fiber loaded transversely.

(3) Why does superposition of a constant load accelerate the dimensional change of a metal matrix composite that is already under thermal cycling?

(4) Describe a possible mechanism for the transverse swelling of an as-thermally cycled unidirectional metal matrix composite.

(5) Discuss the main difference between the Garmong and the Yoda et al. thermal cycling damage models. What is the main difference in the predicted results between them?

(6) What would be the most effective reinforcing geometry that will give rise to better creep resistance for the loading in all directions?

(7) What would be an appropriate creep model for a particulate metal matrix composite?

(8) Discuss the effect of fiber volume fraction V_f and fiber aspect ratio l/d on the degree of strain-rate sensitivity of a metal matrix composite.

(9) What would be the difference in the fatigue behavior between polymer and metal matrix composites?

(10) If one wants to apply the method of the analysis by Williams and Fine, that was developed for a short fiber metal matrix composite, to the case of a continuous fiber metal matrix composite, what would be the modification of the Williams and Fine model?

(11) Beside the key parameters used in eq. (4.67), what would be additional parameters that influence the wear resistance of a metal matrix composite?

(12) Describe the micro-wear modes for short fiber and continuous fiber metal matrix composites. What would be the major difference between them?

(13) From the oxidation resistance viewpoint, identify a few examples of coated fiber metal matrix composite systems that have better oxidation resistance at high temperature in dry flowing air condition.

(14) When intermetallic compounds are formed at the matrix–fiber interfaces of a unidirectional metal matrix composite, where would be possible sites of cracking due to the formation of the compounds? Explain the reasons for the cracking.

(15) What would be other possible severe environments that have not been discussed in this chapter, yet will cause the degradation of a metal matrix composite?

(16) Why does thermal exposure reduce the strength of a SiC/Al composite, whereas neutron irradiation increases it?

(17) In addition to a SiC/Al composite, what would be another good candidate as a structural material for the first wall of a fusion reactor?

References

1. Winsa, E. A. *Mechanical Behavior of Metal Matrix Composites*, edited by J. E. Hack and M. F. Amateau, TMS of AIME, 1983, pp. 283–299.
2. Metcalfe, A. G. and Klein, M. J., *Composite Materials*, Vol. I, edited by Broutmann and Krock, Academic Press, 1972, pp. 125–168.
3. Kim, W. H., Koczak, M. J. and Lawley, A., *Proc. ICCM-II*, AIME, edited by B. Norton et al. 1978, pp. 487–505.
4. Kyono, T., Hall, I. W. and Taya, M., *J. Mater. Sci.*, Vol. 21, 1986, pp. 1879–1888.
5. DiCarlo, J. A., *Mechanical Behavior of Metal Matrix Composites*, edited by J. E. Hack and M. F. Amateau, AIME, 1983, pp. 1–14.
6. Okura, A. and Motoki, K., *Composites Science and Technology*, Vol. 24, 1985, pp. 243–252.
7. Rosen, B. W., *AIAA J.*, Vol. 2, 1964, pp. 1985–1991.
8. DiCarlo, J. A. and Smith, R. J., NASA TM-82806, 1982.
9. Wright, M. A. and Intwala, B. D., *J. Mater, Sci.*, Vol. 8, 1973, pp. 957–963.
10. Olsen, G. C. and Tomplins, S. S., NASA Tech. Paper 1063, 1977.
11. Kyono, T., Hall, I. W. and Taya, M., *J. Mater. Sci.*, Vol. 21, 1986, pp. 4269–4280.
12. Hunn, D. L., *Mechanical Behavior of Metal Matrix Composites*, edited by J. E. Hack and M. F. Amateau, AIME, 1983, pp. 83–94.
13. Kashiwaya, H., Morita, M., Nagata, K., Nishimura, T., Miyamoto, T. and Ishiwata, Y., *Composites, 86: Recent Advances in Japan and the United States*, edited by K. Kawata et al., Japan Society for Composite materials, 1986, pp. 529–536.
14. Lilholt, H. and Taya, M., *Proc. ICCM-6/ECCM-2*, Vol. 2, edited by F. Matthews et al., Elsevier, 1987, pp. 2.234–2.244.
15. Mclean, M., *Proc. R. Soc. Lond.*, Vol. A371, 1980, pp. 279–294.
16. Mclean, M., *Proc. R. Soc. Lond.*, Vol. A373, 1980, pp. 93–109.
17. Bullock, E., Mclean, M. and Miles, D. E., *Acta Metal.*, Vol. 25, 1977, pp. 333–344.
18. Kelly, A. and Tyson, W. R., *J. Mech. Phys. Solids*, Vol. 14, 1966, pp. 177–186.
19. DeSilva, A. R. T. *J. Mech. Phys. Solids*, Vol. 16, 1968, pp. 169–186.
20. Mileiko. S. T., *J. Mater. Sci.*, Vol. 5, 1970, pp. 254–261.
21. Mclean, D., *J. Mater. Sci.*, Vol. 7, 1972, pp. 98–104.
22. Kelly, A. and Street, K. N., *Proc. R. Soc. Lond.*, Vol. A328, 1972, pp. 283–293.
23. Garmong, G., *Metal. Trans.*, Vol. 45, 1974, p. 1257–1258.
24. Lilholt, H., *Comp. Sci. Tech.*, Vol. 22, 1985, pp. 277–294.
25. Taya, M. and Lilholt, M., *Advances in Composite Materials and Structures*, ASME bound volume, edited by S. S. Wang and Y. Rajapakse, in press.
26. Kelly, A. and Street, K. N., *Proc. R. Soc. Lond.*, Vol. A328, 1972, pp. 267–282.
27. Nieh, T. G., *Metal. Trans.*, Vol. 15A, 1984, pp. 139–146.
28. Min, B. K., *J. Mech. Phys. Solids*, Vol. 29, 1981, pp. 327–352.
29. Crossman, F. W. and Min, B. K., ASME WAM, 1981.
30. Johnson, A. F., *J. Mech. Phys. Solids*, Vol. 25, 1977, pp. 117–126.
31. Yoda, S., Kurihara, N., Wakashima, K. and Umekawa, S., *Metal. Trans.*, Vol. 9A, 1978, pp. 1229–1236.
32. Yoda, S., Takahashi, R. Wakashima, K. and Umekawa, S., *Metal. Trans.*, Vol. 10A, 1979, pp. 1796–1798.
33. Petrasek, D. W. and Signorelli, R. A., *Ceramic Eng. Sci. Proc.*, Vols . 7 and 8, July–August, 1981, pp. 739–786.
34. Grimes, H. H., Lad, R. A. and Maisel, J. E., *Metal. Trans.*, Vol. 8A, 1977, pp. 1999–2005.
35. Olsen, G. C. and Thompkins, S. S., *Failure Modes in Composites*, Vol. IV, edited by J. A. Cornie and F. W. Crossman, TMS of AIME, 1977, pp. 1–21.
36. Wright, M. A., *Metal. Trans.*, Vol. 6A, 1975, pp. 129–134.
37. Kim, W. H., Koczak, M. J. and Lawley, A., *New Developments and Applications in Composites*, edited by D. Kuhlmann-Wilsdorf and W. C. Harrigan, Jr, TMS of AIME, 1979, pp. 40–53.
38. Bhatt, R. T. and Grimes, H. H., *Mechanical Behavior of Metal–Matrix Composites*, edited by J. E. Hack and M. F. Amateaux, TMS of AIME, 1983, pp. 51–64.
39. Park, Y. H. and Marcus, H. L., ibid., pp. 65–75.
40. Patterson, W. G. and Taya, M., *Proc. ICCM-V*, edited by W. C. Harrigan, Jr, et al., TMS of AIME, 1985, pp. 53–66.

41. Kyono, T., Hall, I. W. and Taya, M., *Composites '86: Recent Advances in Japan and the United States*, edited by K. Kawata et al. Japan Society for Composite Materials, 1986, pp. 553–561.
42. Wu, M. Y. and Sherby, O. D. *Scripta Metal.*, Vol. 18, 1985, pp. 773–776.
43. Garmong, G., *Metal. Trans.*, Vol. 5, 1974, pp. 2183–2190.
44. Garmong, G. ibid., pp. 2199–2205.
45. Derby, B., *Scripta Metal.*, Vol. 19, 1985, pp. 703–707.
46. Taya, M. and Mori, T., *Proc. IUTAM Symposium: Thermo-Mechanical Couplings in Solids*, Elsevier, Science edited by H. D. Bui and Q. S. Nguyen, 1987, pp. 147–162.
47. Min, B. K., *Advances in Composite Materials and Structures*, ASME bound volume, edited by S. S. Wang and Y. Rajapakse, in press.
48. Mukherjee, A. K., Bird, J. E. and Dorn, J. E., *Trans. ASM*, Vol. 62, 1969, pp. 155–179.
49. Manson, S. S., *Thermal Stress and Low-Cycle Fatigue*, McGraw-Hill, 1966.
50. Mori, T., Okabe, M. and Mura, T., *Acta Metal.*, Vol. 28, 1980, pp. 319–325.
51. Frost, H. J. and Ashby, M. F., *Deformation–Mechanism Maps*, Pergamon Press, 1982.
52. Gittus, J., *Creep, Viscoelasticity and Creep Fracture in Solids*, Applied Science Publishers, 1975.
53. Schuster, D. M. and Reed, R. P., *J. Comp. Mater.*, Vol. 3, 1969, pp. 562–576.
54. Gray, T. D., *Fatigue of Composite Materials*, ASTM STP 569, 1975, pp. 262–279.
55. Marchand, A., Duffy, J., Christman, T. A. and Suresh, S., Brown University Tech. Report to ONR, December 1986.
56. Harding, J., Taya, M., Derby, B. and Pickerd, S., *Proc. ICCM-6/ECCM-2*, edited by F. L. Matthews et al. Elsevier Applied Science, Vol. 2, 1987, pp. 2.224–2.233.
57. Tsangarakis, N., Slepetz, J. M. and Nunes, J., *Recent Advances in Composites in the United States and Japan*, ASTM STP 864, edited by J. R. Vinson and M. Taya, 1985, pp. 131–152.
58. Nunes, J., Chin, E. S. C., Slepetz, J. M. and Tsangarakis, N., *Proc. ICCM-5*, edited by W. C. Harrigan, Jr, et al., TMS of AIME, 1985, pp. 723–745.
59. Chamis, C. C. and Sullivan, T. L., *Fatigue of Composite Materials*, ASTM STP 569, 1975, pp. 95–114.
60. Stinchcomb, W. W., Reifsnider, K. L., Marcus, L. A. and Williams, R. S. *Fatigue of Composite Materials*, ASTM STP 569, 1975, pp. 115–129.
61. Dvorak, G. J. and Johnson, W. S., *Int. J. Fract.*, Vol. 16, 1980, pp. 582–602.
62. Gouda, M., Prewo, K. M. and McEvily, A. J., *Fatigue of Fibrous Composite Materials*, ASTM STP 723, 1981, pp. 101–115.
63. Williams, D. R. and Fine, M. E., *Proc. ICCM-5*, edited by W. C. Harrigan, Jr, et al., TMS of AIME 1985, pp. 639–670.
64. Halling, J., *Principles of Tribology*, Macmillan Press, 1978.
65. Hosking, F. M., Portillo, F. F., Wunderlin, R. and Mehrabian, R., *J. Mater. Sci.*, Vol. 17, 1982, pp. 447–498.
66. Surappa, M., Prasad, S., Rohatgi, P., *Wear*, Vol. 77, 1982, pp. 295–302.
67. Banerji, A., Prasad, S., Surappa, M. Rohatgi, P., *Wear*, Vol. 82, pp. 141–151.
68. Anand, K. and Kishore, *Wear*, Vol. 85, 1983, pp. 163–169.
69. Donomoto, T., Funatani, K., Miura, N. and Miyake, N., SAE Tech. paper 830252, 1983.
70. Akutagawa, K., Ohtsuki, H., Hasegawa, J. and Miyazaki, M., SAE Tech. paper 870441, 1987.
71. Grant, N. J., *Refractory Metals and Alloys*, Proceedings of the Metallurgical Society Conference, Detroit, Michigan, 25–26 May, 1960, Vol. 11, 1961, pp. 119–168.
72. Veltri, R. D. and Galasso, F. S., *J. Amer. Ceram. Soc.*, Vol. 54(6), 1971, pp. 319–320.
73. Essock, D. M., *Engineered Materials Handbook*, Vol. 1, ASM International, 1987, pp. 878–888.
74. El-Dahshan, M. E., Whittle, D. P. and Stringer, J., *Oxidation of Metals*, Vol. 9(1), 1975, pp. 45–67.
75. Kvernes, I. and Kofstad, P., *Scand. J. Metal.*, Vol. 2, 1973, pp. 291–297.
76. Trzaskoma, P. P., *J. Electrochem. Soc.*, Vol. 129(7), 1982, pp. 1398–1402.
77. Trzaskoma, P. P., *Corrosion*, Vol. 42(10), 1986, pp. 609–613.
78. Aylor, D. M. and Kain, R. M., *Recent Advances in Composites in the United States and Japan*, ASTM STP 864, edited by J. R. Vinson and M. Taya, 1985, pp. 632–647.
79. Paciej, R. C. and Agarwala, V. S., *Corrosion*, Vol. 42(12), 1986, pp. 718–729.
80. Aylor, D. M., Ferrara, R. J. and Kain, R. M., *Materials Performance*, Vol. 23, No. 7, 1984.
81. Vassilaros, M. G., Davis, D. A., Steckel, G. L. and Gudas, J. P., *Mechanical Behavior of*

Metal Matrix Composites, edited by J. E. Hack and M. F. Amateau, TMS of AIME, 1983, pp. 335–352.
82. Hopkins, G. R. and Price, R. J., *Nuclear Eng. Design/Fusion*, Vol. 2, 1985, pp. 111–143.
83. Armstrong, H. H., *Proc. SAMPE National Symp.*, San Francisco, CA, 1979.
84. Okamura, K., Matsuzawa, T., Sato, M., Higashiguchi, Y. and Morozumi, S., *J. Nucl. Mater.*, 1985, *pp.* 705–708.
85. Okamura, K., Matsuzawa, T., Sato, M., Higashiguchi, Y., Morozumi, S. and Kohyama, A., *J. Nucl. Mater.*, 1986, pp. 102–109.
86. Kohyama, A., Tezuka, H., Igata, N., Imai, Y., Teranichi, H. and Ishikawa, T., *J. Nucl. Mater.*, 1986, pp. 96–101.
87. Kohyama, A., Tesuka, H. and Igata, N., *Proc. ICCM-6/ECCM-2*, edited by F. L. Matthews et al., Elsevier, Vol. 2, 1987, pp. 2.245–2.254.

CHAPTER 5

Thermal Behavior

5.1 Introduction

The thermal behavior of metal matrix composites has not been considered at the same level as the mechanical behavior. This is because the mechanical behavior of the composite must be shown to be significantly better than that of the matrix; if so then studies of thermal behavior will be undertaken. One of the use environments for most of the metal matrix composite systems is the thermal environment, particularly in high temperatures. Thus the mechanical behavior of metal matrix composites in high temperatures has been studied extensively (see Section 4.2 in Chapter 4). In the analysis of the mechanical behavior of metal matrix composites in thermal environments, one must determine the temperature distribution first, which will then be used to study the mechanical behavior. Determination of the temperature is obtained by solving the heat conduction equation in a composite where the specific heat (c) and thermal conductivity (K_{ij}) of the composite must be known. Then the analysis of stresses in a composite subjected to the temperature field so obtained will follow. Namely, the thermoelasticity equations, eqs. (2.3) and (2.5), will be solved where coefficient of thermal expansions (CTE) of the composite, α_{ij}, are to be found. Thus, the analysis of the mechanical behavior of a metal matrix composite requires the analysis of the thermal behavior as a prerequisite step. The basic thermal properties are: CTEs, K_{ij}, and c, which will be discussed in the following sections.

5.2 Coefficient of Thermal Expansion

Of the possible thermal behavior characteristics, the behavior of thermal expansion of metal matrix composites has been the characteristic most extensively studied, since it affects the mechanical behavior of the composites in severe thermal environments, especially the application of composites in engine components and space structures. In these applications the stability of components and structures made of metal matrix composites over a long period of time becomes the critical design consideration. The stability can be described in two aspects, geometrical changes and mechanical property changes. In the former case the coefficient of thermal expansion (CTE) of composite structures and components plays a key role, while in the latter case the mismatch of CTEs between the metal matrix and fibers has a dominant effect. In this section we will discuss CTEs of metal matrix composites which can be either obtained experimentally or can be predicted by analytical models.

There exist two different CTEs, linear and volumetric ones. The linear CTE is the more commonly measured one. The measurement of the linear CTE (CTE, hereafter) is usually made by a quartz tube dilatometer [1, 2]. A typical quartz tube dilatometer consists of a heating unit, a quartz tube connecting rod, a displacement measurement unit where the change in the length of the specimen is measured through the quartz tube in contact with the specimen, which in turn is measured by a displacement measurement unit. Two types of displacement measurement unit are popular: the linear variable differential transformer (LVDT) and the optical device. A schematic view of the dilatometer with the optical displacement measurement unit is shown in Fig. 5.1. With the dilatometer the length change ΔL of the specimen of length L is measured. Figure 5.2 shows the results of $\Delta L/L$ vs. T (K) curves that were measured by the quartz tube dilatometer for the longitudinal direction of boron fiber/aluminum $(B/Al)_0$, its transverse direction $(B/Al)_{90}$ [3]. The slope of $\Delta L/L$ vs. T curve is the CTE, which is usually a function of T. Hence one must define the CTE value for a given temperature (instantaneous CTE) or for a given range of temperature (average CTE). Christian and Campbell [2] measured the average CTEs of several types of aluminum-based composites and the results are tabulated in Table 5.1. It is seen from Table 5.1 that the average CTEs of B/Al or BORIC/Al systems increase slightly with T.

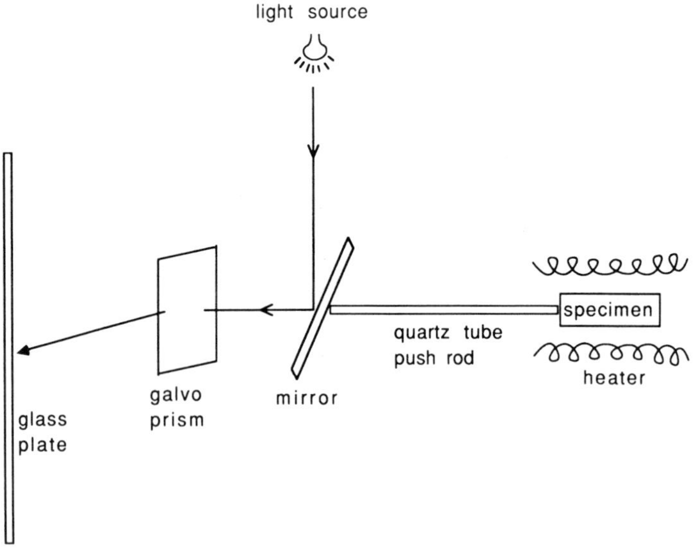

FIG. 5.1 A dilatometer with quartz tube and optical device.

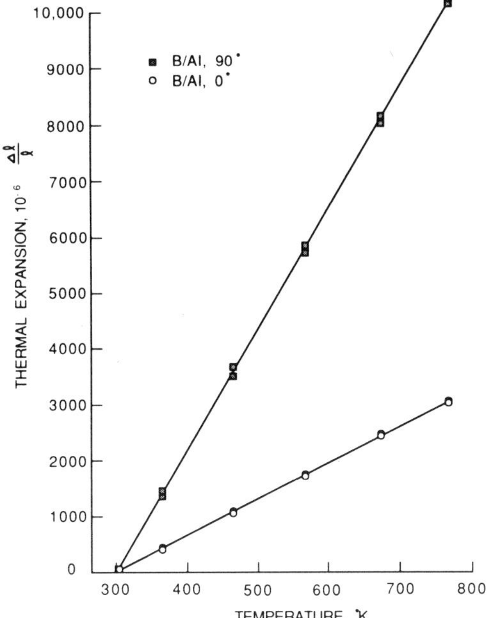

FIG. 5.2 Thermal expansion ($\Delta l/l$) vs. temperature for boron fiber/aluminum composite [3].

The longitudinal (axial) CTEs of unidirectional metal matrix composites are small due to the small axial CTE value of continuous fibers which are either refractory metals or ceramics. The measurement of such CTEs by the quartz tube dilatometers (which require some kind of mechanical contacts) would give rise to crude values of CTEs [4]. To avoid this, one must account for the deformation of specimen ends [4] or use more precise measurement techniques, for example, a non-contact dilatometer [5–7]. An example of a non-contact dilatometer is the two-channel Michelson laser interferometer [7] which would induce the error in CTE of order $1.2 \times 10^{-9}\,K^{-1}$. Wolff measured CTEs of several classes of unidirectional metal matrix composites by using both a contact dilatometer and a non-contact dilatometer with laser interferometer [5–7].

Most of the experimental data of CTEs are on unidirectional metal matrix composites where both longitudinal (along the fiber axis) and transverse (perpendicular to the fiber axis) CTEs were measured [1–4, 7, 8]. Some of these experimental data have been compared with the prediction by earlier analytical models. Kreider and Patarini [1] measured both longitudinal (α_0) and transverse CTEs (α_{90}) of unidirectional BORSIC fiber/aluminum composites which were then compared with law of mixtures values, Schapery's model [9], and Kreider and Patarini's model [1]. The results are shown in

Table 5.1 *Average thermal expansion of aluminum based composites* [2]

Material	Direction	CTE (20^{-6}/K) 77–144 K	144–297 K	297–450 K	450–644 K
B/Al*	0	0.65	1.17	1.86	1.87
	90	2.78	4.61	5.56	7.37
BORSIC/Al	0	0.81	1.39	1.73	1.85
	90	2.68	4.64	5.04	6.08
BORSIC/Ti/Al	0	0.72	1.16	1.70	2.21
	90	1.61	3.23	3.94	4.20
BORSIC/SS/AL	0	0.96	1.30	1.83	1.69
	90	2.78	4.85	5.86	7.59
6061-T6		3.70	6.06	8.12	8.39

* Al here denotes 6061.

Table 5.2 *Comparisons of thermal conductivities* [25]

Materials	Conductivities (BTU/(h·ft °F))			
	K_L		K_T	
	Analytical	Experimental	Analytical	Experimental
Composites				
Gr(V50054)/ Al(6061-T6)	143 ($V_f = 36\%$)	142 ($V_f = 36\%$)	45.28 ($V_f = 36\%$)	44 ~ 53 ($V_f = 30 \sim 37\%$)
		130 ($V_f = 37\%$)		59 ($V_f = 30\%$) 46 ($V_f = 38 \sim 41\%$)
Gr(V50054)/ Mg(AZ91C)	112 ($V_f = 38\%$)	106 ($V_f = 38\%$) 107 ($V_f = 37\%$)	18.6 ($V_f = 38\%$)	20 ($V_f = 38\%$) 17 ~ 18 ($V_f = 37 \sim 38\%$)
Constituents				
6061-T6 Al		96		96
AZ91C Mg		41		41
V50054 Gr		229		0.2

Fig. 5.3 where (a) and (b) denote the results of α_0 and α_{90}, respectively. In both figures the subscripts L and U for CTEs denote the upper and lower bounds proposed by Levin [7, 10].

$$\alpha_{0L} = \frac{\overline{E\alpha}}{\overline{E}} < \alpha_0 < \frac{\overline{K\alpha}}{\overline{K}} = \alpha_{0U} \tag{5.1}$$

where the bar denotes the volume average quantity, E is the Young's modulus and K is the bulk modulus and related to E by $K = E/3(1 - 2v)$ and where v is Poisson's ratio. Levin's model given by eq. (5.1) is valid only for isotropic constituents. Similarly, the upper (α_{90U}) and lower bounds (α_{90L}) on the transverse CTE have been proposed by Schapery [9] and they are given by

$$\begin{aligned} \alpha_{90L} &= \bar{\alpha} = \alpha_m V_m + \alpha_f V_f \\ \alpha_{90U} &= (1 + v_f)\alpha_f V_f + (1 + v_m)\alpha_m V_m - \alpha_0(v_f V_f + v_m V_m) \end{aligned} \tag{5.2}$$

where α_0 is the exact (or experimental) value of the longitudinal CTE of the composite. The analytical results of Schapery's bounds are shown in

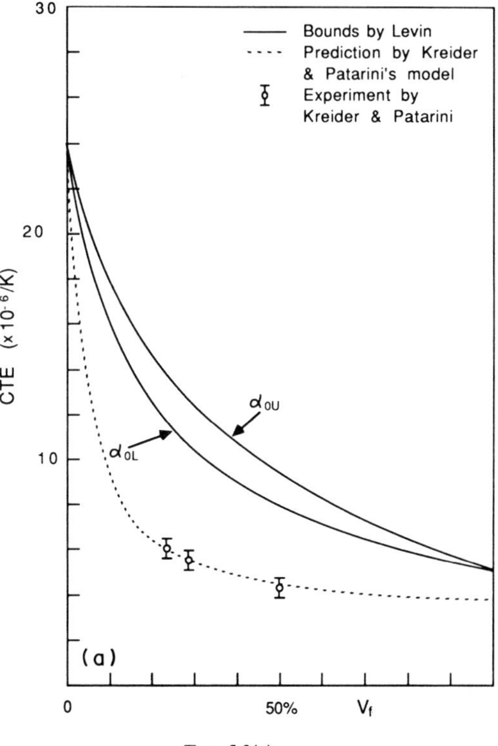

FIG. 5.3(a)

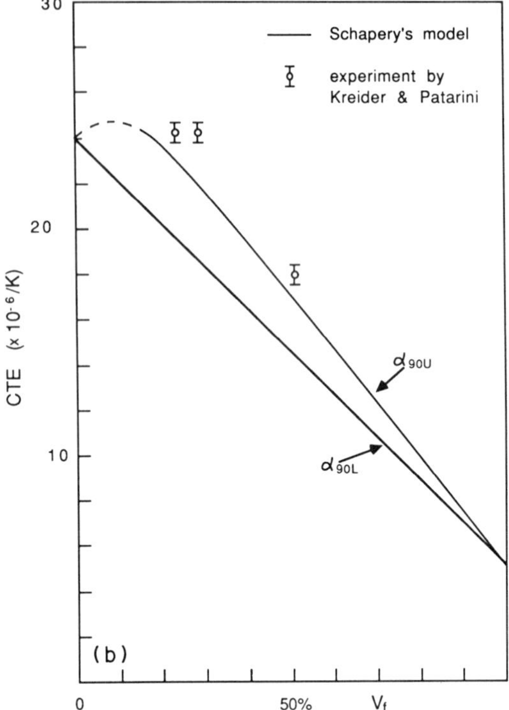

FIG. 5.3 CTEs of BORSIC fiber/aluminum composite; (a) longitudinal (along fiber axis) and (b) transverse direction (perpendicular to the fiber axis).

Fig. 5.3(b). The dashed line in Fig. 5.3(a) is the prediction based on a simple model proposed by Kreider and Patarini [1], who accounted for the yielded matrix metal in the composite which underwent a temperature change ΔT, cooling down from its fabrication temperature to room temperature. The formula to predict α_0 is thus given by

$$\alpha_0 \Delta T = \alpha_f \Delta T + \frac{A_m \sigma_{my}}{A_f E_f} \tag{5.3}$$

where α_f is fiber axial CTE, A_i is the cross-sectional area of the ith phase with $i = \text{m}$ (matrix) of f (fiber), E_f is the Young's modulus of fiber, and σ_{my} is the matrix yield stress. Although this model predicts well the experimental results, it should be valid only for intermediate fiber volume fractions. Namely, as A_f (which is equivalent to fiber volume fraction V_f) goes to zero, $\alpha_0 \Delta T$ goes to infinity, which is not acceptable.

As short fiber metal matrix composites have become increasingly popular, a number of analytical models to predict CTEs have also been proposed

[11–16]. Short fibers are usually not aligned, but randomly distributed in a metal matrix due to processing. Some of the above models have dealt with such a random short fiber system [12, 13, 15]. Among these models, Eshelby's model [11, 15, 16] is considered to be applicable to the most general type of short fiber metal matrix composite, i.e., a semi-randomly distributed short fiber composite system which includes, as special cases, aligned short fiber and completely randomly distributed short fiber systems. The Eshelby model that was originally developed for dilute suspension [17] has been modified for finite fiber volume fractions (see for examples Ref. 18 and subsection 2.4.4), and therefore can account for interaction between short fibers and fiber misorientation. The application of the Eshelby model for the prediction of the CTEs will be briefly reviewed for a specific case of an aligned short fiber composite [11, 16]. The analytical model for prediction of CTEs of an aligned short fiber composite is the same as that used to calculate the thermal stress field in the composite, i.e., Fig. 3.52 where the entire composite is subjected to a uniform temperature change ΔT. Then, the governing equations are given by eqs. (3.93) to (3.95) where $\boldsymbol{a}$ is given by

$$\boldsymbol{a} = (\boldsymbol{a}_{\mathrm{f}} - \boldsymbol{a}_{\mathrm{m}})\Delta T \tag{5.4}$$

where $\boldsymbol{a}_{\mathrm{f}}$ and $\boldsymbol{a}_{\mathrm{m}}$ are the CTE tensors of fiber and matrix, respectively. The **tildes beneath** symbols denote tensorial quantity (index forms are replaced by **these tildes** for simplicity). By eliminating $\mathbf{e}$ and $\bar{\mathbf{e}}$ from eqs. (3.93) to (3.95), we have

$$\{-V_{\mathrm{f}}(\mathbf{C}_{\mathrm{f}} - \mathbf{C}_{\mathrm{m}})(\mathbf{S} - \mathbf{I}) + (\mathbf{C}_{\mathrm{f}} - \mathbf{C}_{\mathrm{m}})\mathbf{S} + \mathbf{C}_{\mathrm{m}}\}\mathbf{e}^* = \mathbf{C}_{\mathrm{f}}\boldsymbol{a} \tag{5.5}$$

where $\mathbf{C}_{\mathrm{m}}$ and $\mathbf{C}_{\mathrm{f}}$ are stiffness tensors of the matrix metal and fiber, respectively, $\mathbf{S}$ is the Eshelby's tensor and $\mathbf{I}$ is the identity tensor. The eigenstrain $\mathbf{e}^*$ will be solved from eq. (5.5). The volume average of the strain induced by ΔT in the entire composite (domain D) γ_{D} is given by

$$\gamma_{\mathrm{D}} = \frac{1}{V_{\mathrm{D}}}\int_{\mathrm{D}} \{\bar{\mathbf{e}} + \mathbf{e}\}\,\mathrm{d}V \tag{5.6}$$

where the above integral is over the entire composite body D with its volume V_{D}. Noting that $\bar{\mathbf{e}}$ is originally defined by the average strain disturbance in the matrix, we obtain

$$\int_{\mathrm{D}-\Omega} \{\bar{\mathbf{e}} + \mathbf{e}\}\,\mathrm{d}V = V_{\mathrm{D}-\Omega}\bar{\mathbf{e}} \tag{5.7}$$

where Ω denotes the fiber domain, thus $\mathrm{D} - \Omega$ is the matrix domain with its volume given by $V_{\mathrm{D}-\Omega}$. From eqs. (3.95) (5.6) and (5.7), we have

$$\gamma_{\mathrm{D}} = V_{\mathrm{f}}\mathbf{S}\mathbf{e}^* \tag{5.8}$$

The CTEs of the composite, $\boldsymbol{\alpha}_c$, are the sum of $\boldsymbol{\alpha}_m$ and $\gamma_D/\Delta T$, and are obtained as

$$\boldsymbol{\alpha}_c = \boldsymbol{\alpha}_m + V_f \mathbf{S} \mathbf{e}^*/\Delta T \tag{5.9}$$

A substitution of $\mathbf{e}^*$ from eq. (5.5) into eq. (5.9) yields the CTEs of the composite. By using eq. (5.9) Takao [15] has calculated CTEs of short fiber and particulate SiC/aluminum (SiC/Al) composites with various V_f and fiber aspect ratios and the results are plotted as a function of fiber volume fraction V_f in Fig. 5.4, where α_0 and α_{90} denote the CTEs of the composite along the fiber axis and its perpendicular direction, respectively. The thermomechanical properties of Al and SiC are given in Appendix B where those of other fibers and metals are also given. It is noted in Fig. 5.4 that the CTE of a composite with fiber aspect ratio $l/d = 1$ gives rise to an isotropic composite, $\alpha_0 = \alpha_{90}$, thus both solid and dashed lines coincide. Takao and Taya [16] have examined the effect of other material parameters on the CTEs of an aligned short carbon fiber/aluminum (C/Al) composite. The effect of the fiber aspect ratio l/d, the matrix Poisson's ratio ν_m and the longitudinal stiffness of carbon fiber C_{33} are shown in Fig. 5.5(a), (b), and (c), respectively, where α_{cL} and α_{cT} denote α_0 (longitudinal) and α_{90} (transverse) of C/Al composite, respectively. It follows from Fig. 5.5(a) that the maximum peak of α_{CT} is more

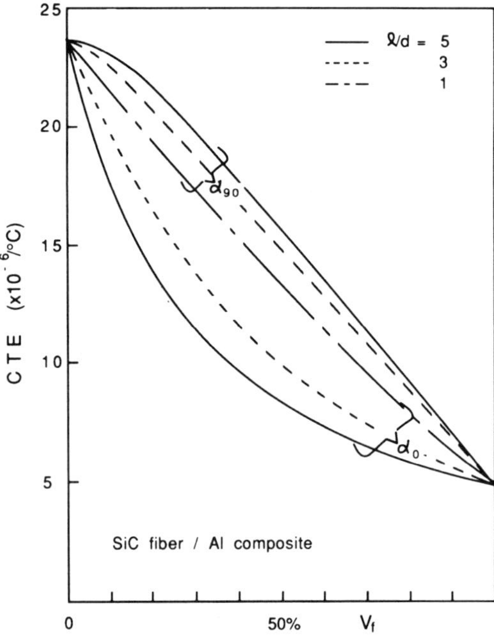

FIG. 5.4 CTEs vs. fiber volume fraction V_f for aligned short fiber composite [15].

1.5
$\frac{\alpha_c}{\alpha_m}$
α_{CT}
ℓ/d
100 ~ 1000
10
1.0
0.5
α_{CL}
ℓ/d
10
100 ~ 1000
(a)
0.0
0.0 0.1 0.2 0.3 0.4 0.5
V_f

1.5
$\frac{\alpha_c}{\alpha_m}$
α_{CT}
ν_m
0.4
0.33
0.2
1.0
0.5
α_{CL}
0.2 ~ 0.4
(b)
0.0
0.0 0.1 0.2 0.3 0.4 0.5
V_f

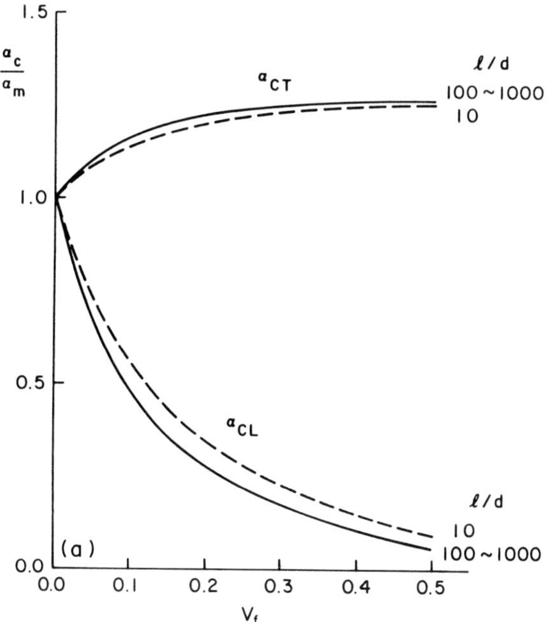

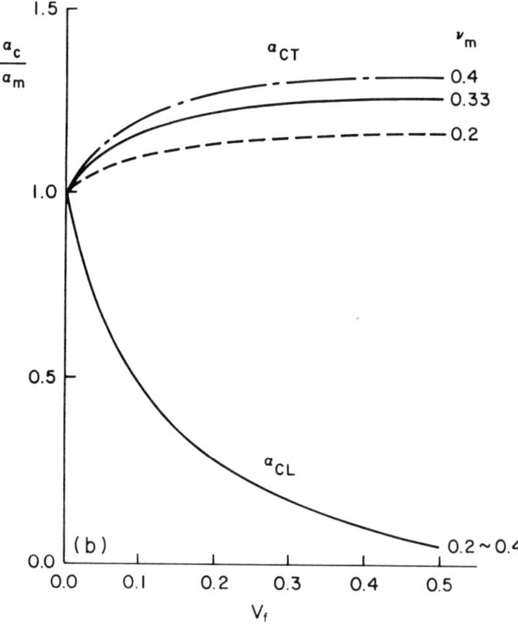

FIG. 5.5(a, b)

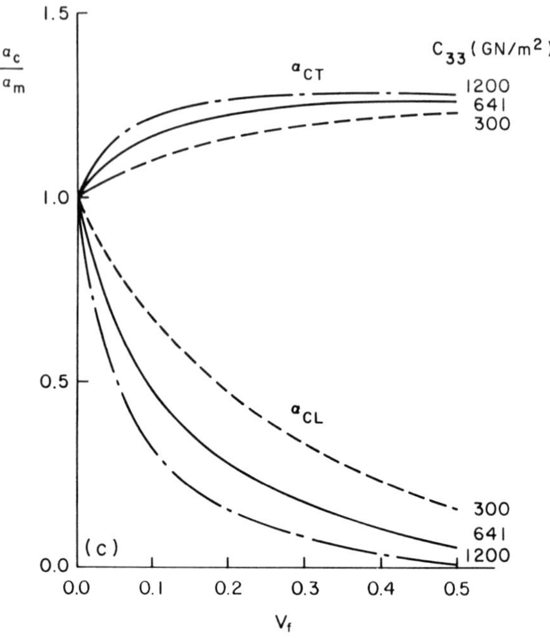

FIG. 5.5 CTEs of C/Al composite predicted by Takáo and Taya [16]; (a) fiber aspect ratio l/d being a parameter, (b) matrix Poisson's ratio ν_m and axial stiffness of carbon fiber C_{33}.

exaggerated for a C/Al as compared with a SiC/Al composite (Fig. 5.4). The effect of ν_m on α_{cT} is visible, but that on α_{cL} is not. Figure 5.5(c) indicates that C_{33} has measurable impact on both CTEs of the composite. The results of Fig. 5.5 suggest that a metal matrix composite with zero CTE (along the fiber axis) can be easily tailored. If both matrix and reinforcing phase are isotropic and the geometry of the reinforcing phase becomes simplified, the formulae to predict CTEs of such composites are given by simple equations [11]. For spherical reinforcement, the composite CTE (α_c) is given by

$$\alpha_c = \bar{\alpha} + V_f(1 - V_f)(\alpha_f - \alpha_m)\frac{K_f - K_m}{(1 - V_f)K_m + V_f K_f + (3K_m K_f)/(4\mu_m)} \tag{5.10}$$

where

$$\bar{\alpha} = (1 - V_f)\alpha_m + V_f \alpha_f \tag{5.11}$$

and where K and μ are the bulk and shear moduli, α_c is isotropic and coincides with Kerner's prediction [19], and also with Schapery's upper bound [9]. For a disc-shaped reinforcement the composite CTE parallel to

($\alpha^c_{11} \equiv \alpha^c_{22}$) and perpendicular to disc plane (α^c_{33}) are given by

$$\alpha^c_{11} = \alpha^c_{22} = \bar{\alpha} + \bar{\bar{\alpha}} \cdot \frac{2(\nu_f E_m - \nu_m E_f)}{(1 - V_f)(1 - \nu_f)E_m + V_f(1 - \nu_m)E_f}$$
$$\alpha^c_{33} = \bar{\alpha} + \bar{\bar{\alpha}} \cdot \frac{(1 - \nu_m)E_f - (1 - \nu_f)E_m}{(1 - V_f)(1 - \nu_f)E_m + V_f(1 - \nu_m)E_f} \tag{5.12}$$

where

$$\bar{\bar{\alpha}} = V_f(1 - V_f)(\alpha_f - \alpha_m) \tag{5.13}$$

For the long fiber reinforcement the composite CTE along the fiber axis (x_3-axis) and perpendicular to it (x_1–x_2 plane) are given by

$$\alpha^c_{11} = \alpha^c_{22} = \bar{\alpha} + \bar{\bar{\alpha}}\left\{\frac{(C - D)E_f}{(AC - BD)} - 1\right\}$$
$$\alpha^c_{33} = \bar{\alpha} + \bar{\bar{\alpha}}\left\{\frac{(A - B)}{(AC - BD)} E_f - 1\right\} \tag{5.14}$$

where

$$A = \frac{(1 - V_f)}{(1 - \nu_m)}\{2\nu_m \nu_f \mu_f + 2(1 - 2\nu_f)\mu_m\} + 2V_f(1 - \nu_f)\mu_f$$
$$B = \frac{(1 - V_f)}{(1 - \nu_m)}\{\nu_m \mu_f + (1 - 2\nu_f)\nu_m \mu_m\} + 2V_f \nu_f \mu_f \tag{5.15}$$
$$C = \frac{(1 - V_f)}{(1 - \nu_m)}\{\mu_f + (1 - 2\nu_f)\mu_m\} + 2V_f \mu_f$$
$$D = \frac{(1 - V_f)}{(1 - \nu_m)}\{2\nu_f \mu_f + 2(1 - 2\nu_f)\nu_m \mu_m\} + 4V_f \nu_f \mu_f$$

The foregoing Eshelby model is based on the assumption that both matrix and reinforcing phase deform elastically during temperature change ΔT. For a large ΔT the matrix metal is expected to deform plastically. Wakashima et al. [11] considered the case of a plastically deforming matrix in an aligned short (or continuous) fiber metal matrix composite. They assumed that a large temperature change ΔT induces uniform plastic strain $\mathbf{e}^p$ in the matrix while the fibers remain elastic. Then the thermal expansions of the composite per unit temperature change, $\boldsymbol{\alpha}_c$ are given

$$\boldsymbol{\alpha}_c = \boldsymbol{\alpha}_m + (V_f \mathbf{S}\mathbf{e}^* + \mathbf{e}^p)/\Delta T \tag{5.16}$$

where $\mathbf{e}^*$ is solved from eqs. (3.93) to (3.95) except that $\boldsymbol{\alpha}$ defined by (5.4) is now replaced by $\boldsymbol{\alpha}^*$ given by

$$\boldsymbol{\alpha}^* = \boldsymbol{\alpha} - \mathbf{e}^p \tag{5.17}$$

The solution for $\mathbf{e}^*$ still contains unknown $\mathbf{e}^{\mathrm{p}}$, which can be solved from the energy balance equation, i.e., the change in the total potential energy being equal to the plastic work in the matrix [11, 20]. The analytical results of thermal expansion of a continuous fiber metal matrix composite based on the above elastic–plastic Eshelby model have been obtained explicitly by Wakashima et al. [11] and they are given by

$$e_{33}^{\mathrm{c}} = \alpha_{33}^{\mathrm{c}} \Delta T + e_{\mathrm{p}} \left\{ 1 - V_{\mathrm{f}} \cdot \frac{\mu_{\mathrm{f}}(1-2\nu_{\mathrm{f}})(2C+D)}{(AC-BD)} \right\}$$

$$e_{11}^{\mathrm{c}} = e_{22}^{\mathrm{c}} = \alpha_{11}^{\mathrm{c}} \Delta T - \frac{1}{2} e_{\mathrm{p}} \left\{ 1 - V_{\mathrm{f}} \cdot \frac{2\mu_{\mathrm{f}}(1-2\nu_{\mathrm{f}})(A+2B)}{(AC-BD)} \right\} \tag{5.18}$$

with

$$e_{\mathrm{p}} = \frac{2(1+\nu_{\mathrm{m}})E_{\mathrm{f}}}{E^*}(\alpha_{\mathrm{f}} - \alpha_{\mathrm{m}})\Delta T \pm \frac{(1-\nu_{\mathrm{m}})(AC-BD)\sigma_{\mathrm{y}}}{V_{\mathrm{f}}(1-2\nu_{\mathrm{f}})\mu_{\mathrm{m}}\mu_{\mathrm{f}}E^*} \tag{5.19}$$

and

$$E^* = 2(1+\nu_{\mathrm{m}})E_{\mathrm{f}} + 3(1-V_{\mathrm{f}})(1-2\nu_{\mathrm{f}})E_{\mathrm{m}} + 3V_{\mathrm{f}}(1-2\nu_{\mathrm{m}})E_{\mathrm{f}} \tag{5.20}$$

where e_{11}^{c} $(= e_{22}^{\mathrm{c}})$ and e_{33}^{c} are the thermal expansion of the composite along and perpendicular to the fiber axis (x_3-axis), α_{ij}^{c}, A, B, C and D are given by eq. (5.15), and the + and − signs in eq. (5.19) correspond to $(\alpha_{\mathrm{f}} - \alpha_{\mathrm{m}})\Delta T < 0$ and $(\alpha_{\mathrm{f}} - \alpha_{\mathrm{m}})\Delta T > 0$, respectively. Wakashima et al. also conducted experiments on a continuous tungsten fiber/copper (W/Cu) composite with $V_{\mathrm{f}} = 0.32$ and the results of the axial strain vs. temperature relationship are plotted as solid lines in Fig. 5.6 [11] where the cases of Cu and W fiber are also plotted as dash–dot and dash lines, respectively. It follows from Fig. 5.6 that the hysteresis loop shown consists of two regions, one corresponding to the solid lines with a steep slope (Region I) and the other with a less steep slope (Region II). They found that Regions I and II can be predicted by the elastic analysis, eq. (5.14) and the elasto-plastic analysis, eq. (5.18), respectively. Namely, they plotted the experimental results of α_{33}^{c} in Regions I and II as a function of fiber volume fraction V_{f} by open circles in Fig. 5.7 where the analytical results by eqs. (5.14) for Region I and eq. (5.18) for Region II are also plotted by solid lines. Figure 5.7 indicates that the two regions are distinguishable from each other and the analytical model can predict the CTEs in these regions well. The α_{33}^{c} dependence on V_{f} when the matrix plastic deformation is expected to occur during temperature change ΔT was also studied by Kreider and Patarini [1], as discussed earlier. The analytical prediction of Wakashima et al., eq. (5.18) is considered to be more accurate in comparison to that of Kreider and Patarini, eq. (5.3). Wolff et al. [21] have also reported such a hysteresis of thermal strain vs. temperature for a unidirectional graphite fiber/magnesium composite where they attempted a comparison between the experimental and analytical results, obtaining reasonably good agreement for thermal cycles with small temperature difference.

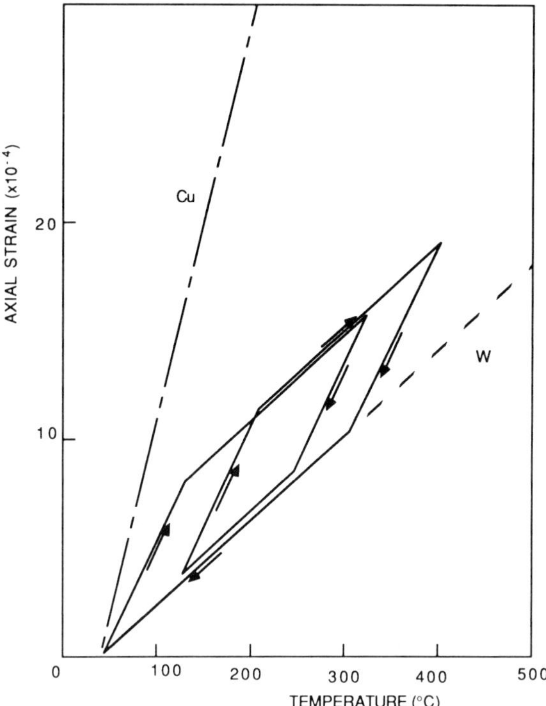

FIG. 5.6 Axial strain vs. temperature for continuous tungsten (W) fiber/copper (Cu) composite where the hysteresis shown by solid lines shows experimental results and dash and dash–dot lines are the predicted results of W and Cu, respectively [11].

The analytical model used in the above study is based on the plane stress numerical model to account for elastic/plastic matrix and elastic fiber [22].

In most of the short fiber metal matrix composites, short fibers are misoriented either two-dimensionally or three-dimensionally depending on the processing and the geometry of composites. Then the prediction of the CTEs of such composites must be made based on the model which can account for fiber misorientation. Takao [15] has recently constructed an analytical model based on Eshelby's method. Hence, the foregoing formulation, eqs. (5.4) to (5.9) was also used in his model except for eqs. (5.5) and (5.6) where all variables are referred to the local coordinates (x') associated with a misoriented fiber while the global coordinates are referred to by x (Fig. 5.8). The misorientation of short fibers is characterized by a density function ρ which is a function of misorientation angles θ and ϕ based on the spherical coordinate system (see Fig. 5.8). Takao calculated two cases of fiber misorientation, (1) $\rho = \rho(\theta)$ and ϕ is uniform over 0 to 2π, (ii) and $\rho = \rho(\phi)$ and

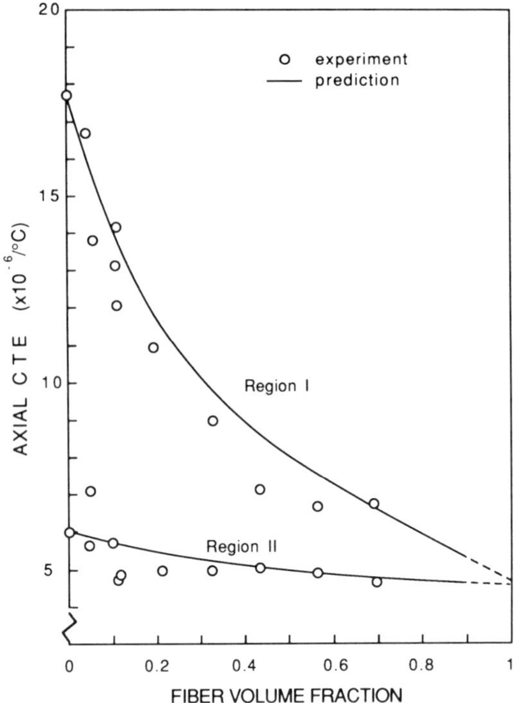

FIG. 5.7 A comparison between the experimental (solid lines) and analytical results (open circles) of axial CTE for Regions I and II (see definition in the text) [11].

$\theta = \pi/2$. The first case corresponds to the metal matrix composite made by extrusion process and the properties becomes transversely isotropic. The second case represents a two-dimensionally misoriented short fiber metal matrix composite usually in plate form. In both cases a density function can take any form to simulate the actual distribution of short fibers. Takao's model includes, as an extreme case, the case of a three-dimensional random composite ($\rho(\theta) =$ constant over $-\pi/2$ to $\pi/2$) and a two-dimensional random short fiber metal matrix composite ($\varphi(\theta)$ = constant over 0 to 2π). Figure 5.9 illustrates the effect of a limit angle β* of a cosine type density function, $\rho(\theta)$ (corresponding to Fig. 3.6(b)) on the longitudinal (α_0) and transverse (α_{90}) CTEs of a misoriented short SiC fiber/Al composite with $l/d = 10$. The larger the limit angle β, the more randomly the short fibers are distributed in a two-dimensional space, thus α_0 and α_{90} become closer, but they never

* The limit angle defines the maximum misalignment angle (see Fig. 3.6).

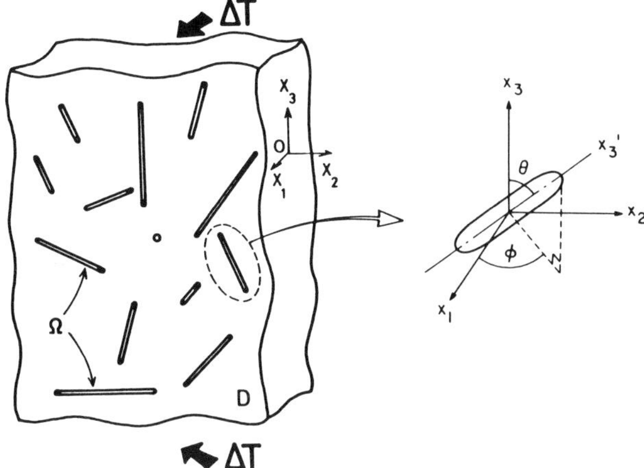

FIG. 5.8 Analytical model used by Takao [15] to predict CTEs of a misoriented short fiber composite.

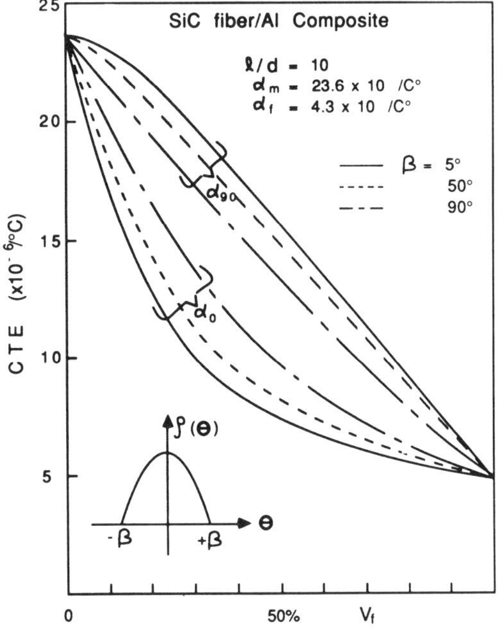

FIG. 5.9 CTEs of a misoriented short fiber SiC/Al composite as a function of fiber volume fraction V_f for several cases of β [15].

coincide even for $\beta = 90°$ since the distribution function is of a cosine type (note they coincide for uniform distribution; Fig. 3.6(a) with $\beta = 90°$).

5.3 Thermal Conductivity

The thermal conductivity of metal matrix composites has been recognized as one of the important thermomechanical properties since metal matrix composites are considered as a strong candidate for structural materials in space application and high-performance engines. For example, space structures require high specific stiffness (E/ρ) and high thermal conductivity to CTE ratio (K_T/α_L) and metal matrix composites usually satisfy these requirements where subscripts L and T denote the longitudinal and transverse properties of a unidirectional metal matrix composite system [8]. Figure 5.10 illustrates a performance map of structural metals, polymeric composites and several metal matrix composites. The best material for space structures should possess high E/ρ and K_T/α_L, thus occupying the upper right corner in the graph. It is clear from Fig. 5.10 that a graphite fiber (V50054)/magnesium alloy (AZ31) exhibits the best performance.

Standard methods of measurement of thermal conductivity of a composite are guarded hot plate (ASTM Standard (C177–76)) or the cut-bar method

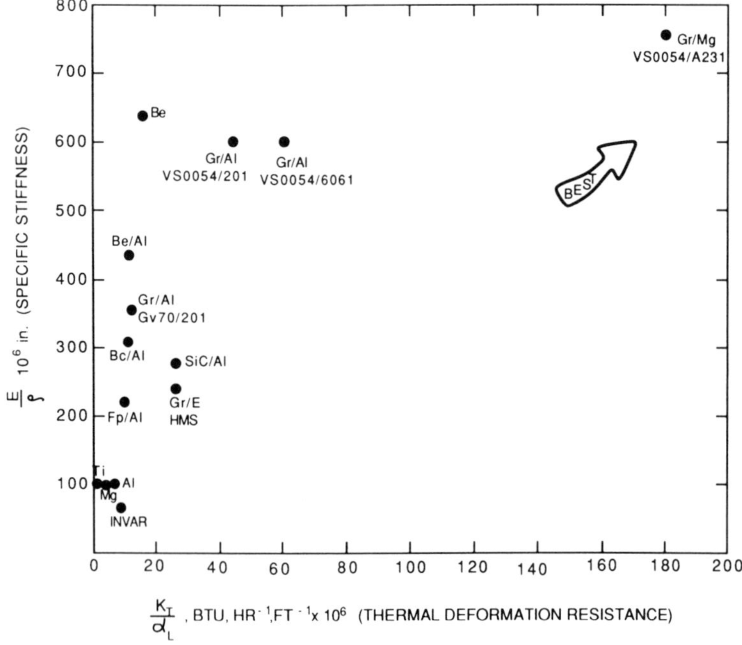

FIG. 5.10 Specific stiffness (E/ρ) vs. thermal deformation resistance (K_T/α_L) of various metals and metal matrix composites [8].

[23]. The former and latter were used to measure the thermal conductivity of unidirectional composite (graphite/polymid) along the fiber direction (K_L) and its perpendicular direction (K_T), respectively. The former method requires a symmetric arrangement of composite and reference material plates, thus it is a more accurate measurement than the latter. The thermal conductivity of a composite (K_c) can be obtained based on the cut-bar method [23]

$$K_c = K_r \frac{\Delta T_r}{\Delta T_c} \cdot \frac{X_c}{X_r} \cdot \frac{A_r}{A_c} \tag{5.21}$$

where:

K_i = thermal conductivity of ith phase material.
ΔT_i = temperature difference across ith phase material.
X_i = distance between thermocouples on the ith phase material.
A_i = cross-sectional area of the ith phase material.
i = r (known reference material) or c (composite).

In obtaining ΔT_i one must wait for the steady-state condition to be established. Although these methods have been used only for polymeric composites, they also should be applicable to metal matrix composites. Kelly [24] measured thermal conductivities of several metal matrix composites by using a Dynatech axial rod conductivity measurement device. This device requires that the composite specimen is of a rod type. Figure 5.11 shows some of Kelly's experimental results of the conductivities as a function of temperature T (°F) where (a) and (b) denote the longitudinal (K_L) and transverse conductivity (K_T) of unidirectional boron/aluminum (B/Al) composite. The volume fraction of the boron fiber in the B/Al composite is 35%. It can be seen from Fig. 5.11 that K_L is larger than K_T, though not by a large margin. This is due to the fact that the thermal conductivity of boron ($K_B = 20$ BTU/(h·ft°F)) is smaller than that of aluminum ($K_{Al} = 107$ BTU/(h·ft°F)), thus the heat flow along the transverse direction is disturbed by boron fibers. Kelly [24] attempted to explain the experimental results by a simple analytical model which yields upper and lower bounds on K. Kelly's bounds seem to be too broad. Gurtman et al., on the other hand, developed a more rigorous model to predict K_L and K_T of unidirectional metal matrix composites [25]. The Gurtman et al. model is based on an unit cell which represents a binary unidirectional composite with a periodic array of anisotropic constituents and where the method of multi-variable asymptotic expansion was used. The analytical results of K_L and K_T are given by

$$K_L = V_f K_{fL} + V_m K_{mL} \tag{5.22}$$

$$K_T = V_f K_{fT}(1 + V_m A) + V_m K_{mT}(1 - V_f A) \tag{5.23}$$

where

$$A = \frac{K_{mT} - K_{fT}}{\{K_{mT} + K_{fT} + V_f(K_{mT} + K_{fT})\}} \tag{5.24}$$

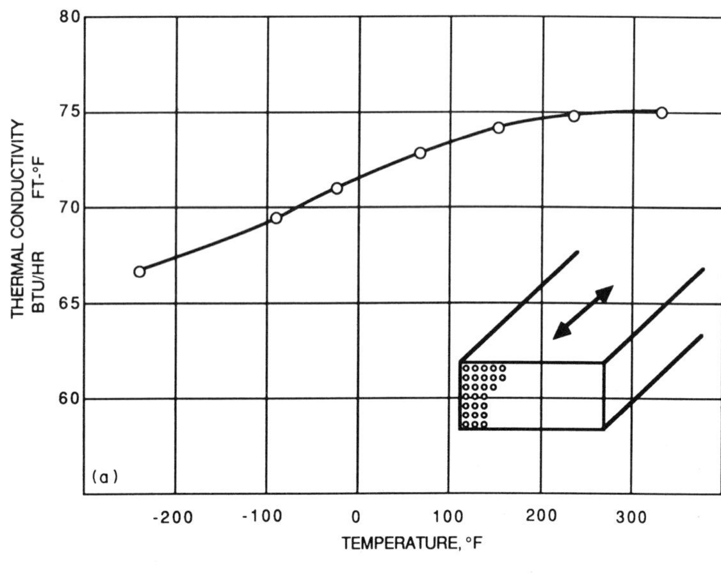

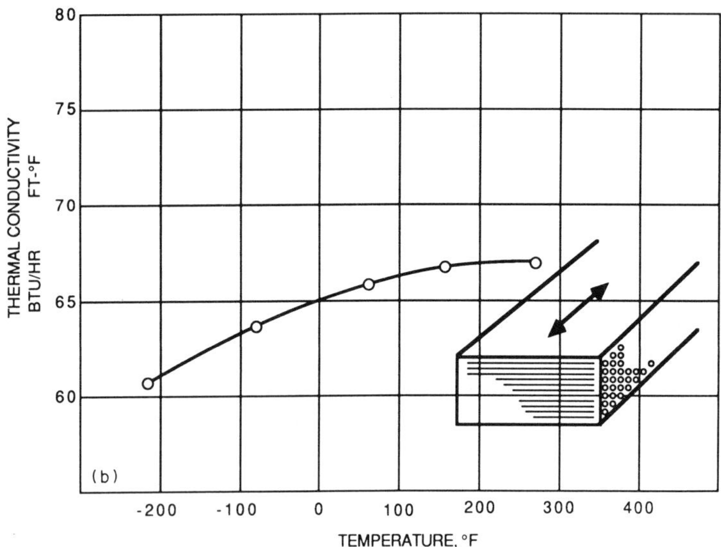

FIG. 5.11 Thermal conductivity K vs. temperature of a B/Al composite; (a) K_L (longitudinal, along fiber axis), (b) K_T (transverse, perpendicular to the fiber axis) [24].

and where V_i is the volume fraction of the ith phase with i being m (matrix) and f (fiber), K_{iL} and K_{iT} are the thermal conductivities of the ith phase along the longitudinal and transverse directions, respectively. Equation (5.22) coincides with the prediction based on the law of mixtures. Gurtman et al. compared the conductivities predicted by eqs. (5.22) and (5.23) with the measured values of two types of unidirectional metal matrix composites, graphite (VS0054)/aluminum (6061-T6) and graphite (VS0054)/magnesium (AZ91C)s. A comparison between the analytical and experimental results is given in Table 5.2, where the thermal conductivities of constituents are also given.

Springer and Tsai proposed a model to predict K_T of a unidirectional composite system [26]. However, their model is designed for the case where $K_{fT}/K_m > 1$ while in most of the metal matrix composite systems, $K_{fT}/K_m < 1$. Thus the model by Springer and Tsai is not applicable to most of the metal matrix composite systems.

Hatta and Taya have recently proposed a model to predict the thermal conductivity of a composite [27]. The Hatta–Taya model takes advantage of the Eshelby's method developed for elasticity problems [17] where stress (σ_{ij}) strain (e_{ij}) and stiffness tensor (C_{ijkl}) in the elasticity problem correspond to heat flax (q_i), temperature gradient ($T,_i$) and thermal conductivity (K_{ij}), respectively (see subsection 2.4.4). This model can be applied to cases of more complicated morphologies of reinforcement such as a misoriented short fiber composite [28]. Based on this model they derived a simple formula to predict the transverse thermal conductivity K_T of a unidirectional composite system

$$K_T = K_m + \frac{V_f(K_{fT} - K_m)K_m}{0.5(1 - V_f)(K_{fT} - K_m) + K_m} \tag{5.25}$$

and the longitudinal thermal conductivity predicted by Hatta–Taya model coincides with the law of mixtures given by eq. (5.22).

The K_T predicted by eq. (5.25) for a 36% V_f graphite/Al composite is 45.33 BTU/(h·ft °F), which is close to the experimental results and the analytical prediction by eq. (5.23) (see Table 5.2). Thus both models (Gurtman et al. and Hatta–Taya) seem to predict well the K_T of unidirectional metal matrix composites. Of course, the K_L of unidirectional metal matrix composites predicted by eq. (5.22) agrees well with the experiment, as seen from Table 5.2.

For $K_{fT}/K_m \ll 1$, which is the case of some of the metal matrix composites, the Gurtman et al. model (eq. (5.23)) and the Hatta–Taya model yield basically the same result, i.e.

$$\frac{K_T}{K_m} = \frac{1 - V_f}{1 + V_f} \tag{5.26}$$

However, for the ratio K_{fT}/K_m becomes comparable to 1, the Hatta–Taya model gives slightly larger values of K_T than the Gurtman et al. model. This is

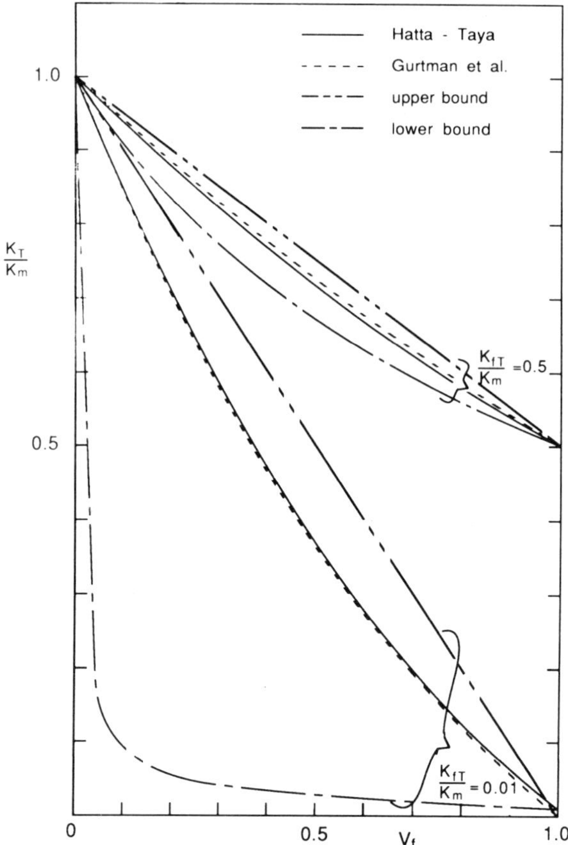

FIG. 5.12 Transverse thermal conductivity of a unidirectional metal matrix composite; K_T as a function of fiber volume fraction V_f predicted by several models.

illustrated in Fig. 5.12 where the predictions by Hatta–Taya and Gurtman et al. models and the upper (eq. (5.27)) and lower bounds (eq. (5.28)) are shown by solid, dash, dash–two dots and dash–dot curves, respectively, for two cases of K_{fT}/K_m, 0.01 and 0.5. The upper (K_{TU}) and lower bounds (K_{TL}) on K_T are given by

$$K_{TU} = V_f K_{fT} + V_m K_m \tag{5.27}$$

$$K_{TL} = \frac{1}{V_f K_m / K_{fT} + V_m} \tag{5.29}$$

It follows from Fig. 5.12 that for small K_{fT}/K_m values the bounds on K_T deviate appreciably from the predictions by Gurtman et al. and Hatta and Taya.

Unlike unidirectional metal matrix composites, investigation of the thermal conductivity of a short fiber metal matrix composite has been limited to the special geometry of short fiber morphology, for example, aligned short fibers [14]. The short fibers in a metal matrix composite are often misoriented either two- or three-dimensionally. The Hatta–Taya model can predict the thermal conductivity of such a misoriented short fiber composite [28]. The analytical model of Hatta–Taya is shown in Fig. 5.13 where a misoriented short fiber composite is subjected to uniform heat flux (q_0) along the x_3-axis and the misorientation of a short-fiber is defined by two angles based on the spherical coordinates system, θ and ϕ. They considered two cases of a misoriented short fiber composite: (1) a three-dimensional misorientation with θ as a variable and no dependence on ϕ, i.e., $\rho(\phi)$ being uniform over 0 to 2π, and where ρ is the density function; and (2) a two-dimensional misorientation with $\theta = \pi/2$ and ϕ as a variable. The first case corresponds to a transversely isotropic composite with the x_1–x_2 plans as an isotropic plane. The first case corresponds to a short fiber metal matrix composite that is processed by extrusion (with the extrusion axis given by the x_3-axis (Fig. 5.13), whereas in the metal matrix composite belonging to the second case, all short fibers are misoriented in-plane (the x_1–x_2 plane in Fig. 5.13). Hence, the distribution functions in the first and second cases are defined by $\rho = \rho(\theta)$ and $\rho = \rho(\phi)$, respectively. They considered two types of distribution

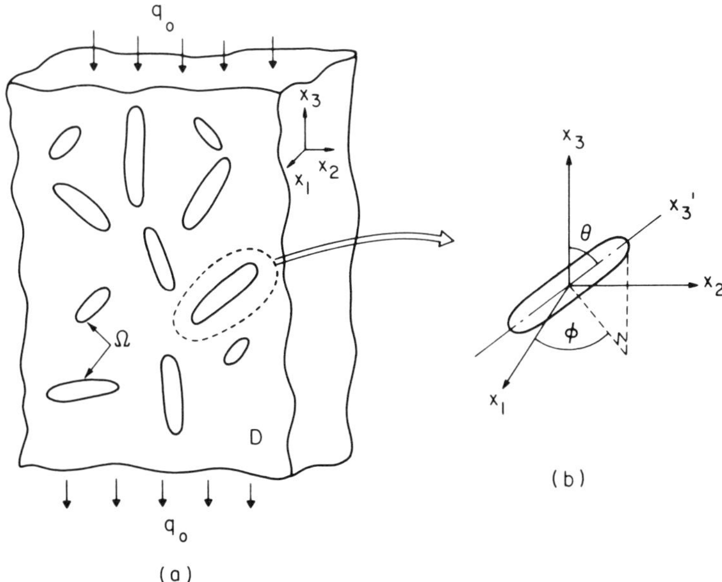

FIG. 5.13 Analytical model to predict thermal conductivities of a misoriented short fiber composite used by Hatta and Taya [28].

function, the uniform and cosine types which are also used to predict the stiffness of a misoriented short fiber composite [29] and given by Fig. 3.6(a) and (b), respectively, where β is a limit angle and can take values of 0 to $\pi/2$. Uniform type $\rho(\theta) = \rho_0$ with $\beta = \pi/2$ corresponds to a 3D completely random distribution, thus its conductivity becomes isotropic. In this case Hatta and Taya obtained a simple formula to predict the thermal conductivity $K_c (= K_{11} = K_{22} = K_{33})$ of a short fiber composite:

$$\frac{K_c}{K_m} = 1 - \frac{V_f(K_m - K_f)[(K_f - K_m)(2S_{33} - S_{11}) + 3K_m]}{3(K_f - K_m)^2(1 - V_f)S_{11}S_{33} + K_m(K_f - K_m)R + 3K_m^2} \quad (5.30)$$

where

$$R = 3(S_{11} + S_{33}) - V_f(2S_{11} + S_{33})$$

and where S_{ij} are the second-order tensor similar to the Eshelby's fourth-order tensor S_{ijkl} (see Chapter 2), and they are functions of a fiber aspect ratio and its detailed expressions for various geometries of fiber are given in Appendix D. K_m and K_f are the thermal conductivities of the matrix and fiber, both of which are assumed to be isotropic (but the model can treat the case of an anisotropic fiber and matrix). Similarly the thermal conductivity $K_c (= K_{11} = K_{33})$ of two-dimensionally (i.e. x_1–x_3 plane) complete random short fiber composite can be obtained as

$$\frac{K_c}{K_m} = 1 + V_f(K_f - K_m)\{(K_f - K_m)(S_{11} + S_{33}) + 2K_m\}/A \quad (5.31)$$

where

$$A = 2(K_f - K_m)^2(1 - V_f)S_{11}S_{33} + K_m(K_f - K_m)(2 - V_f)(S_{11} + S_{33}) + 2K_m^2$$

In order to see clearly the effect of various distributions of the fiber orientation angle, Hatta and Taya determined the thermal conductivity (K_{33}) of a short fiber composite with four different types of distribution as a function of fiber volume fraction V_f (Fig. 5.14). The four types of distribution function used are explained in Table 5.3, where the limit angle β is set equal to $\pi/2$. Though the parameters chosen, $K_f/K_m = 20$ and fiber aspect ratio $(a_3/a_1) = 100$, do not necessarily represent a typical case of a short fiber metal

Table 5.3 *The notations in Figs. 5.14 and 5.15*

Notation	Type of composite	Orientation distribution, $\rho(\theta)$
AL	aligned short fiber	$\rho_0\delta(\theta)$
3-UNI	three-dimensionally misoriented	ρ_0
3-COS	three-dimensionally misoriented	$\rho_0 \cos a\theta$
2-UNI	two-dimensionally misoriented	ρ_0
2-COS	two-dimensionally misoriented	$\rho_0 \cos a\theta$

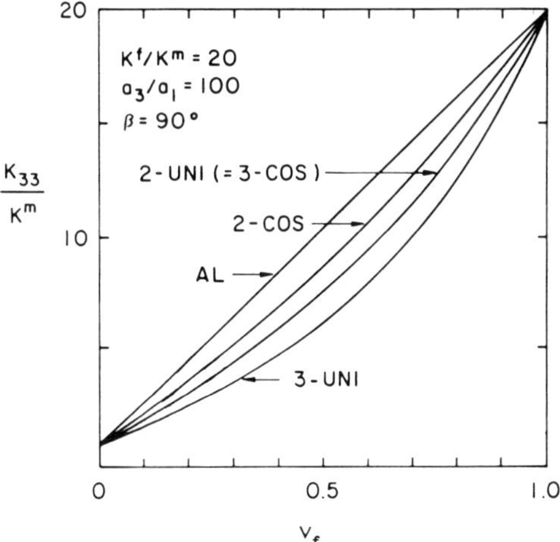

FIG. 5.14 Thermal conductivity of a short fiber composite, K_{33} as a function of fiber volume fraction V_f where various distributions of short fiber orientation angles are used (see Table 5.3 for the definition of symbols) [28].

matrix composite, Fig. 5.14 illustrates the sensitivity of distribution shape on the conductivity. Namely, for a given fiber volume fraction, an aligned short fiber composite gives rise to the largest conductivity along the fiber axis (x_3-axis in this case). Then it is followed by a two-dimensional cosine type, a two-dimensionally uniform type (which is equal to a three-dimensional cosine) and a three-dimensionally uniform type, in this case given by eq. (5.30) which yields the smallest conductivity. The fiber aspect ratio (a_3/a_1) has a strong effect on the thermal conductivity as shown in Fig. 5.15 where the case of $\beta = 90°$, $V_f = 0.4$ and $K_f/K_m = 20$ was used as a demonstration. It should be noted in Fig. 5.15 that the thermal conductivity along the x_3-axis, K_{33}, increased with the fiber aspect ratio, a_3/a_1 of an aligned short fiber composite with the fiber axis being the x_3-axis. This trend is true for a misoriented short fiber composite system with the reference axis (or the axis of the majority of fibers aligned) being the x_3-axis, including the cases of two-dimensionally (2-UNI with $\beta = \pi/2$) and three-dimensionally random (3-UNI with $\beta = \pi/2$) short fiber composites. The gain in K_{33}, however, becomes saturated as the a_3/a_1 reaches the higher i.e., 50.

Thermal conductivitiy of a composite can usually be predicted once those of the constituent phases, the relevant phase geometry, K_f/K_m, a_3/a_1 and V_f, and also the distribution function, are given. However, the conductivity of a composite can depend also on the properties of the matrix interface, for example a debonded interface and thin coating on fiber. Coated fibers are sometimes used as a reaction barrier between fiber and the metal matrix

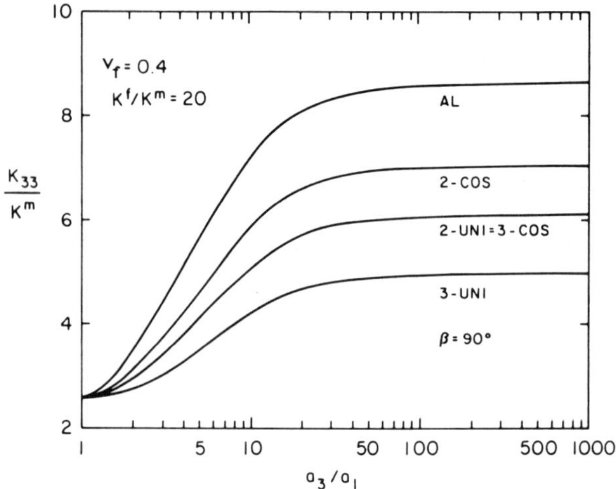

FIG. 5.15 Thermal conductivity of a short fiber composite, K_{33} as a fiber aspect ratio a_3/a_1 [28].

which would otherwise react chemically leading to the degradation of the interfaces, and as the matrix–fiber interfaces of an uncoated fiber metal matrix composite when subjected to severe thermal environment which may lead to debonding. Hence it is important to assess the effect of the coating or debonding at the interface on the thermal conductivity of a metal matrix composite. Hatta and Taya studied analytically the effect of coating on the conductivity of a coated fiber composite [30]. The analytical model by Hatta and Taya is an extension of their model for a short fiber composite (Fig. 5.13) in that short fibers are coated. The thermal conductivities (K_{33}) of a coated aligned short fiber composite with different K_c/K_m ratios are shown as solid curves against the fiber volume fraction V_f in Fig. 5.16, where (a) and (b) denote the cases of highly conductive coating and highly resistive coating, respectively (and the K_{33} of a non-coated fiber composite are also shown by dashed curves). In Fig. 5.16 K_m, K_f and K_c are the thermal conductivity of the matrix, fiber and coating, respectively, w/a_1 is the ratio of the coating thickness to fiber radius, and a_3/a_1 is the fiber aspect ratio. The fiber axis is taken along the x_3-axis.

It is obvious from Fig. 5.16 that the use of a relatively thin coating which is either highly conductive or resistive has a strong effect on the overall conductivity of the composite. The effect of the coating thickness (w) on K_{33} is demonstrated in Fig. 5.17 where the ratio of the coating thickness to fiber radius, w/a_1, is the key parameter ranging from 0.001 to 0.4. Hasselman and Johnson studied analytically the thermal conductivity of a composite with various interfacial thermal barrier resistances [31]. The model by Hasselman and Johnson is a modification of the model used by Maxwell [32] in that

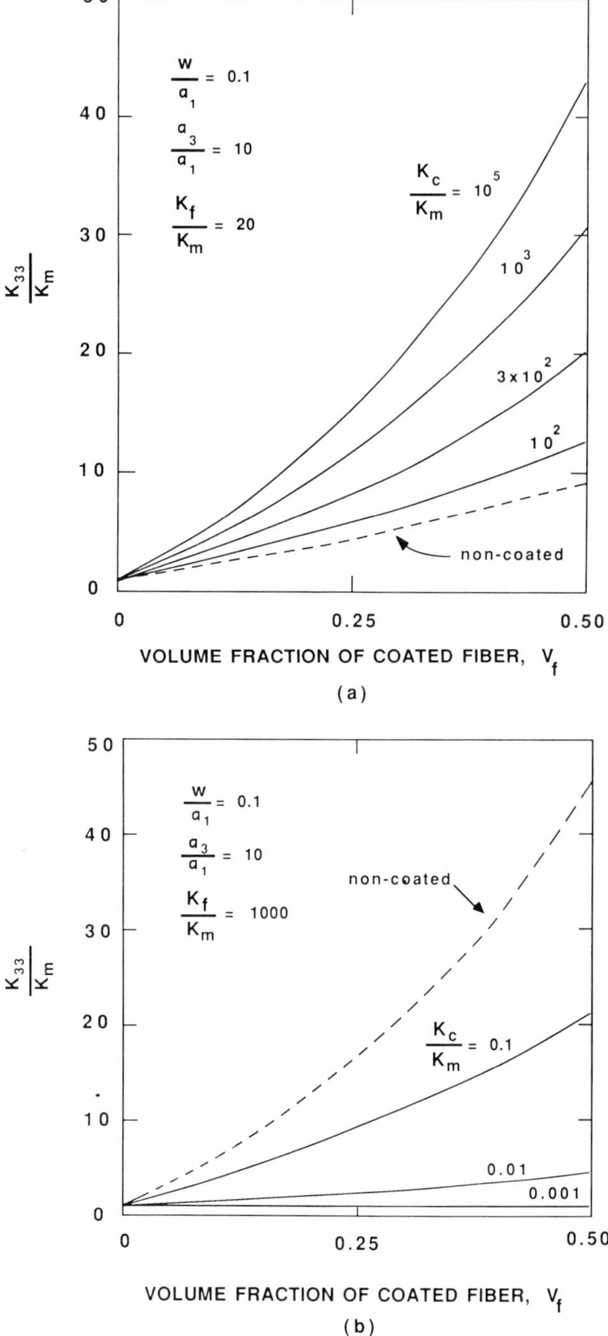

FIG. 5.16 Thermal conductivity K_{33} of a coated short fiber composite as a function of fiber volume fraction V_f where fiber aspect ratio (a_3/a_1) is 10 and the ratio of coated thickness to fiber radius is 0.1; (a) for highly conductive coating, (b) for very low conductive coating. As a comparison, K_{33} of non-coated fiber composite also is shown by a dashed line [30].

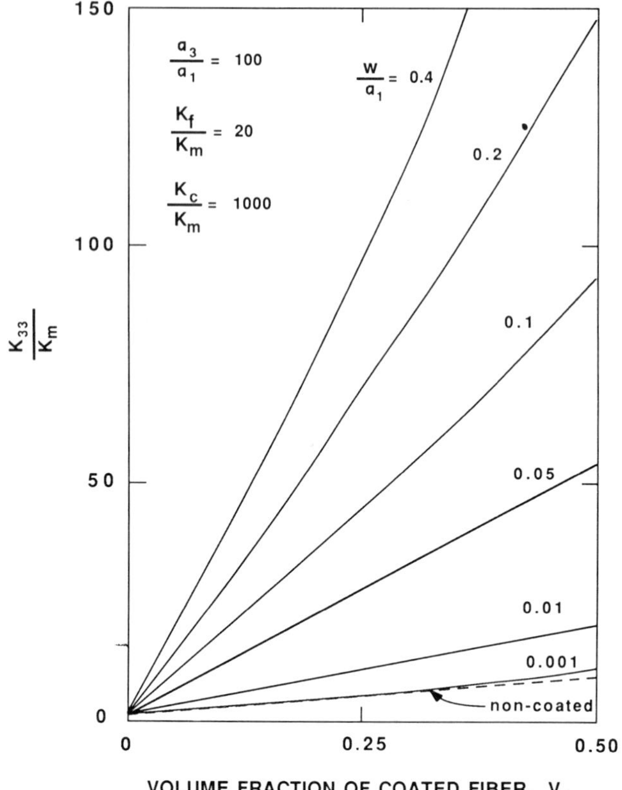

FIG. 5.17 Thermal conductivity K_{33} of a coated short fiber composite as a function of fiber volume fraction V_f for various ratios of coating thickness to fiber radius (w/a_1) [30].

fillers are distributed dilutely in the matrix (thus valid only for small fiber volume fractions) and the matrix–fiber interface acts as a thermal barrier resistance with the heat transfer coefficient h_c. Namely, at the interface the following equation must be satisfied

$$T_f - T_m = -(K_f/h_c)(\partial T_f/\partial r) \tag{5.32}$$

where T_f and T_m are the temperatures in the fiber and matrix, respectively, K_f is the thermal conductivity of the fiber and r is the radial coordinate normal to the interface. The values of $h_c = 0$ and ∞ correspond to no heat flow and perfect heat flow, respectively. The former case represents approximately that of a debonded interface, while the latter represents perfectly bonded interface resulting in $T_f = T_m$, which is the assumption used in the Hatta and Taya model [30].

The conductivity of a composite with an interfacial thermal barrier has also been studied analytically by Benveniste and Miloh [33], who used the interfacial boundary condition given by eq. (5.32) and discussed several cases of reinforcement geometry. More recently, Benveniste [34] has extended this mode to the case of finite volume fractions of fiber by use of the Mori–Tanaka theory [18]. Benveniste's model [34] is thus very similar to the model developed by Hatta and Taya [16] for prediction of the effective thermal conductivity of a composite with perfect interfacial bonding, i.e., $h_c \to \infty$ in eq. (5.32). As to the effective thermal conductivity of a composite with an interfacial thermal barrier resistance for finite volume fractions, the Hatta and Taya model [30] for a coated fiber composite can still be used by considering a coating with conductivity K_c and thickness w as an interface barrier with conductance h_c. Namely, the concept of coating (K_c, w) is related to that of the interface barrier (h_c), as

$$h_c = \frac{K_c}{w} \tag{5.33}$$

A careful examination of the above models has revealed that the prediction by Hasselman and Johnson coincides with that by Benveniste [35]. For example, the formulae to predict the thermal conductivity k_p of a composite with spherical filler and the transverse conductivity k_T of a composite with cylindrical filler are given by

$$k_p = \frac{(2-2V_f)+h_c\{1+2V_f+(2-2V_f)/\lambda\}}{(2+2V_f)+h_c\{1-2V_f+(2+2V_f)/\lambda\}} \tag{5.34}$$

for spherical filler; and

$$k_T = \frac{(1-V_f)+h_c\{1+V_f+(1-V_f)/\lambda\}}{(1+V_f)+h_c\{1-V_f+(1+V_f)/\lambda\}} \tag{5.35}$$

for cylindrical filler, where $\lambda = k_f/k_m$.

In order to see the difference between the above models clearly, let us consider two cases of interface condition; $h_c = 0.1$ (poorly conductive interface), and $h_c = 10$ (highly conductive interface). Figure 5.18 shows the numerical results of the thermal conductivity k_p of a composite with spherical filler as a function of the volume fraction of filler V_f where (a) and (b) denote the cases of $h_c = 0.1$ and 10.0, respectively. Similarly in the case of cylindrical filler, the numerical result are given in Fig. 5.19 where (a) and (b) denote the cases of $h_c = 0.1$ and 10.0, respectively. In Figs. 5.18 and 5.19 the results by the Benveniste–Miloh dilute model are also shown by dash–dot curves. In this calculation the ratio of the conductivity of filler to matrix, k_f/k_m has been set at 10 and $k_m = 1.0$. It follows from Fig. 5.18 that for the case of spherical filler the Hatta–Taya coated fiber composite model gives rise to the same numerical prediction as the Hasselman–Johnson–Benveniste model, eq. (5.34), thus shown as solid curves whereas for cylindrical filler the results based on

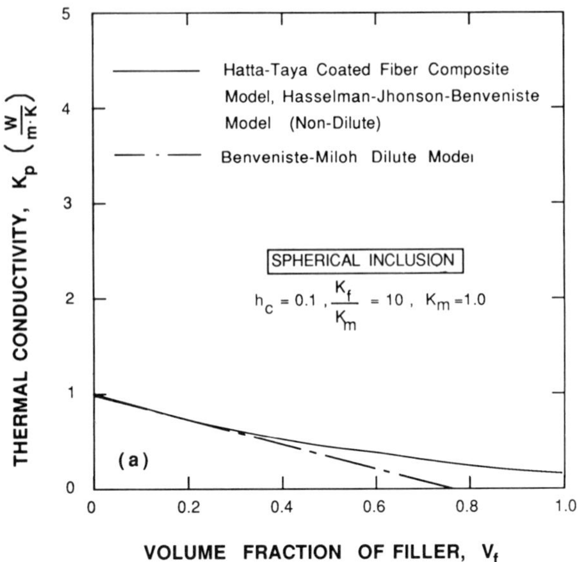

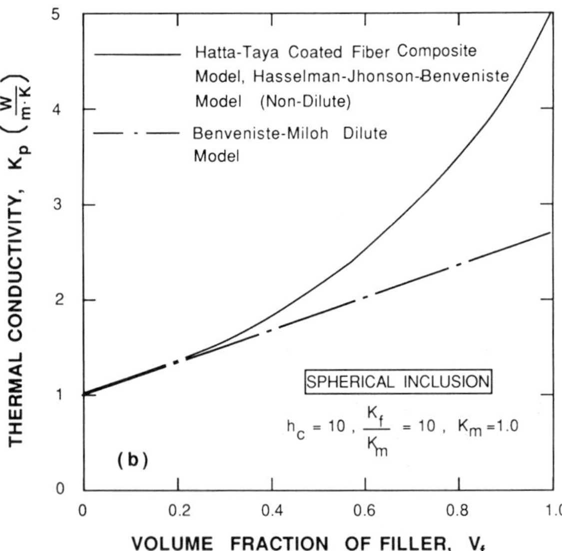

FIG. 5.18 Thermal conductivities k_p of a composite with spherical filler and interface thermal barrier of conductance h_c, predicted by various models; (a) for $h_c = 0.1$ (poorly conductive interface), and (b) for $h_c = 10$ (highly conductive interface).

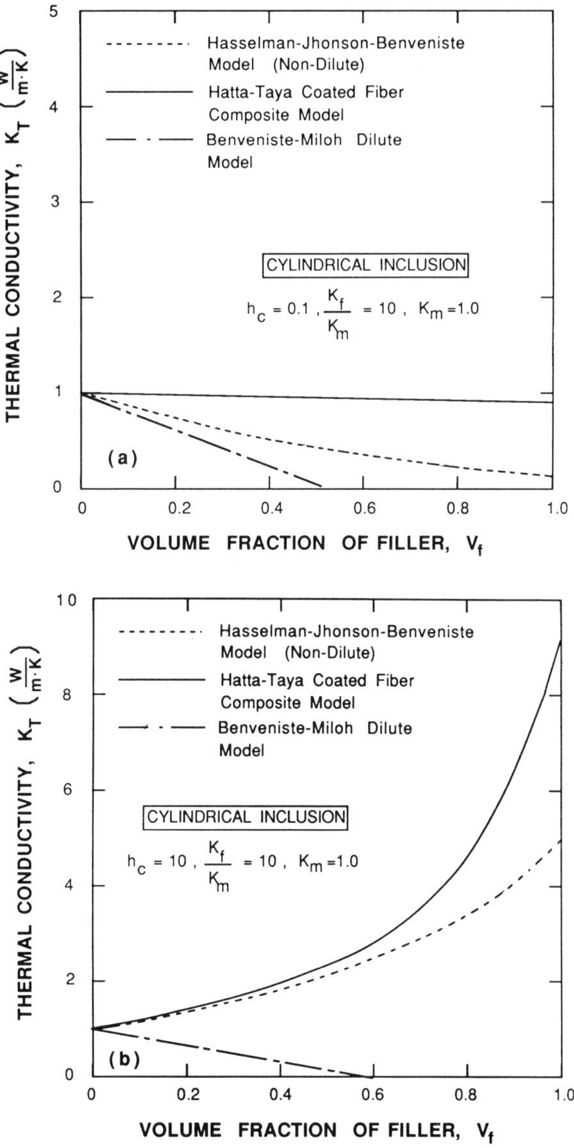

FIG. 5.19 Transverse thermal conductivities k_T of a composite with cylindrical filler and interface thermal barrier of conductance h_c, predicted by various models; (a) for $h_c = 0.1$ (poorly conductive interface), and (b) $h_c = 10$ (highly conductive interface).

the Hatta–Taya model exceed those by the Hasselman–Johnson–Benveniste model. It is also seen from these figures that the Benveniste–Miloh dilute model predicts the conductivities which are lower than those by the others, especially for higher volume fractions of filler. It is also clear from these figures that the interfacial condition which affects interfacial thermal resistance (or conductance) h_c, has a strong impact on the overall thermal conductivity of a composite.

5.4 Specific Heat

Specific heat (C) of a body is defined by

$$\Delta Q = C \Delta T \tag{5.36}$$

where ΔQ is the change in heat content per unit volume for uniform temperature change within the body ΔT. Thus C is the heat per unit volume for unit temperature change, and is generally a function of temperature T. Specific heat is either C_v or C_p under conditions of constant volume or constant pressure. C_v and C_p are related by

$$C_p - C_v = C_{ijkl} \alpha_{ij} \alpha_{kl} T \tag{5.37}$$

where C_{ijkl} is the elastic stiffness tensor, α_{ij} is the CTE tensor and T is the absolute temperature of the body. For an isotropic body, eq. (5.37) is simplified as

$$C_p - C_v = K \xi^2 T \tag{5.38}$$

where K is the bulk modulus and ξ is the bulk CTE. Specific heat C (C_v or C_p) appears in the transient heat conduction equation, eq. (2.50). Thus, determination of C becomes important if one wants to calculate the transient temperature distribution within a body. Despite the above importance, only a limited number of studies of specific heat of metal matrix composites have been made.

The standard method for measuring specific heat is to use a drop calorimeter. A specimen of known mass is heated to a known temperature and it is transferred to the calibrated calorimeter, from which the enthalpy vs. temperature curve is obtained. The slope of the curve at a given temperature is the specific heat. The specific heat of a specimen at temperatures below ambient temperature can be measured by a dynamic heating method [2]. Christian and Campbell [2] measured the specific heat of various aluminum matrix composites for the ranges of cryogenic to elevated temperatures. Except for the specific heat at 88 K, those measured are close to the value predicted by the law of mixtures, i.e.

$$C = V_m C_m + V_f C_f \tag{5.39}$$

where V_i and C_i are the volume fraction and specific heat of the i-phase with i being m (matrix) of f (fiber). Budiansky [36] proposed a model to predict the constant volume specific heat (C_v) of isotropic composites where both matrix and fiber, and also the composite, are assumed to be isotropic. Budiansky's formula is given by

$$C_v = \sum_{i=1}^{N} V_{fi}(C_v)_i + aK\xi^2 TA \tag{5.40}$$

where

$$A = \sum_{i=1}^{N} V_{fi}\left(\frac{K_i}{K}\right)^2\left(\frac{\xi_i}{\xi}\right)^2\left\{1 - a + a\left(\frac{K_i}{K}\right)\right\}^{-1} - 1$$

$$a = \frac{(1+\nu)}{3(1-\nu)}$$

In the above equations, V_{fi}, $(C_v)_i$, K_i and ξ_i are the volume fraction, the constant-volume specific heat, the bulk modulus and bulk CTE of the interphase, with i being the multi-reinforcing phases and the matrix, and K and ξ are the bulk modulus and bulk CTE of the composite. The constant-pressure specific heat of the composite can be easily obtained from eq. (5.38) once C_v is calculated by eq. (5.40). Usually the second term in eq. (5.40) is much smaller than the first term, and thus can be neglected. Budiansky's results, eq. (5.40), are reduced to the law of mixtures prediction, eq. (5.39). The law of mixtures formula seems to be accurate enough for predicting the C_v of most of the composite systems including the short and continuous fiber composite which usually exhibits strong anisotropic behavior. Aboudi [14] confirmed in his parametric study that the fiber aspect ratio has little effect on the overall specific heat of the composite. He also confirmed that the term on the right-hand side in eq. (5.37) is much smaller than the C_v terms. Hence

$$C_v \cong C_p \tag{5.41}$$

The case of anisotropic constituents and composites was discussed rigorously by Rosen and Hashin who obtained bounded solutions to specific heat [37].

5.5 Problems

(1) Discuss the sources of error in measurement of CTEs of a composite specimen by a standard dilatometer as shown in Fig. 5.1, and then suggest possible improvement in the dilatometer.
(2) Explain why the bounded solutions by Levin [7, 10] are not in agreement with the experimental results of unidirectional metal matrix composites (see Fig. 5.3(a)).
(3) Compare the Eshelby type model (eq. (5.14)) with Levin's (eq. (5.1)) for prediction of the longitudinal CTE of a unidirectional metal matrix composite.
(4) Discuss the difference in CTEs of a composite predicted by the model which can account for the plastic deformation of the metal matrix (eqs. (5.16–5.20)) and that without such plastic deformation of the matrix (eqs. (5.10–5.15)).
(5) Compare the Springer and Tsai model [26] with eq. (5.25) for prediction of the transverse thermal conductivity (K_T) of a unidirectional metal matrix composite. Plot the results of K_T as a function of V_f for two cases of metal matrix composites, G_r/Al and G_r/Mg (see the input data in Table 5.2).

(6) Walpole [38] proposed a simple model to predict the temperature field within a thin coating in a coated fiber composite. Discuss the validity of the Walpole model for prediction of the temperature gradient within a thin coating in a coated short fiber composite as the thickness of the coating increases, particularly in comparison with the Hatta and Taya model [30].

(7) When the matrix–fiber interfaces in a particulate metal matrix composite are all deboned, how do you estimate the effective thermal conductivity of the composite? (See Refs [31] and [34].

References

1. Kreider, K. G. and Patarini, V. M. *Metal. Trans.*, Vol. 1, 1970, pp. 3431–3435.
2. Christian, J.L. and Campbell, M. D., *Proceedings of the 1972 Cryogenic Engineering Conference, Advances in Cryogenic Engineering*, Vol. 18, edited by K. Timmerhaus, Plenum Press, 1972, pp. 175–183.
3. Christian, J. L. and Adsit, N. R., in *Hybrid and Selected Metal Matrix Composites—A State of the Art Review*, edited by W. J. Renton, AIAA, New York, 1977, pp. 67–97.
4. Kural, M. H. and Ellison, A. M., *SAMPE J.*, September/October 1980, pp. 20–26.
5. Wolff, E. G. and Eselun, S. A., *J. Comp. Mater.*, Vol. 11, 1977, pp. 30–32.
6. Wolff, E. G., *9th National SAMPE Tech. Conf.*, Ser. 9, 1977, pp. 57–72.
7. Wolff, E. G., *Thermal Expansion of Metal Matrix Composites*, Tech. Report to Air Force Systems Command, TR-0081 (6950-01)-1, 1981.
8. Armstrong, H. H., *SAMPE National Symp.*, Vol. 2, San Francisco, CA, 1979, pp. 1250–1264.
9. Schapery, R. A., *J. Comp. Mater.* Vol, 2(3), 1968, pp. 380–404.
10. Levin, V. M., *Meckanica Tverdogo Tela*, Vol. 2, 1967, pp. 88–94.
11. Wakashima, K., Otsuka, M. and Umekawa, S., *J. Comp. Mater.*, Vol. 8, 1974, pp. 391–404.
12. Craft, W. J. and Christensen, R. M., *J. Comp. Mater.*, Vol. 15, 1981, pp. 2–20.
13. Nomura, S. and Chou, T. W., *Int. J. Eng. Sci.*, Vol. 19, 1981, pp. 1–9.
14. Aboudi, J., *Fibre Sci. Tech.*, Vol. 20, 1984, pp. 211–225.
15. Takao, Y., in *Recent Advances in Composites in the United States and Japan*, ASTM STP 864, edited by J. R. Vinson and M. Taya, American Society for Testing Materials, Philadelphia, 1985, pp. 685–699.
16. Takao, Y. and Taya, M., *J. Appl. Mech.*, Vol. 107, 1985, pp. 806–810.
17. Eshelby, J. D., *Proc. R. Soc. London*, Ser. A, Vol. 241, 1957, pp. 376–396.
18. Mori, T. and Tanaka, K., *Acta Metal.*, Vol. 21, 1973, pp. 571–574.
19. Kerner, E. H., *Proc. Phys. Soc. B Sec.*, Vol. 69, 1956, pp. 808–813.
20. Arsenault, R. J. and Taya, M., *Acta Metal.*, Vol. 35, No. 3, 1987, pp. 651–659.
21. Wolff, E. G., Min, B. K. and Kural, M. H., *J. Mater. Sci.*, Vol. 20, 1985, pp. 1141–1149.
22. Min, B. K. and Flaggs, D. L., *ASME Conf. Volume*, 1985.
23. Cambell, M. D. and Burleigh, D. D., *Composites for Extreme Environments*, ASTM STP 768, edited by N. R. Adsit, American Society for Testing and Materials, 1982, pp. 54–72.
24. Kelly, K. P., *Evaluation of Metal Matrix Composites*, Vought Miss. Space Comp. Report 001471 to NASA, 1971.
25. Gurtman, G. A., Rice, M. H. and Maewal, A., *Thermomechanical Analysis of Graphite/Metal Matrix Composites*, Final Report (SSS-R-81–4862) to DARPA (# 3788), February 1981.
26. Springer, G. S. and Tsaï, S. W., *J. Comp. Mater.*, Vol. 1, 1967, pp. 166–173.
27. Hatta, H. and Taya, M., *Int. J. Eng. Sci.*, Vol. 24(7), 1986, pp. 1159–1172.
28. Hatta, H. and Taya, M., *J. Appl. Phys.* Vol. 58(7). 1985, pp. 2478–2486.
29. Takao, Y. Chou, T. W. and Taya, M., *J. Appl. Mech.*, Vol. 49, 1982, pp. 536–540.
30. Hatta, H. and Taya, M., *J. Appl. Phys.* Vol. 59(6), 1986, pp. 1851–1860.
31. Hasselman, D. P. H. and Johnson, L. F., *J. Comp. Mater.*, Vol. 21, 1987, pp. 508–515.
32. J. C. Maxwell, *Treatise on Electricity and Magnetism*, 3rd edn, Oxford University Press, 1904.
33. Benveniste, Y., and Miloh, T., *Int. J. Eng. Sci.*, Vol. 24, 1986, pp. 1537–1552.
34. Benveniste, Y., *J. Appl. Phys.*, Vol. 61, 1987, pp. 2840–2843.
35. Taya, M. and Fadale, T. Unpublished work.
36. Budiansky, B., *J. Comp. Mater.*, Vol. 4, 1970, pp. 286–295.
37. Rosen, B. W. and Hashin, Z., *Int. J. Eng. Sci.*, Vol. 8, 1970, pp. 1567–173.
38. Walpole, I. J., *Proc. Camb. Phil. Soc.*, Vol. 83, 1978, pp. 495–506.

CHAPTER 6

Engineering Problems

6.1 Introduction

In order to characterize the thermomechanical behavior of a metal matrix composite, the metal matrix composite must be first fabricated by an appropriate processing route, followed by machining it into a specimen geometry. Then the machined metal matrix composite specimen will be subjected to thermal and mechanical tests to determine its thermomechanical behavior, so that the experimental thermal–mechanical properties can be compared with those predicted by analytical models.

In this chapter some of the engineering problems relevant to the thermomechanical behavior of a metal matrix composites will be discussed. These include fabrication and machining. Of these, fabrication appears to be the most popular, as evidenced by the large number of papers written on this subject. However, quite recently fabrication has been studied rigorously from the scientific viewpoint, i.e., in terms of simulation by an appropriate thermomechanical model. A brief review of the recent fabrication studies, with a scientific approach, will be made here. Unlike fabrication studies, there has not been much focus on machining of metal matrix composites despite the fact that all the metal matrix composites fabricated must be machined for use as structural components or characterization by thermal and mechanical tests. Also, the machining cost often represents a large portion of the total cost of a metal matrix composite component. Hence, machinability studies of metal matrix composites should not be overlooked.

6.2 Fabrication

Fabrication methods for the production of metal matrix composites have been revived recently as one of the thrust areas of composite technology, and a number of books have focused on the fabrication methods for metal matrix composites; for example, books by Kelly and Mileiko [1] and Schwartz [2]. The first book [1] is extensive in describing some of the scientific aspects relevant to fabrication of metal matrix composites, and the second one [2] provides a list of various fabrication methods and routes shown schematically by illustrations. In addition to the above books, a number of technical papers on fabrication of metal matrix composites have recently been published in journals and conference proceedings. Based on these, one can summarize the popular fabrication methods and their applications to various metal matrix composites as listed in Table 6.1. The illustrative figures corresponding to the fabrication methods listed in Table 6.1 are shown in Figs. 6.1 through 6.8. It should be noted that these fabrication methods are not mutually exclusive, but some of them can be combined into another fabrication route as demonstrated in Fig. 6.6, where HIPing and powder metallurgical methods are combined into another processing route.

Among these fabrication methods, squeeze and pressure casting methods have been studied relatively well, and will be reviewed first. Clyne [6] studied the processability of 2D misoriented short fiber metal matrix composites by squeeze casting. The basic principle used in Clyne's study is Kelvin's equation,

$$\Delta P = \sum_{i} \frac{\gamma}{r_i} \tag{6.1}$$

where ΔP is the pressure change required for the infiltration of molten metal into an array of fibers, γ is the surface energy at the melt–atmosphere interface, and r_i are the principal radii of curvature. If eq. (6.1) is applied to the

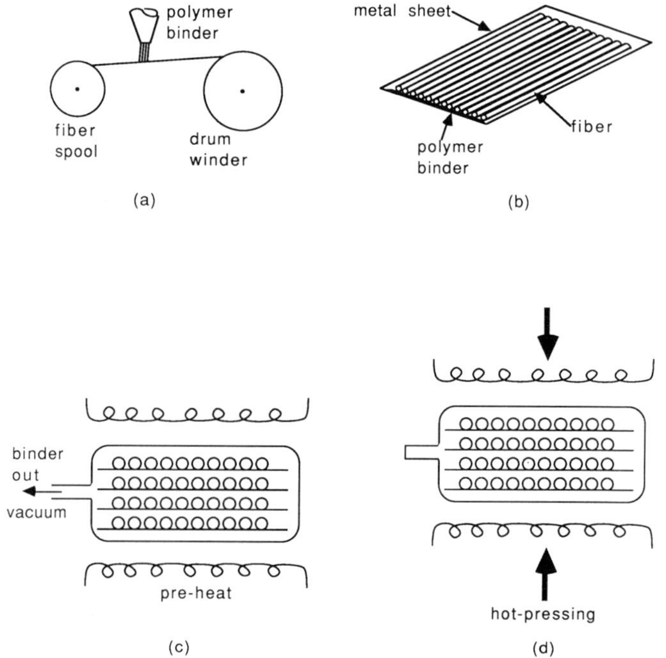

FIG. 6.1 Diffusion bonding processing route; (a) preform fabrication, (b) cut out preform sheet, (c) burn off binder and vacuuming and (d) hot pressing.

Table 6.1 *Various fabrication methods*

Fabrication method	Reinforcing morphology	Examples of MMC systems	References	Figures
Diffusion bonding (vacuum, or air hot press)	Laminated composites	B/Al	[3, 4]	6.1
		B-SiC/Ti	[5]	
Squeeze casting	Generally applicable to any type of reinforcement	δ-Al_2O_3/Al	[6, 7]	6.2
		C/Mg	[8]	
		SiCw/Al	[9]	
		Si_3N_4w/Al	[10]	
Pressure casting	Generally applicable to any type of reinforcement	SiC/Al	[11]	6.3
		γ-Al_2O_3/Al	[12]	
Hot isostatic pressing (HIPing)	Generally applicable to any type of reinforcement	W/Fe-alloy	[13]	6.4
		α-Al_2O_3(FP)/FeCrAlY	[14]	
Reocasting (compocasting)	Particulates, short fibers composites	Al_2O_3/Al	[15]	6.5 (a), (b)
		SiCw/Al	[16]	
Powder metallurgy	Particulates, short fibers composites	SiCw/Al	[17]	6.6
		K_2O–6TiO_2w/Al	[18]	
Arc spray forming	Unidirectional and woven composites	W/super-alloy	[19, 20]	6.7
Superplastic forming	Particulates short fibers and unidirectional composites	SiCp, SiCw/Al	[21]	6.8
		Stainless Steel/Al	[22]	

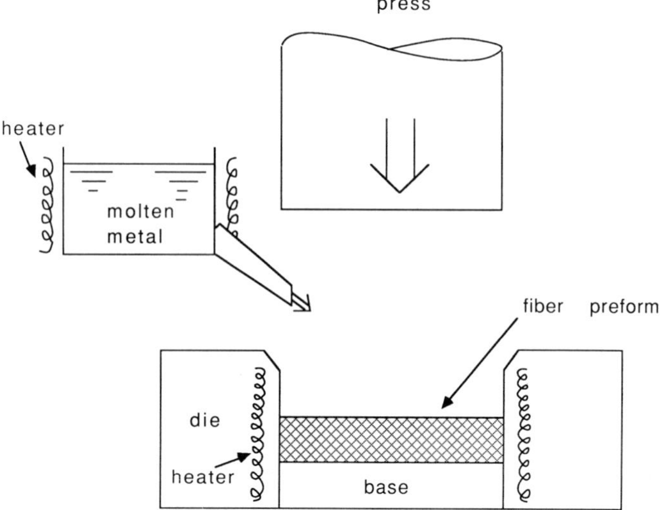

FIG. 6.2 Squeeze casting [6].

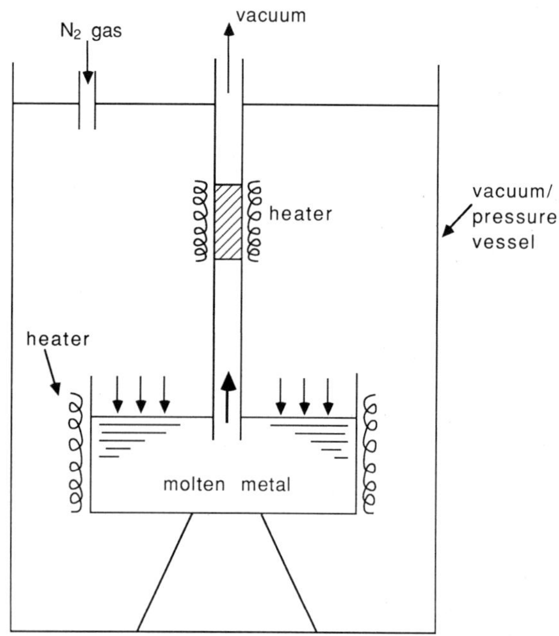

FIG. 6.3 Pressure casting.

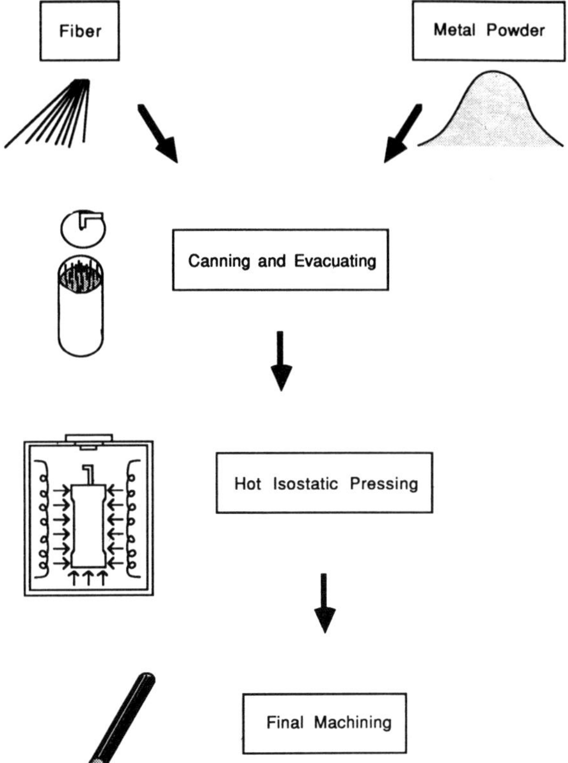

FIG. 6.4 Hot isostatic pressing [14].

case of pure aluminum infiltration of melt along the axis of unidirection (parallel) fiber bundle where the fiber diameter is 15 μm and the fiber array is hexagonal, then ΔP becomes approximately 1 MPa (10 atm.), which is considered modest. The case of a 2D misoriented fiber network, which is more convenient for squeeze casting than that of parallel fiber bundle, requires a more rigorous analysis for ΔP. Clyne [6] proposed a simple model to predict the melt infiltration pressure for a given 2D misoriented short fiber preform which was assumed to consist of planar fiber network and cross-linking fibers (Fig. 6.9). This model also can predict the infiltration pressure for the onset of fiber fracture of cross-linking fibers. One of the important conclusions in this study is that short fibers of smaller diameter (submicron) require much greater infiltration pressure, which results in possible fracture of such submicron fibers. Clyne also found that squeeze casting of aluminum-based composites gives rise to finer grain size in the matrix, due primarily to fast solidification of the melt contacted with fibers, and also to oxidation-free interfaces leading to good bonding of the matrix–fiber interface [23]. In

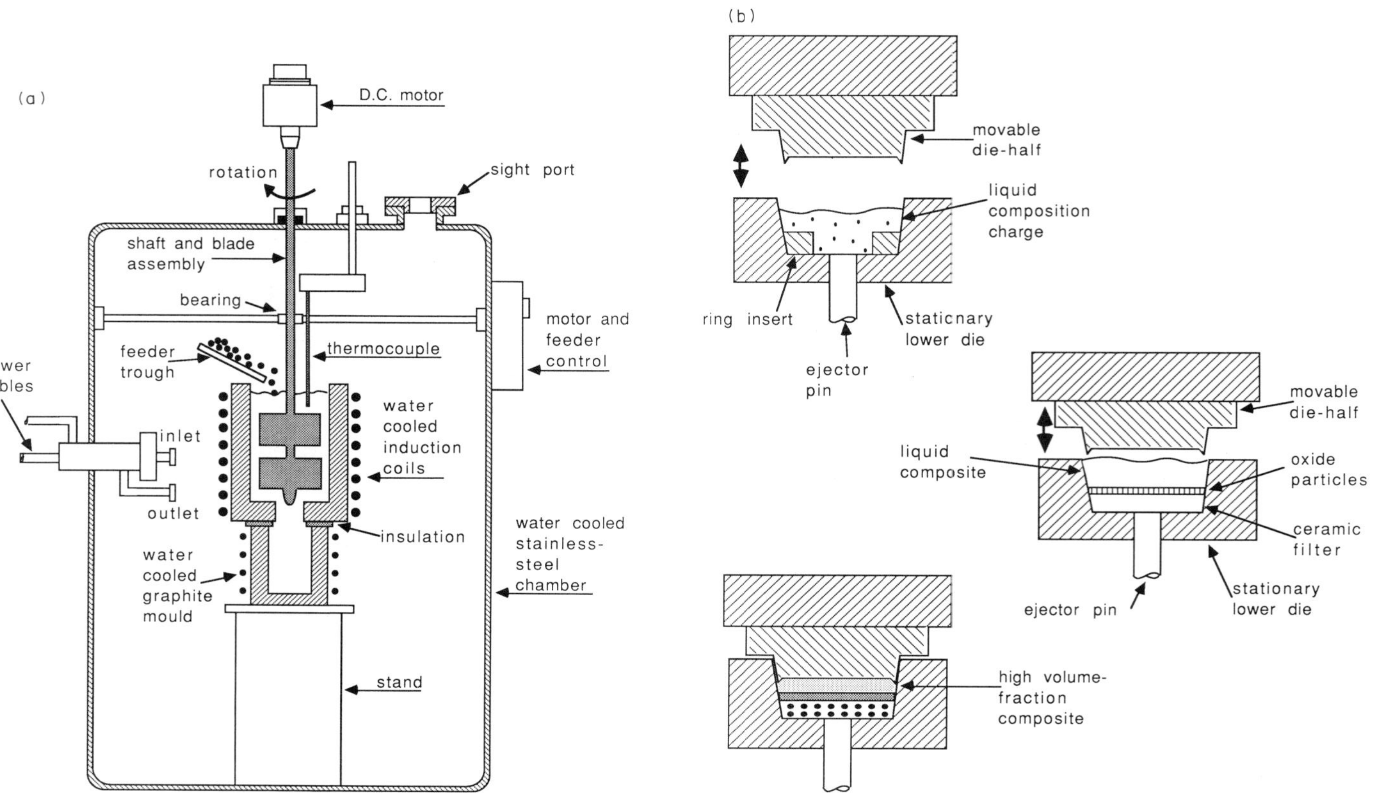

FIG. 6.5 Reocasting; (a) mixing fibers (or particulates) with metal, followed by (b) die casting [15].

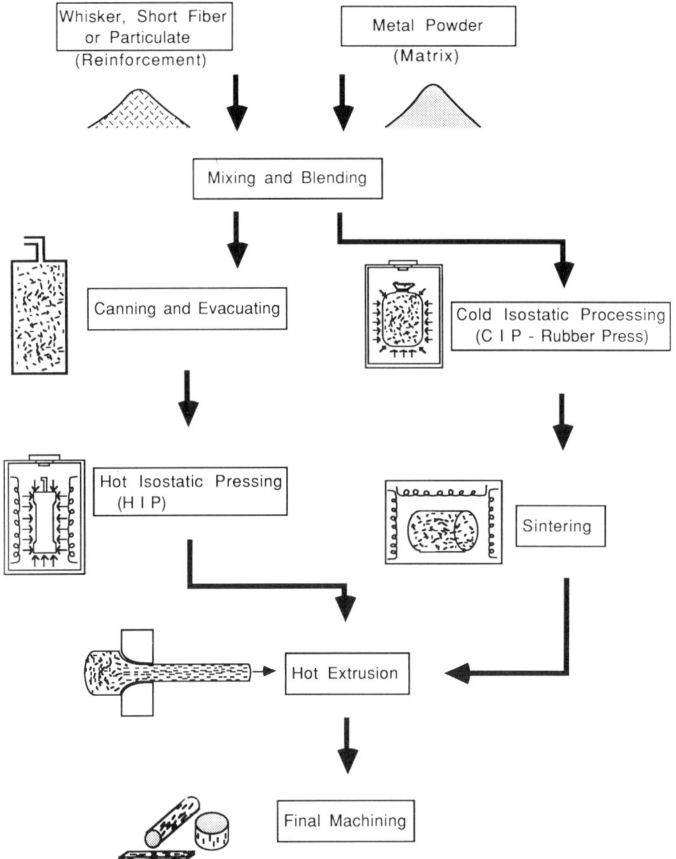

FIG. 6.6 Powder metallurgy processing route where two intermediate steps, HIP and CIP, are shown.

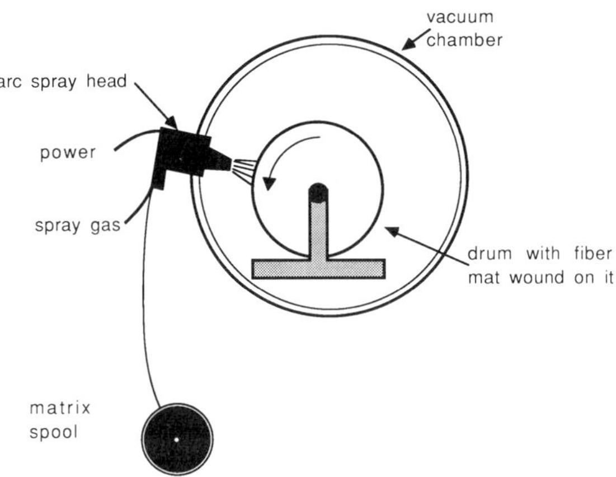

FIG. 6.7 Arc spray forming.

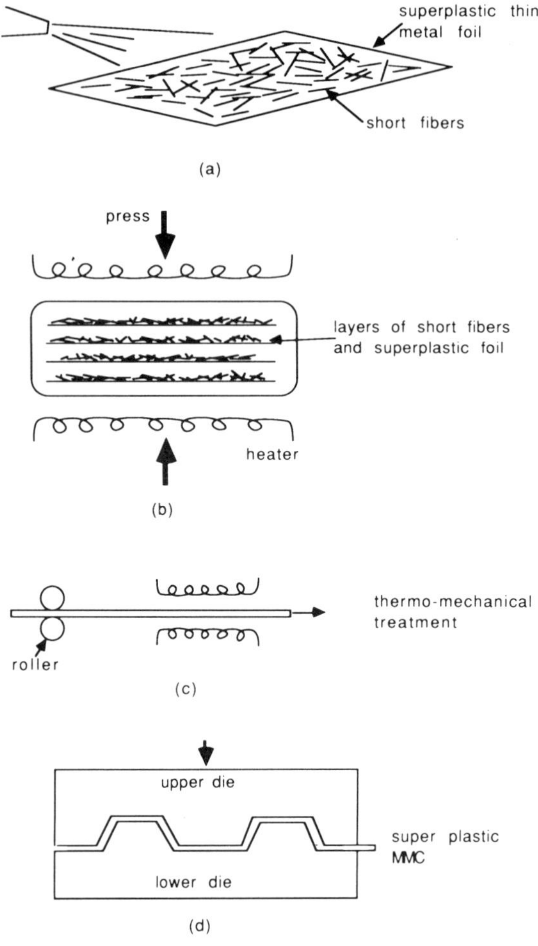

FIG. 6.8 Superplastic forming [22].

contrast, the powder metallurgical method tends to yield more oxide films at the interfaces, which can be eliminated by applying shearing (through extrusion).

The MIT group [11, 24] has recently studied the effect of several processing parameters on the performance of pressure cast aluminum matrix composites where the reinforcements are SiC and B_4C particulates. The processing parameters examined are pressure (P), temperature of molten metal (T_m), that of fiber preform (T_f), fiber volume fraction (V_f) and percentage of alloying element (V_{al}). A summary of this study [11] is shown in Fig. 6.10, where the order of parameters influencing the performance is found to be V_f,

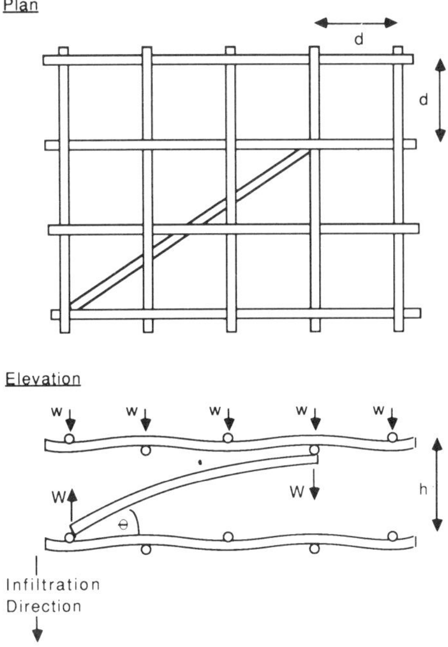

FIG. 6.9 2D misoriented short fiber preform is assumed to consist of planar fiber network and cross-linking fibers in Clyne's model [23].

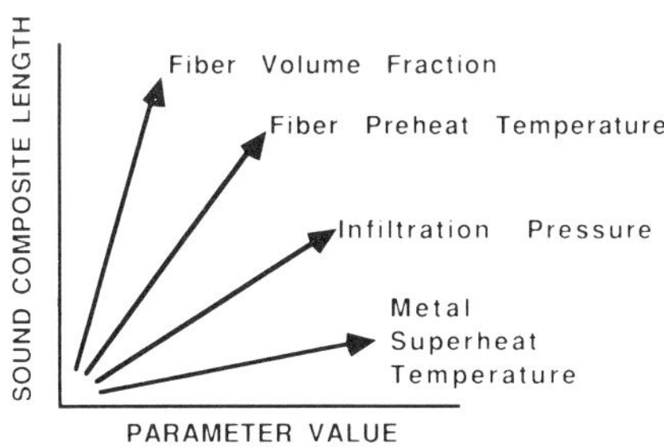

FIG. 6.10 Effect of several processing parameters on infiltration length where steeper slope denotes a more significant effect [11].

T_f, p, T_m, and V_{al}. In a related investigation [24], they re-examined the performance of infiltration in terms of specific permeability k defined by

$$k = \mu \frac{dL}{dt} \cdot \frac{L}{P} \tag{6.2}$$

where μ is the viscosity of the melt, L is the infiltration length by the melt, t is

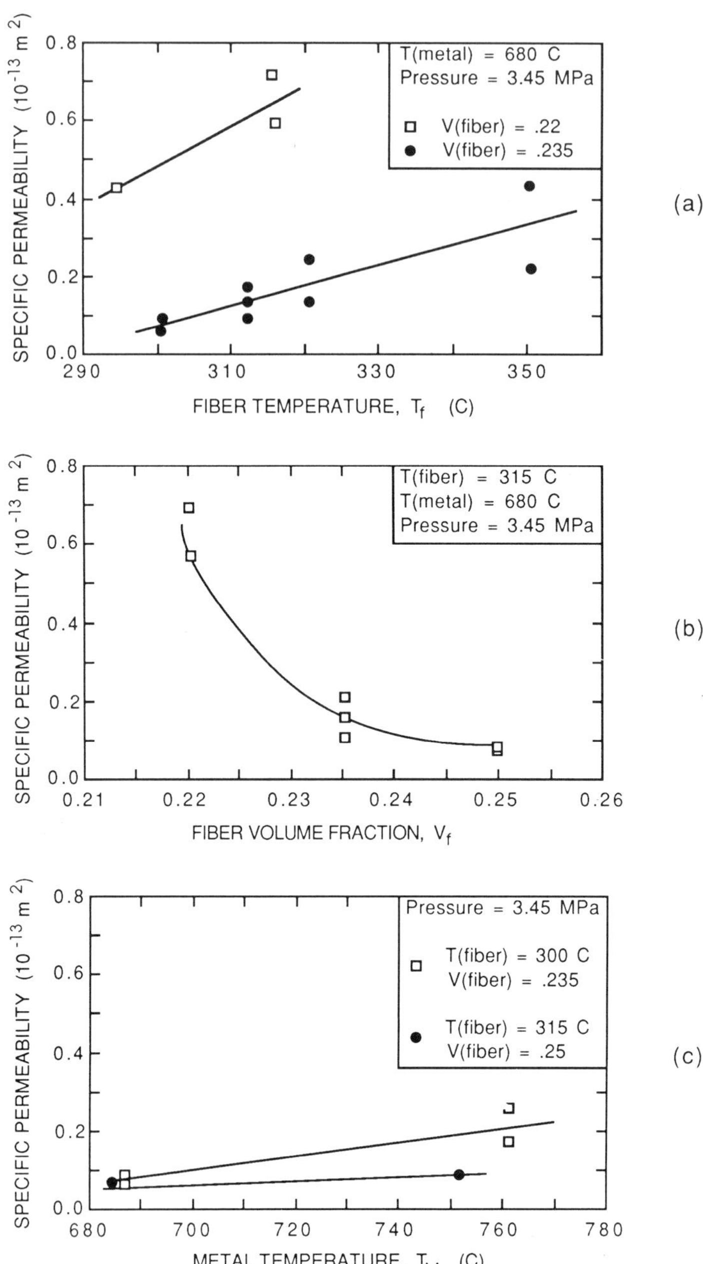

FIG. 6.11 Effect of processing parameters on specific permeability of a composite; (a) fiber temperature, (b) fiber volume fraction, and (c) metal temperature [24].

time and P is pressure. The measured values of k of misoriented short fiber (Saffil)/pure aluminum are shown in Fig. 6.11 where (a), (b), and (c) denote dependence of k on fiber preform temperature (T_f), fiber volume fraction (V_f) and metal temperature (T_m). The results of Fig. 6.11 agree with the general trend of Fig. 6.10, i.e. V_f and T_m are the most and least influential parameters, respectively.

Other research groups have studied the effect of inhomogeneous distribution of fiber preform [16, 25, 26]. Girot et al. studied the mechanical behavior of short SiC fiber/Al composites that were fabricated by compocasting [16], and found that the stiffness, hardness and dynamic toughness of the composites increased compared with the unreinforced metal, but the yield stress (σ_y) ultimate strength (σ_B) and rupture strain (ε_f) decreased compared with the unreinforced metal. This reduction in σ_y, σ_B, and ε_f was attributed to the misorientation and uneven distribution of short fibers. Fujita et al. [25] investigated the effect of homogeneity of fiber preform on the mechanical properties of squeeze cast SiCw/Al composite, and concluded that the composites with homogeneous fiber preform exhibited a good mechanical behavior (for example σ_B of 20% SiCw/Al = 500 MPa), while those with inhomogeneous fiber preform showed relatively low σ_B (300 MPa).

Lloyd [26] investigated the effect of some processing parameters on the degree of homogenization in SiC particulate/aluminum cast composites. First he examined the relationship between cooling rate and cell size of 6061 aluminum and SiC particulate/6061 aluminum (SiCw/6061/Al) composites, the results of which are shown in Fig. 6.12. Figure 6.12 indicates that the faster cooling rate results in the smaller cell size, and under the same cooling rate the composite gives rise to smaller cell size than the unreinforced aluminum. Lloyd found strong dependence of the homogenization of SiC particulates in the composite on the cell size, i.e., the smaller cell size resulted in a more uniform distribution of SiC particulates which, in turn, leads to better mechanical properties of the composite. In the case of the larger cell size, SiC particulates tend to be segregated into the intercellular regions as shown in Fig. 6.13 where the cooling rate of 1 °C/s was used, thus resulting in large cell size (approximately 100 μm, see Fig. 6.12). The distribution of SiC particulates in as-cast composites can be altered by the extrusion process. Lloyd examined the microstructures of as-extruded SiCp/6061 Al composites which were first processed by casting. Figure 6.14 shows the typical microstructure of a 15% V_fSiCp/6061 Al composite where (a) and (b) denote the microstructure along the extrusion direction and along its perpendicular direction, respectively, some SiC particulates are clustered as inherited from the as-cast microstructure. A similar study on the microstructure of as-extruded SiCp/Al composites was also conducted by Sun and Greenfield [27] who examined the effect of extrusion ratio (measured in terms of the reduction of the cross-section) on the size distribution and orientation of SiC particulates. They also discussed the mechanism by which particulates are broken into smaller pieces.

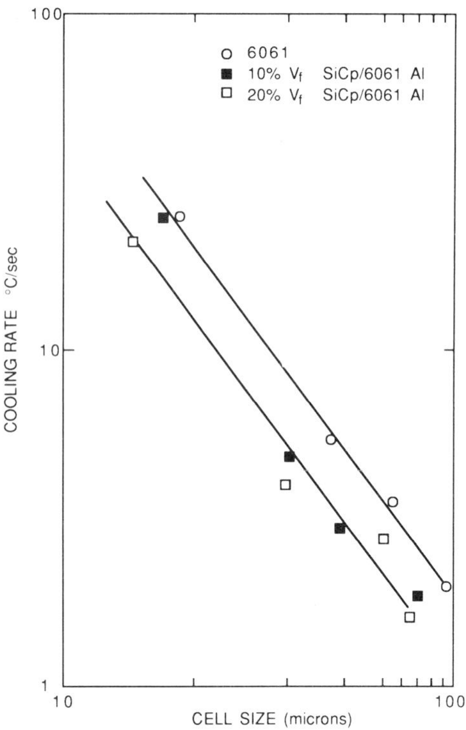

FIG. 6.12 Relationship between cooling rate and cell size in 6061 Al and SiCp/6061 Al composite [26].

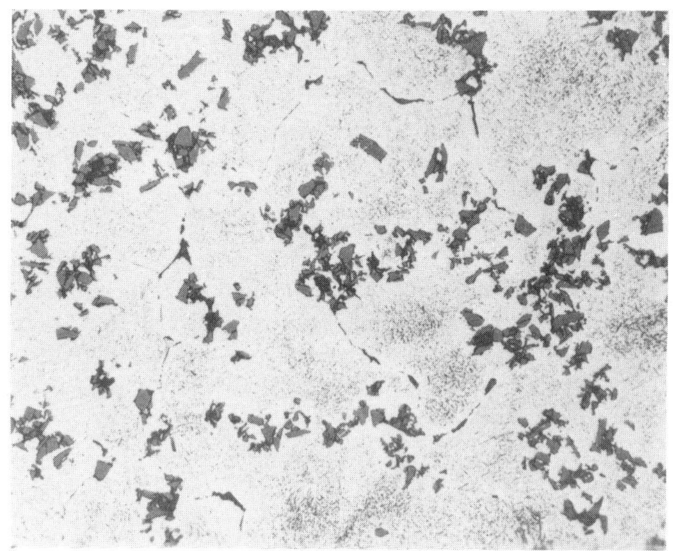

FIG. 6.13 Microstructure of the 10% V_f SiCp/6061 Al composite that was cooled at the rate of 1 °C/s during its fabrication [26].

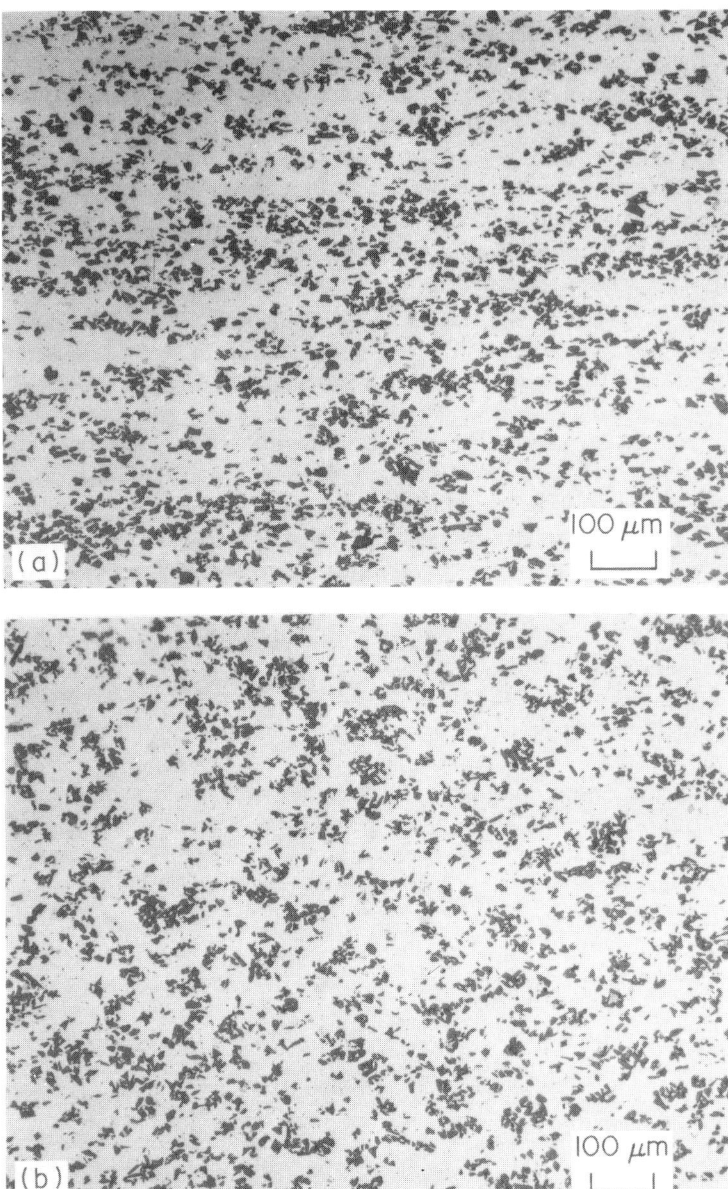

FIG. 6.14 Microstructure of the as-extruded 15% SiCp/6061 Al composite, (a) along the extrusion direction and (b) along its perpendicular direction [26].

6.3 Machining

Development of cost-effective machining methods will reduce the total cost of a metal matrix composite component, thus increasing the competitiveness of metal matrix composites over other high-temperature material systems [28]. In this respect other fabrication methods—joining, welding, riveting and adhesive bonding—where machining constitutes the initial processing step must also be studied.

In most metal matrix composites, fibers and particulates are ceramics or refractory metals, which are considered 'hard to machine' material. This hard reinforcement sometimes makes machining of metal matrix composites difficult. In this respect ceramic matrix composites share similar machining difficulties, while the machining of polymer matrix composites is much easier than metal and ceramic matrix composites.

Most of the machining methods that have been used for metals are basically applicable to metal matrix composites if the volume fractions of fibers or particulates (V_f) are not too large, although use of highly wear-resistant cutting tools is often required. For metal matrix composites with large V_f, the machining methods used for ceramics are equally applicable. Machining methods can be categorized into two groups, traditional and non-traditional. The first group includes turning, milling, drilling, grinding and sawing by mechanical machines. The second that has proved to be applicable to machining of metal matrix composites includes electro-discharge machining (EDM), abrasive water (AWJ) and laser beam machinings.

Machinability investigation of metal matrix composites has been limited because the main thrust of the research in metal matrix composites has still been on its processing and characterization. Thus, information on machinability is lacking in the composite community. Based on these limited studies, and the information collected from machinists, we have summarized the machinability of a selected metal matrix composite, which will be given in the following subsection, using traditional and non-traditional machinings.

6.3.1 Traditional machining

Metal matrix composites with relatively small volume fractions (V_f) of fiber and particulate can be machined by traditional methods such as turning by lathe. Traditional machining methods (turning, milling and drilling) are basically applicable to most of the metal matrix composites provided that the reinforcement is particulate or short fiber or even for continuous fiber of small diameter. McGinty and Preuss [29] investigated the machinability of FP fiber/aluminum (FP/Al) and FP fiber/magnesium (FP/Mg) composites where they used turning, milling and drilling. The metal matrix composites used are FP/Al of 12.7 mm in thickness for drilling and milling processes, and FP/Al rod of 50.8 mm in diameter for the turning process. The volume fraction of these continuous FP fiber metal matrix composites is 55%. Some of the findings in this study are that drilling of FP/Al composite plates can be performed with solid carbide high-helix drills at a speed of 30.48 m/min, and turning of the composite plates can also be performed with C-2 uncoated carbide or ceramic-coated carbide at a speed of 30.48 m/min; also in turning, large nose radius to round inserts should be used to enhance surface finish at high feed rates. Similar tools can be used for milling the composite. Tool life in the above machining process is short compared with machining other materials. In contrast to carbide tools, polycrystalline and natural diamond tool materials are not applicable to FP/Al or FP/Mg composites. As for particulate or short fiber metal matrix composites, for example SiC whisker/aluminum [30], the above machining with carbide tools is also applicable without any difficulty. Matsubara et al. [10] have reported that SiCw/Al is still harder to machine than other short fiber metal matrix composites, Si_3N_4 and K_2O–TiO_2 whisker/Al composites.

For metal matrix composites with continous fiber of large diameter such as boron fiber, machining with diamond is recommended, although it is time-consuming.

6.3.2 Non-traditional machining

Non-traditional machining is becoming increasingly popular since it is possible to machine complex-shaped parts or hard-to-machine materials, including metal matrix composites, at relatively high speed. Non-traditional machining includes:

(1) ultrasonic machining (USM)
(2) abrasive jet machining (AJM)
(3) electron chemical machining (ECM)
(4) chemical milling (CHM)
(5) electro-discharge machining (EDM)
(6) electron beam machining (EBM)
(7) laser beam machining (LBM)
(8) plasma arc machining (PAM)

Performance of the above eight non-traditional machinings has been well documented for application to materials of various shapes [31]. However, application of these non-traditional machining processes to metal matrix composites has been limited. We will review below the results of a limited number of studies on machinability of metal matrix composites. The non-traditional machining processes reviewed here are AJM [32, 33], EDM [34], and LBM [35, 36].

Abrasive jet machining, which is also called abrasive water jet (AWJ) machining, is insensitive to materials and free of thermal effects with acceptable surface finish. AWJ has recently been applied to B_4C particulate/magnesium (B_4C/Mg) [32] and SiC whisker/aluminum (SiCw/Al) [33] composites. The material removal mechanism in AWJ machining is erosion of material by high-speed water jet with abrasives, and its schematic view is shown in Fig. 6.15. Hashish [32] has investigated the machinability (turning and milling) of a B_4C/Mg composite into a tensile specimen rod by using the AWJ system similar to Fig. 6.15. Hashish [32] found that AWJ machining can machine any materials into complex shapes at very high material removal rates, but the AWJ machined surface is rough, requiring grinding to obtain a smooth surface finish. Figure 6.16 shows the surface texture of the AWJ machined B_4C/Mg composite rod. Savrun and Taya [33] have studied the machinability of SiCw/Al composite plate (of 6.3 mm in thickness) by AWJ where three different machining speeds (V) were used, $V = 127$, 381 and 635 mm/min. In AWJ machining, water pressure of 275.6 MPa, and 80-mesh garnet particles as abrasive, with a flow rate of 0.67 kg/min, were used. These parameters are not necessarily optimized. Figure 6.17(a), (b), and (c) show the AWJ machined surfaces of a 25% V_f SiCw/Al composite at cutting speeds of $V = 127$, 381 and 635 mm/min, respectively, where white arrows denote the direction of the abrasive waterjet. Savrun and Taya measured the surface finish of the AWJ machined surface of SiCw/Al composites by profilometry and the results are shown in Table 6.2. They have also measured the hardness of the AWJ machined composite at three different depths by the knoop

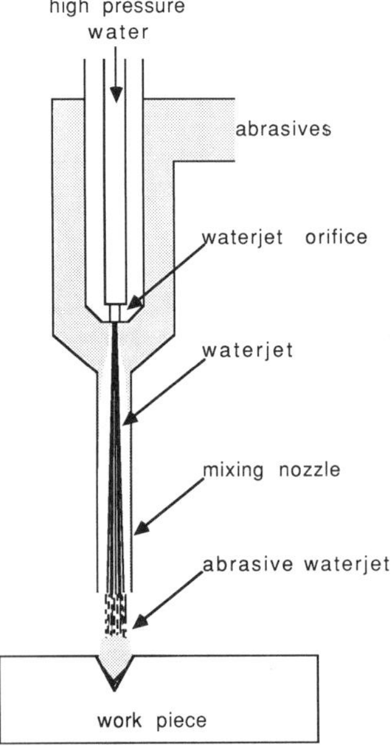

FIG. 6.15 Schematic of an abrasive water jet machining [32].

FIG. 6.16 Surface texture of the AWJ machined B_4C particulate/Mg composite [32].

Table 6.2 *Surface finish of the AWJ machined SiCw/Al composite [33]*

Materials	Cutting speed (mm/min)	Surface finish
25% V_f	127	2.54
SiCw/2124Al	381	2.79
	635	3.81

Table 6.3 *Microhardness of the AWJ machined SiCw/Al composite at different depths [33]*

Distance from surface (mm)	Microhardness knoop (mm)
80	220
180	220
3000	220

indentation test, indicating no change in the microhardness knoop, as shown from Table 6.3.

Electro-discharge machining (EDM) is an electrothermal process where material removal takes place between an anode (tool electrode) and a cathode (workpiece), submerged in a dielectric fluid. The workpiece in the EDM

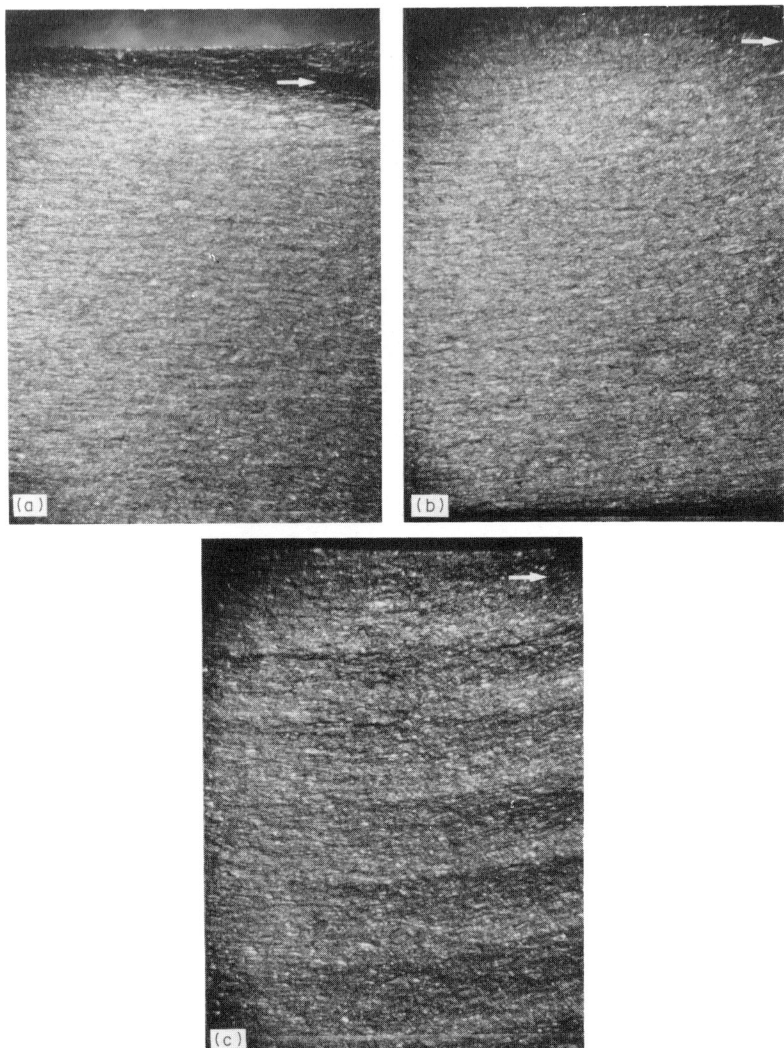

FIG. 6.17 Surfaces of the AWJ machined 25% V_f SiC whisker/2124 aluminum at various cutting speeds (V) (a) $V = 127$, (b) $V = 381$ and (c) $V = 635$ mm/min where white arrows denote the cutting direction [33].

process must be at least semi-conductive. A typical setup for EDM is shown in Fig. 6.18. Two types of EDM processing are commonly used, die-sinking and wire-cutting. The former is suited for machining holes or shaping complex components, the latter for straight two-dimensional line cutting. Ramulu and Taya [34] studied the EDM machinability of SiCw/2124 Al composites with $V_f = 15$ and 25% and their thicknesses 6.3 mm, where the

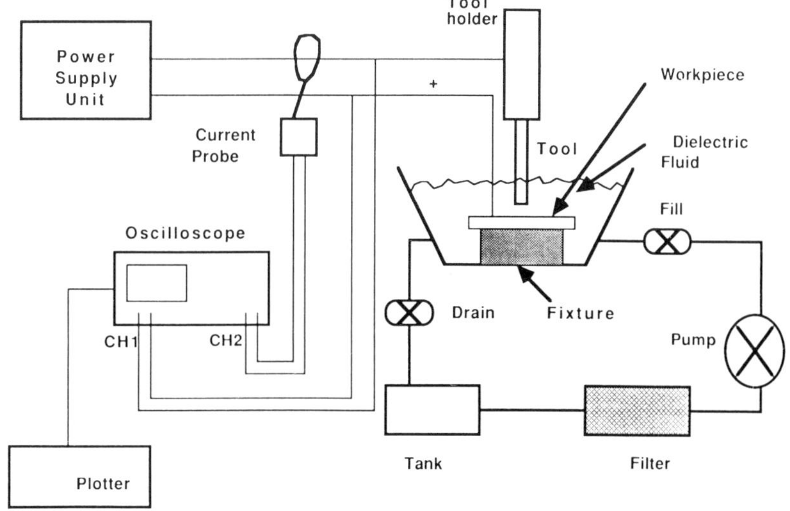

FIG. 6.18 Schematic diagram of EDM [34].

EDM setup of Fig. 6.18 type was used and the tool electrodes are brass and copper.

The conditions for the EDM process used in Ramulu and Taya's experiment are given in Table 6.4.

Table 6.4 *EDM conditions [34]*

Rate of cutting	Process conditions	Power (amp)	Frequency (Hz)
Very slow	Fine	1/3	600
Medium	Medium	1	300
Rapid	Coarse	6	50

The results of he EDM machined SiCw/Al composites are shown in Table 6.5.

The EDM machined surfaces of 15% SiCw/Al composite were examined by SEM. The SEM study revealed that the surface damage and crater size increased with cutting speed. To investigate the damage caused by EDM, these authors sectioned the EDM machined composite by diamond saw, an example of which in the case of 25% SiCw/Al composite machined by rapid EDM process is shown in Fig. 6.19(a), revealing several large voids and micro voids near the EDM machined surface. Then a part of Fig. 6.19(a) in tilted condition was examined by SEM at high magnification (Fig. 6.19(b)). It can

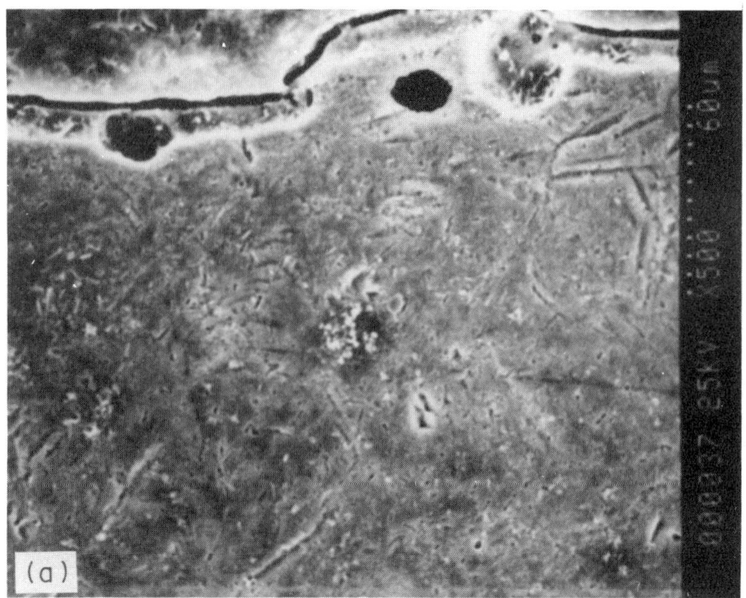

FIG. 6.19 Surface damage in the EDM processed 25% V_f SiC_w/2124 Al composite; (a) profile of the AWJ machined surface (the upper part) and (b) the enlarged view of (a) at tilted condition where the upper portion denotes the EDM processed surface and the lower part is the cross-section machined by diamond saw [34].

Table 6.5 *Results of EDM machined SiCw/Al composites [34]*

Materials	Rate of cutting	Process conditions	Cutting time (min/mm)	Relative electrode wear (%)	Surface finish (μm)
15% SiCw/Al	Very slow	Fine	5.3	28	3.56
	Medium	Medium	2.5	29	3.68
	Rapid	Coarse	1.2	32	6.99
25% SiCw/Al	Very slow	Fine	8.5	44.5	3.05
	Medium	Medium	5.5	30.49	3.73
	Rapid	Coarse	2.5	39.10	6.99

be concluded from Fig. 6.19(a) and (b) that the EDM at higher cutting speed can damage the composite severely. In order to study the effect of heat along the thickness direction they also conducted microhardness knoop tests on several points in the subsurface of Fig. 6.19(a) and the results are given in Table 6.6.

Table 6.6 *Results of microhardness knoop of SiCw/Al composites machined by EDM [34]*

Materials	Distance from surface (μm)	Microhardness knoop
15% SiCw/Al	50	180
	230	180
	3000	180
25% SiCw/Al	50	178
	206	206
	3000	220

Microhardness knoop tests were also taken of as-received SiCw/Al composites and the results are 180 for 15% SiCw/Al and 220 for 25% SiCw/Al. This, combined with the results of Table 6.3, leads to the conclusion that the EDM applied to the higher V_f SiCw/Al composite did change the mechanical properties near the surface while AWJ did not. It can also be concluded from the AWJ and EDM studies [33, 34] that both EDM and AWJ on 25% SiCw/Al composite tend to yield the same degree of surface finish, although AWJ is a faster machining process.

Laser beam machining (LBM) is possible because of the distinguishing features of laser light; monochromaticity, coherence, less divergence and as a result of the high energy intensity, being transmitted to a target material which will then melt and evaporate. Figure 6.20 shows a schematic LBM

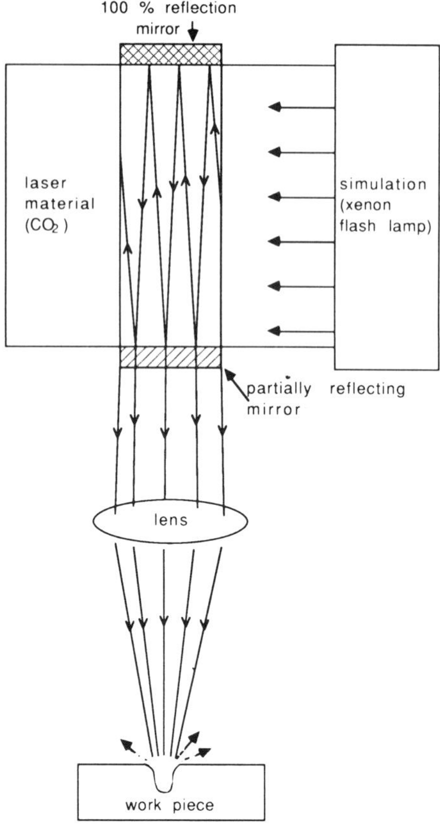

FIG. 6.20 Schematic of laser beam machining.

setup. Although LBM has been used for polymer matrix composites, application of LBM to metal matrix composites is still in its infant stage. The only investigations reported to date of LBM are CO_2–LBM on continuous fiber metal matrix composites [35, 36].

Utsunomiya et al. [35] conducted a LBM study on several types of aluminum-based composites, carbon/1100 Al and carbon/6061 Al composites. The LBM system used in this study was CO_2 gas laser with a maximum power output of 1.5 kW made by Mitsubishi Electric (MLIS) where the wavelength is 10.6 μm and the wave mode is TEM_{00}. Additional features of this LBM include the energy density of Gaussian type with a beam diameter of 80 μm, focal distance of the lens of 127 mm, and the duration of pulse is 0.1 ms. The laser beam was kept perpendicular to the surface of a metal matrix composites plate and also to fibers which are located in plane of the plate. Some of the results obtained in terms of weight loss (mg) vs. power

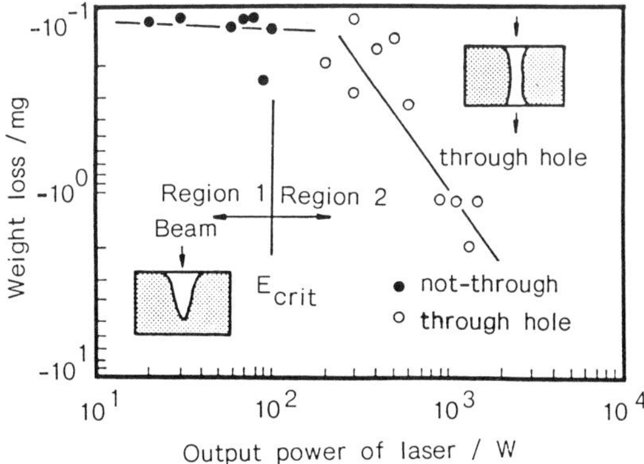

FIG. 6.21 Weight loss vs. laser output power [35].

(W) for carbon fiber/1100 Al composite are shown in Fig. 6.21. One can observe two regions in Fig. 6.21—lower power and higher power regions. The former tends to give rise to pit-like holes which do not penetrate through the plate thickness, resulting in smaller weight loss, while in the high-power region the composite plates are successfully perforated, leading to larger weight loss. The carbon fiber/1100 Al composites machined by laser beam were sectioned for SEM examination. The SEM photos of the cross-section of the composite, with pit-like hole and through hole, are shown in Fig. 6.22(a) and (b) respectively. The power used for (a) was 20 W and that for (b) was 1.2 kW. In Fig. 6.22(a), three zones are distinguishable: zone 1 (innermost zone) consisting of resolidified aluminum rich in Al_4C_3 reactant but without carbon fiber; zone 2 (adjacent to zone 1) consisting of carbon fibers and aluminum and numerous cavities; and zone 3 consisting of carbon fibers, aluminum and cavities, but without Al_4C_3 reactant. In the case of through hole, Fig. 6.22(b), zones 1 and 2 are merged into a narrow zone next to the machined surface which does not appear to be straight. When fibers of larger diameters are used, i.e., in the case of SCS-2/6061 Al composite, the micro morphology near the LBM processed surface appears to be different from the case of carbon fiber/1100 Al composite. Figure 6.23 illustrates this situation where the distribution of the energy intensity is shown along with an SEM photo of the laser beam machined through-hole. Figure 6.23 suggests that the matrix metal melts and evaporates away at some lower energy (E^{m}_{crit}). Thus the surface machined by laser beam is not so flat microscopically.

Lee [36] has recently investigated the machinability of continuous fiber metal matrix composites by AWJ, laser beam and diamond saw machinings. The continuous fiber composites used by Lee are graphite/aluminum (Gr/Al) and SCS–6/Ti–15–3 alloy (SCS6/Ti) composites. The laser beam used is 1 kW continuous-wave type with helium gas as a shielding. As to SCS6/Ti

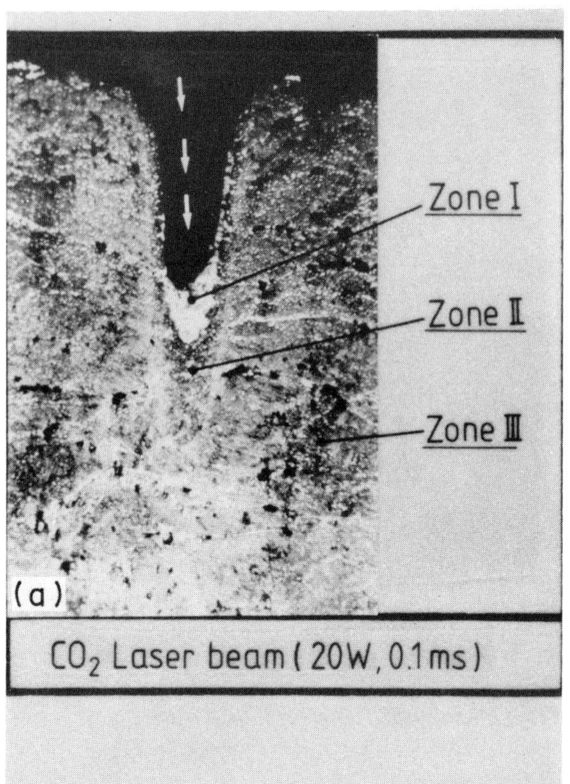

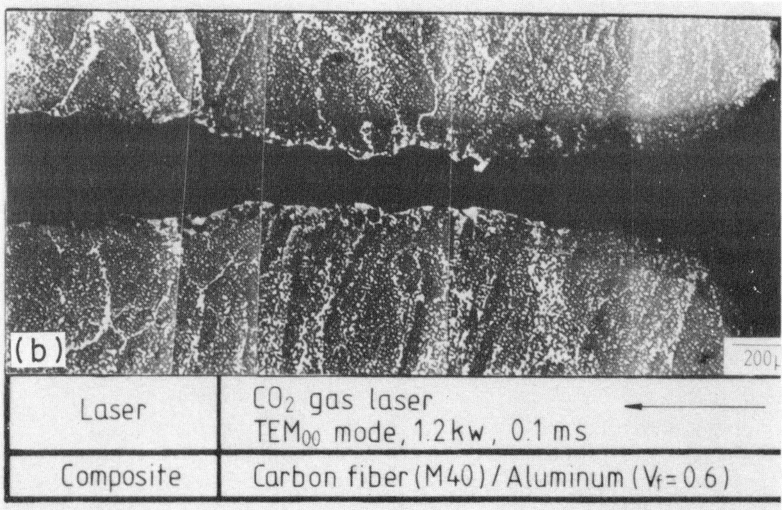

FIG. 6.22 Profile of the laser beam machined holes of carbon fiber/1100 aluminum composite with laser beam power of (a) 20 W and (b) 1.2 kW [35].

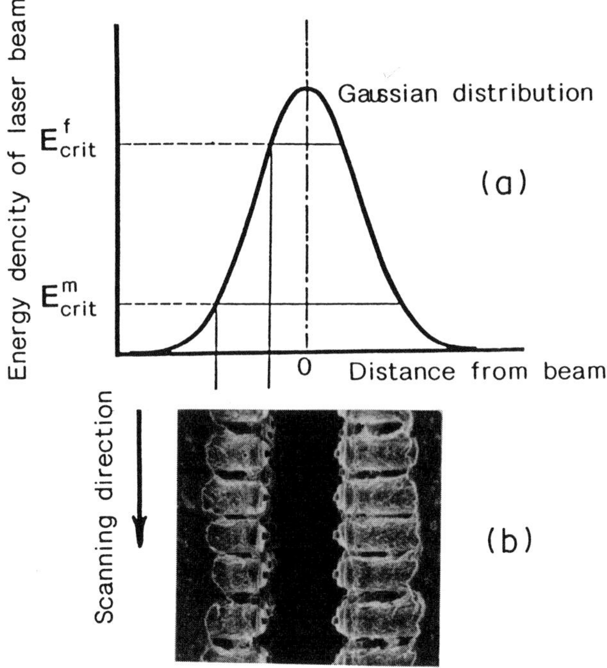

FIG. 6.23 Relationship between (a) laser beam energy distribution, and (b) the profile of laser beam machined surface in the SiC (SCS-2)/6061 Al composite [35].

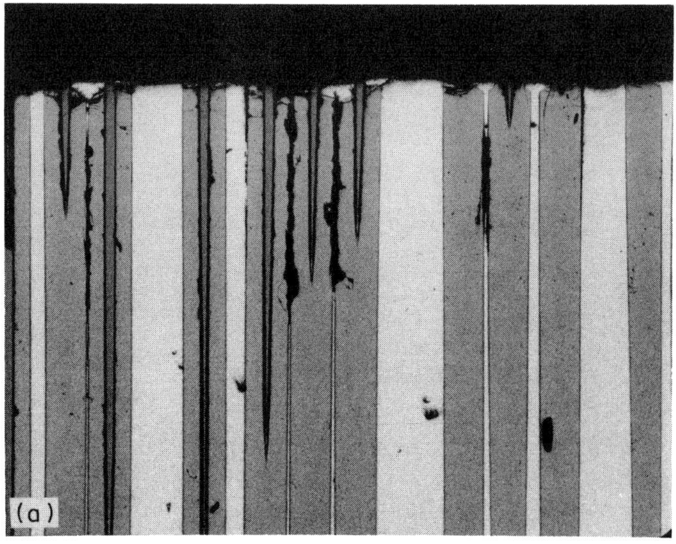

FIG. 6.24(a)

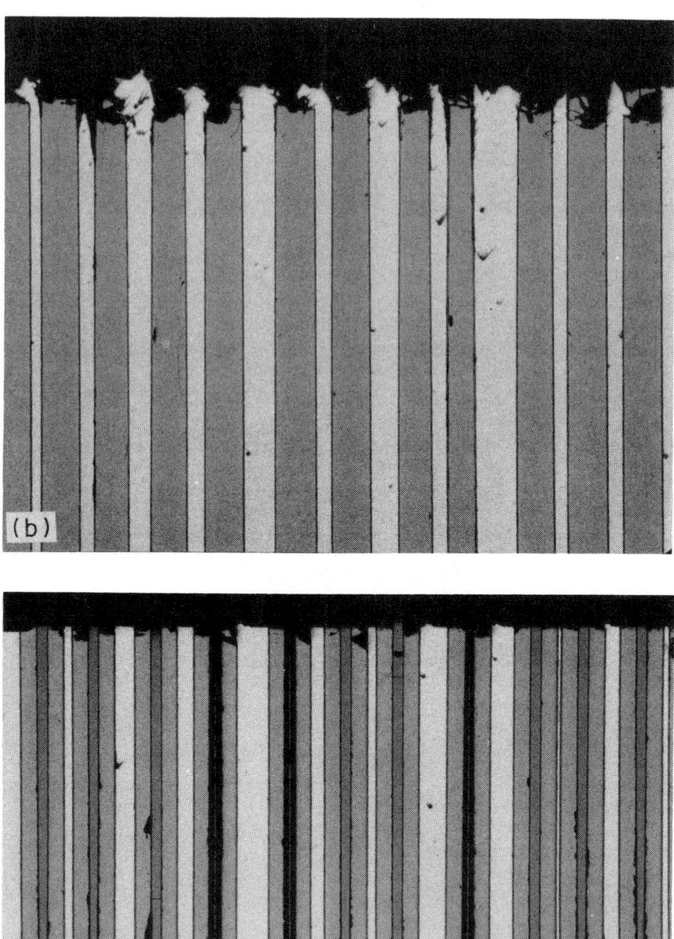

FIG. 6.24 Cross-section of SCS6/Ti-alloy composite machined by (a) laser beam, (b) AWJ and (c) diamond saw [36].

composite, diamond saw cutting gives rise to a smooth cut surface, although it is most time-consuming while AWJ and laser beam cutting tend to induce damage which often extends deeper into the composite. Figure 6.24 shows typical machined surfaces of SCS6/Ti composite where (a), (b), and (c) denote the SEM photos of the longitudinal sections of the composite machined by

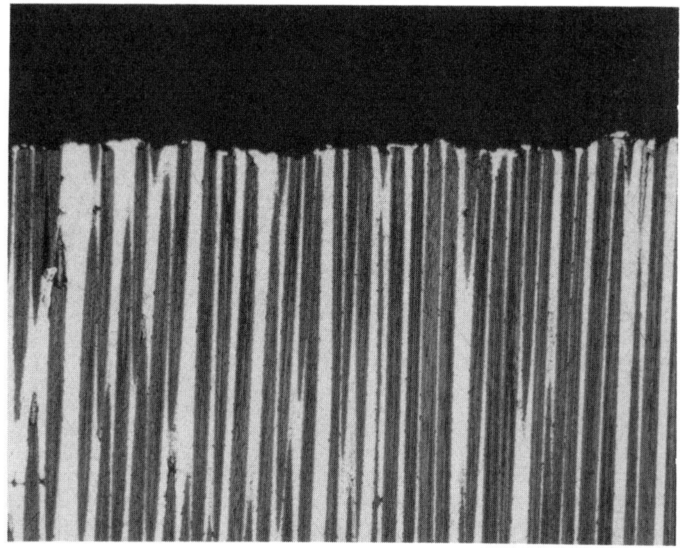

FIG. 6.25 Cross-section of Gr/Al composite machined by AWJ revealing smooth machined surface [36].

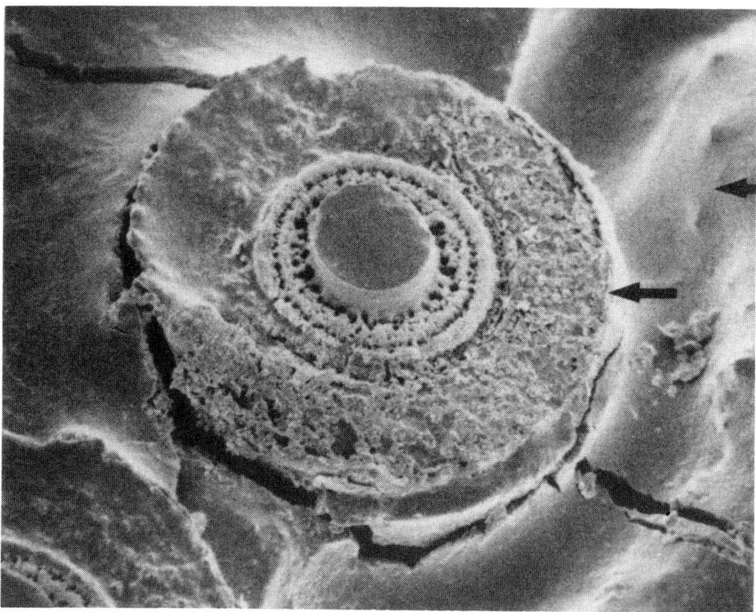

FIG. 6.26 Cracks initiated at the interface of the SiC (SCS-2) fiber/Ti-alloy composite which was machined by 1 kW continuous wave CO_2–laser beam [36].

laser beam, AWJ and diamond saw, respectively. Among these machinings, diamond saw gave the smoothest surface, although it is the most time-consuming and the order of cutting speed from the highest to lowest is laser beam, AWJ, and diamond saw. One of the conclusions Lee arrived at is that AWJ is the best cutting method for Gr/Al composite due to its high speed, thus it is economical and yet keeps the machined surface smooth, as shown in Fig. 6.25. Laser beam is the fastest cutting method, but it tends to induce cracks in the surface due to its high heat flux, as shown in Fig. 6.26. This disadvantage of laser beam machining could be reduced if pulse-type laser beam, combined with an optimum level of power, is employed.

6.4 Problems

(1) Derive the formula to predict the pressure change ΔP that can be observed at infiltration front of unidirection fiber metal matrix composite (see Ref. 6).
(2) Among various fabrication methods, what would be the best suited for processing short fiber metal matrix composite with the aluminum reaction product at the matrix–fiber interfaces?
(3) Discuss the effect of cooling rate on the microstructure of metal matrix composite and its mechanical properties.
(4) Discuss the effect of fiber preform temperature on the mechanical properties of a metal matrix composite.
(5) Explain the mechanism at a microscopic level during superplastic forming of a particulate metal matrix composite.
(6) What type of machining method would you use for straight cutting of a given continuous Nicolon fiber/aluminum composite sheet of thickness 5 mm?
(7) Suppose AWJ was used as a primary machining method for machining a tensile specimen of SCS–6/Ti–alloy composite. What secondary machining methods would you use to machine it into a threaded dog-bone shaped bar?
(8) Explain why laser beam machining appears to be applicable to a graphite fiber/magnesium composite, but not necessarily to a SCS–6 fiber/Ti–alloy composite.
(9) Which type of machining method is best suited for the precision machining assisted by the CNC system?

References

1. Kelly, A. and Mileiko, S. T., *Fabrication of Composites, Handbook of Composites*, Vol. 4, North-Holland, 1983, pp. 221–294.
2. Schwartz, M. M., *Composite Materials Hand Book*, McGraw-Hill, 1984.
3. Prewo, K. M. and Kreider, K. G., *Metal. Trans.*, Vol. 3, 1972, pp. 2201–2211.
4. Metcalfe, A. G., *Composite Materials*, edited by L. J. Broutman and Krack, R. H., Vol. 4, *Metallic Matrix Composites*, edited by K. G. Kreider, Academic Press, New York.
5. Smith, P. R., Froes, F. H. and Cammett, J. T., *Mechanical Behavior of Metal-Matrix Composites*, edited by J. E. Hack and M. F. Amateau, TMS of AIME, 1983, pp. 143–168.
6. Clyne, T. W. and Bader, M. G., *Proc. ICCM-5*, edited by W. C. Harrigan, Jr. et al., TMS of AIME, 1985, pp. 755–791.
7. Donomoto, T., Funatani, K., Miura, N. and Miyake, N., SAE Tech. paper 830252, 1983, Society for Automobile Engineering.
8. Diwanji, A. P. and Hall, I. W., *Proc. ICCM-6/ECCM-2*, edited by F. L. Matthews, et al., Vol. 2, Elsevier, 1987, pp. 2.265–2.274.
9. Fukunaga, H., Goda, H., Kurita, Y., Fujita, Y., ibid., pp. 2.362–2.371.
10. Matsubara, H., Nishida, Y., Yamada, M., Shirayanagi, I. and Imai, T., *J. Mater. Sci. Letts.*, Vol. 6, 1987, pp. 1313–1315.

11. Cornie, J. A., Mortensen, A. and Flemings, M. C., *Proc. ICCM-6/ECCM-2*, edited by F. L. Matthews et al., Vol. 2, Elsevier, 1987, pp. 2.297–2.319.
12. Abe, Y., Horikiri, S., Fujimura, K. and Ichiki, E., *Proc. ICCM-4, Progress in Science and Engineering of Composites*, edited by T. Hayashi et al., Japan Society for Composite Materials, Vol. 2, 1982, pp. 1427–1434.
13. Larsson, L. O. K. and Warren, R., in *Advanced Fibers and Composites for Elevated Temperatures*, edited by I. Ahmad and B. R. Noton, TMS of AIME, 1980, pp. 108–125.
14. Morimoto, T. and Taya, M., Proc. ICF-7, Pergamon Press, In press.
15. Hosking, F. M., Partillo, F. F., Wunderlin, R., and Mehrabian, R., *J. Mater. Sci.*, Vol. 17, 1982, pp. 479–498.
16. Girot, F., Quenisset, J. M., Naslain, R., Coutand, B. and Macke, T., *Proc. ICCM-6/ECCM-2*, edited by F. L. Matthews, et al., Elsevier, Vol. 2, 1987, pp. 2.330–2.339.
17. Rack, H. J., Baruch, T. R. and Cook, J. L., *Proc. ICCM-4, Progress in Science and Engineering of Composites*, edited by T. Hayashi et al., Japan Society for Composite Materials, Vol. 2, 1982, pp. 1465–1471.
18. Imai, T., Nishida, Y., Yamada, M., Shirayanagi, I. and Matsubara, H., *J. Mater. Sci. Letts.*, Vol. 6, 1987, pp. 1257–1258.
19. Westfall, J., U.S. patent Appl. SN 560035, 1985.
20. NASA Cont. Pub. 10003, Aeropropulsion 87, session 1: Aeropropulsion Materials Research, NASA-Lewis Research Center, 17–19 November 1987.
21. Mahoney, M. W., Ghosh, A. K. Bampton, C. C., *Proc. ICCM-6/ECCM-2*, edited by F. L. Matthews et al., Elsevier, Vol. 2, 1987, pp. 2.372–2.381.
22. Lloyd, D. J., *J. Mater. Sci.*, Vol. 19, 1984, pp. 2488–2492.
23. Clyne, T. W., *Proc. ICCM-6/ECCM-2*, edited by F. L. Matthews et al., Elsevier, Vol. 2, 1987, pp. 2.275–2.286.
24. Masur, L. J., Mortensen, A., Cornie, J. A. and Flemings, M. C., ibid., pp. 2.320–2.329.
25. Fujita, Y., Fukumoto, H., Kurita, Y., ibid, pp. 2.340–2.349.
26. Lloyd, D. J., *Canadian Metall. Quarterly*, in press.
27. Sun, J. and Greenfield, I. G., *Proc. ICCM-6/ECCM-2*, edited by F. L. Matthews et al., Elsevier, Vol. 2, 1987, pp. 2.287–2.296.
28. Schoutens, J. E., *J. Metals*, June 1985, TMS of AIME, p. 43.
29. McGinty, M. J. and Preuss, C. W., *High Productivity Machining Materials and Process*, edited by V. K. Sarin, American Society of Metals, 1985, pp. 231–243.
30. Williams, D. R. and Fine, M. E., *Proc. ICCM-5*, edited by W. C. Harrigan, Jr. et al., TMS of AIME, 1985, pp. 639–670.
31. *Nontraditional Machining Processes*, edited by E. J. Weller, 2nd edn, Society of Manufacturing Engineers, 1984.
32. Hashish, M., in *Advances in Non-Traditional Machining*, edited by K. P. Rajurkar et al., ASME, PED-22, 1986, pp. 79–100.
33. Savrun, E. and Taya, M., *J. Mater. Sci.*, Vol. 23, 1988, pp. 1453–1458.
34. Ramulu, M. and Taya, M., *J. Mater. Sci.*, in press.
35. Utsunomiya, S., Kagawa, Y. and Kogo, Y., in *Composites,'86: Recent Advances in Japan and the United States*, edited by K. Kawata et al., Japan Society for Composite Materials, 1986, pp. 589–595.
36. Lee, J. J., Boeing Aerospace Company, Internal Report, 1987.

Appendix A Metal matrix composite systems*

Table A1 *Metal matrix composite systems with metallic fibers*

Fibers	Matrix	Fabrication method	Field of application
Stainless steel	Al	Powder metallurgy, lamination	Aircraft industry Automotive engine
Be ribbons	Al, Ti	Al or Ti clad Be rods are inserted in drilled Al or Ti preforms; mechanical deformation of the preform	Shafts for high-speed rotating engines, e.g., rotor shafts in helicopters and control rods in aircraft
Ta	Mg	Infiltration technique	Aircraft industry
Mo	Ti or Ti alloy	Powder metallurgy, fiber alignment by extrusion rolling, etc.	Supersonic aircraft rocket propulsion
Be or Ti clad Be	Ti alloy, pure Ti	Explosive welding of interposed layers of Ti and Be fibers	Aircraft construction
Be	Ti, Ti–6 Al–4V Ti–6 Al–6V 2Sn Ti–5 Al–5Sn	Hot extrusion of mixed precursor of Ti and Be; the latter can be present as fiber preforms	Aircraft industry

Ni_3Al	Ni, (2–10%) Al	Powder metallurgy; the fiber-like phase is formed *in situ* during mechanical working	Oxidation resistant, high strength at high and low temperatures
Ni–Cr–Al–Y	Ni alloy	Powder metallurgy	Sealing elements in turbines and compressors
W/1% ThO_2	Superalloy	Investment casting	Jet engines
W	W–Ni–Fe alloy	Liquid phase sintering; the W fibers are recrystallized to avoid dissolution	
Stainless steel	Ni alloys	Electroforming	Rocket engines
Mo, Ti, Nb	Ni superalloys	Powder metallurgy	
W	Cu	Melt impregnation	Electrical machinery
Nb filaments	Ni, Cu, Ag	Nb filaments embedded in a Cu, Ni, or Ag matrix are passed through a molten bath of Sn to form Mb_3Sn	Superconductors

* Tables A1 through A3 are taken from a book, *Inorganic Fibre and Composite Materials*, by P. Backe, H. Schurmans, and J. Verhoest, Pergamon Press, Oxford, 1983.

Table A2 *Metal matrix composite with non-oxide ceramic fibers*

Fibers	Matrix	Fabrication method	Field of application
SiC coated B	Al	Powder metallurgy; the composite article is clad with a sheet of Ti by diffusion bonding	Turbine blades
C (graphite, amorphous carbon)	Ni/Co aluminide	Coating C fibers with Ni or Co; mixing with Ni–Co–Al powder; hot pressing	
C coated with boride of Ti, Zr, Hf	Al or Al alloys, Mg, Pb, Sn Cu, Zn	Metal impregnation	
C	Al alloy containing carbide-forming metal, e.g., Ti and Zr	Melt impregnation	
SiC with W core	Al–Cu alloy	Coating with filaments with Cu; passing the Cu-coated filaments through an Al melt	
C	Mg or Mg alloy	Melt impregnation; the molten Mg matrix contains small amounts of magnesium nitride to enhance wetting of the fibers	Turbine fan blades, pressure vessels,
C	Cu	High-energy compaction, diffusion bonding	Electrical brushes, re-entry shielding sheets
SiC	Be or alloys with Ca, W, Mo, Fe, Co, Ni, Cr, Si, Cu, Mg, Zr	Vacuum impregnation with molten Be or plasma spraying fibers with Be and consolidation by metallurgical process	Aerospace and nuclear industries

B + stainless steel; Borsic + Mo fibers	Al, Ti	Impregnation, spraying, etc.; combination of high-strength ductile and brittle fibers	Aerospace industry
SiC	Ti or alloy Ti–3 Al–2.5V	Hot pressing or interposed layers of fibers and matrix sheets; SiC fibers are previously coated with Zr diffusion barrier layer	Compressor blades, airfoil surfaces
Carbides of Nb, Ta, and W	Ni–Co and Fe–Cr alloys	Unidirectional solidification	Aircraft industry
SiC containing 0.01–20% free carbon	Cr-based alloys	Powder metallurgy; the free carbon reacts with the Cr to form carbides, thus improving bonding	High-strength, heat-resistant material, e.g., vanes and blades for turbines, nozzles
SiC containing 0.01–30% free carbon	Co or Co-based alloys	Powder metallurgy or melt impregnation; carbide formation between the fibers and the Co matrix	High-strength, heat-resistant material, e.g., vanes and blades for turbines, rocket nozzles
SiC containing 0.01–20% free carbon	Mo-based alloys	Powder metallurgy	High-strength, heat-resistant material, e.g., vanes and blades for turbines, rocket nozzles
C coated with carbides	Ni or Ni alloys	Melt impregnation	Aeronautical industry
B	Cu–Ti–Sn alloy	Liquid phase sintering	Cutting tools
C	Bronze	Various processes	Bearing materials
C	Cu alloy	Powder metallurgy; the fibers are mixed with a slurry of Cu powder and 2% of a carbide-forming metal powder (Ti or Cr)	High-strength, electrically conductive materials

Table A2 (*continued*)

Fibers	Matrix	Fabrication method	Field of application
C coated with Ti boride	Al, Cu, Sn, Pb, Ag, Zn, and Mg	The matrix contains alloying elements of Ti and B to prevent deterioration of the TiB coating of the fibers	Aeronautical industry
C coated with Ni	Metals with melting point lower than that of Ni	Melt impregnation	Engine components
C coated with $SiO_2 + SiC$	Al, Mg, Ti, and Ni	Melt impregnation, powder metallurgy	Engine components
Monocarbides of Ta, Ti, and W	Al, Al–Si, Ag or Ag alloys, and Cu or Cu alloys	Melt impregnation	Abrasion-resistant materials
β-SiC	Ag or Ag alloys		Electrical conductors, contacts
C	Si	Powder metallurgy	Abrasive materials
SiC_w	Al		
C coated with TiB	Mg, Pb, Zn,	Melt impregnation	

Table A3 *Metal matrix composite systems with oxide ceramic fibers*

Fibers	Matrix	Fabrication method	Field of application
Al_2O_3, SiC, Al oxy-nitride	Al–Cu alloy	Mixing minute filaments in molten matrix; after solidification, the filaments penetrate through grain boundary regions	Aerospace structures, engine components
Al_2O_3	Al–Li alloy	Infiltration with a molten 1–8% Li alloy; reaction occurs between Al_2O_3 fiber and the Li	Aerospace structures, engine components
Al_2O_3–SiO_2, YAl_2O_3	Al, Al–Zn alloy	Melt impregnation, powder metallurgy, etc.	Aeronautical industry
Ni-coated glass ceramic fibers	Al	Powder metallurgy	Sliding parts
Al_2O_3 continuous and polycrystalline	Mg or Mg alloy	Melt impregnation of aligned fibers	Turbine blades, shafts
C, B, glass, ceramic, and metal fibers	Mg alloy	Powder metallurgy; the composite contains two different Mg alloy phases	Multiple applications
Metal-coated glass fibers	Fe, Be, Ti, Al, and Sn	Chopped metal-coated glass fibers are mixed with glass filaments and metal powder; hot pressing	Dimensionally stable machine parts, e.g., friction elements
Glass fibers	Pb	Melt impregnation	Battery plates, bearing materials, acoustic insulation

Appendix B Thermomechanical properties of materials*

Table B1 *Tensile strengths of whiskers and some larger crystals at room temperature*

Material	Maximum tensile strength (GPa)	Young's modulus (GPa)	Specific gravity (ρ)	$\frac{\sigma}{\rho}$ (GPa)	$\frac{E}{\rho}$ (GPa)	Melting temperature (K)
Graphite	19.6	686 [1]	2.2	8.9	312	subl. 3950
$Al_2O_3//\langle 1120 \rangle$	22.3 [2]	420 [3]	4.0	5.6	105	2288
Al_2O_3 large crystal	7.0 [4]	532 (max)	4.0	1.8	133	2288
Fe	12.6 [5]	196	7.8	1.6	25	1808
Si_3N_4	14.0 [6]	385 [8]	3.1	4.5	124	subl. ~2200
	14.7 [7]			4.7		
	13.8 [36]	379	3.18	4.3	119	
SiC	21.0 [9]	700 (max) [10]	3.2	6.6	219	subl. ~3000
B_4C	21.0 [11]	483 [11, 19]	2.5	8.4	193	2623
Si	7.6 [12]	163 [3]	2.3	3.3	71	1683
Si large crystal	4.1	163 [3]	2.3	1.8	71	1683
Ge	4.3 [12]	133 [3]	5.4	0.8	25	1210
BeO	24.8 [13]	133 [3]	3.0	8.3	132	2803
	19.3 [14]			6.4		
AlN	7.0 [15]	350	3.3	2.1	106	subl. ~2300
	7.6 [16]			2.3		
$K_2O \cdot 6TiO_2$	6.9 [37]	274	3.3	2.1	83	

The figures in square brackets refer to specific references.

* Most of Appendix B1–B3 are taken from a book, *Strong Solids*, 3rd edn. by A. Kelly and N. H. Macmillan, Oxford University Press, Oxford, 1986.

Table B2 *Tensile strengths of strong non-metallic fibers at room temperature*

Material	Maximum strength (GPa)	Young's modulus (GPa)	Specific gravity (ρ)	$\frac{\sigma}{\rho}$ (GPa)	$\frac{E}{\rho}$ (GPa)	Melting temperature (K)
Asbestos (crocidolite)	5.9 [17]	189	3.4	1.7	56	Loses water at 573
Mica	3.2 [18]	231	2.7	1.2	86	Losses water at ~673
Etched soda-lime glass	3.0 (mean) [19]	68.6	2.5	1.2	27	—
Drawn silica in air	10.0 (max) [20]	73.5	2.2	4.6	33	—
Drawn silica *in vacuo*	12.0 (max) [20]	73.5	2.2	5.5	33	—
Drawn silica in liq. N_2	14.3 (max) [20]	73.5	2.2	6.5	33	—
Drawn silica in liq. He	15.2 (max) [21]	73.5	2.2	6.9	33	—
B filaments with W core	7.0 [22]	385	2.6	2.7	148	2313
SiC (Nicalon)	8.3 (max) [23]	616 [24]	2.55	3.3	242	Subl. ~3000
Al_2O_3(FP fibers)	2.1 (max) [25]	379	4.0	0.5	95	2288
Al_2O_3(Saffil)	1.5 (mean) [26]	285	4.0	0.4	71	2288
C (graphite)	3.0 [27]	412	~1.9	1.6	217	Sub. 3950
C (graphite)	5.5 (max)	256	~1.9	2.9	135	subl. 3950

The figures in square brackets refer to specific references.

Table B3 *Tensile strengths of metal wires and ribbons of metallic glass (at room temperature except where stated)*

Material	Composition (wt.%)	Diameter (μm)	Tensile strength (GPa)	Young's modulus (GPa)	Specific gravity (ρ)	Specific strength (GPa)	Specific modulus (GPa)	Melting temperature (K)
Patented steel	0.9 C	100	4.2 [28]	220	7.9	0.53	28	1575
Patented steel after 30 min at 533 K	0.9 C	100	3.2 [28]	210	7.9	0.41	27	1575
Maraging steel 4/11	—	—	4.9 [29]	220	7.9	0.62	28	—
Fe	—	—	9.7 [30]	220	7.9	1.23	28	1808
Stainless steel	18 Cr–8Ni–0.8Mo	50	2.1 [28]	224	8.2	0.28	27	—
	Ni–19Cr–11Co–10Mo–3Ti–1.7 Al							
Rene 41 after 30 min at 1073 K	Ni–19Cr–11Co–10Mo–3Ti–1.7 Al	100	1.4 [28]	168	8.2	0.17	20	—
Ni	—	~0.05	10 [31]	210	8.9	1.12	24	1726
β-Ti	13V–11Cr–3Al	150	2.3 [32]	119	4.6	0.50	26	—
W	99.95 W	25	3.9 [28]	405	19.3	0.20	21	3663
Mo	99.9 Mo	150	2.1 [32]	343	10.3	0.20	33	2883
Be	Commercial purity	150	1.3 [33]	315	1.8	0.72	175	1557
Al	High purity	150	0.17 [32]	70	2.7	0.06	26	933
Sb	—	30	0.2 [34]	68	6.7	0.03	10	904
$Fe_{80}B_{20}$ glass	—	—	3.63 [35]	168	6.8	0.53	25	—
$Ni_{49}Fe_{29}P_{14}$	—	—	2.38 [35]	132	7.1	0.34	19	—
B_6Si_2 glass	—	—						
$Pd_{78}Cu_6Si_{16}$	—	—	1.47 [35]	88	10.1	0.15	8.7	—

The figures in square brackets refer to specific references.

Table B4 *Thermal properties of selected fibers at room temperature.**

Fiber	Diameter μm	Specific heat $kJ/(kg \cdot K)$[a]	Thermal conductivity $W/(m \cdot K)$[b]	Coefficient of thermal expansion $10^{-6}/°C$
Graphite				
PAN HM	7	0.71	—	-1.2[c]
PAN HTS (T300)	8	0.71	8	-0.7
Pitch (type P100)	10	0.7	500 [38]	-1.4
Boron on tungsten	102–203	1.3	38	5.0
Borsic	102–203	1.3	38	5.0
Boron on carbon	102–203	1.3	38	5.0
Silicon carbide on tungsten	102–203	1.2	16	4.3
Silicon carbide on carbon	102	1.2	16	4.3
Beryllium	127	1.9	150	11.5
α-Al_2O_3 (FP)	20	—	37.7††	8.3
S glass	9	0.7	13	5.0
E glass	9	0.7	13	5.0
Molybdenum	127	0.3	145	4.9
Steel	127	0.5	29	13.3
Tantalum	508	0.2	55	6.5
Tungsten	381	0.1	168	4.5
Whisker				
Al_2O_3	10–25	0.6	24	7.7
Fe	127	0.5	29	13.3
β-SiC	0.2	0.69 1.19 (1000 K)††	—	2.5** 4.3**†
β-SiC (Nicalon)	10 ~ 15	0.69 1.19 (1000 K)††	25 (473 K)	2.5** 4.3**†
γ-Al_2O_3 (Sumitomo)	17	0.71	—	8.8 (473–773 K)

[a] To convert $kJ/(kg \cdot K)$ to $Btu/(lb°F)$, divide by 4.184
[b] To convert $W/(m°K)$ to $Btu/(ft°h \cdot °F)$, divide by 1.729
[c] CTE values are along the fiber axis
* Some of the data in Table B4 are taken from a book, *Engineers' Guide to Composite Materials*, American Society for Metals, Metals Park, 1987.
** CTE values are taken from Ref. 39.
† CTE was measured for the temperature range 500–1000°k
†† The data are taken from a book, *Thermal Physical Properties of High Temperature Solid Materials*, edited by Y. S. Touloukian, Purdue University, 1966.

Table B5 *Thermal properties of selected matrix metals at room temperatures.**

Matrix Metals	Specific heat kJ/(kg·K)	Thermal conductivity W/(m·K)	Coefficient of thermal expansion $10^{-6}/°C$
Aluminum			
pure	0.9	247	23.6
1100	0.904	230	23.6
2024	0.875	190	22.9
6061	0.896	180	23.6
7075	0.960	130	23.4
Copper			
C10100 (oxygen free, electronic)	.385	391	17.0
C19400 (high strength, modified)	.385	260	16.3
C2200 (commercial bronzc)	.376	189	18.3
C63200 (82 Cu-9Al-5Ni-4Fe)	.439	36	16.2
Magnesium			
AZ 80A (extrusion alloy)	1.05	107	26.6
EZ33 A (sand casting)	1.05	100	26.1
Titanium			
Ti-0.3Mo-6.8Ni	.540	19	9.5
Ti-6Al-6V-2Su	.67	6.6	9.0
Ni-super alloys			
IN-100 (60Ni-10Cr)	.48 (at 540°)	17.3 (at 540°)	13.9 (at 540°C)
In-713C (rotor casting)	.42	10.9	10.6
MAR-M200 (60Ni-9Cr)	.40	12.7	11.9 (at 200°C)
Lead			
Corroding (99.94 + %Pb)	.129	35.0	29.3
50-50 Solder (50 pb-50Su)	.210	46.5	23.4
Silver (comm. high purity)	.234	4187	19.68
Gold (comm. high purity)	.131	300	14.2
Niobium (comm. high purity)	.268	649	7.1
Stainless steel 304	0.5	16.2	17.2
Steel (A151-1008)	.481	57.8	12.6

* The data of Table B5 are taken from a book, *Metals Handbook*, 9th Ed., Vol. 2, American Society for Metals, Metals Park, 1978.

References

1. Measured valued of $1/s_{11}$, see Bacon, R., *J. Appl. Phys.* Vol. 3, 1960, p. 283; and Baker, C. and Kelly, A., *Phil. Mag.*, Vol. 9, 1965, p. 927.
2. Soltis, P. J., *Bull. Am. Phys. Soc.*, Vol. 10, 1965, p. 163.
3. Simmons, G. and Wang, H., *Single Crystal Elastic Constants and Calculated Aggregate Properties*, MIT Press, 1971.
4. Morley, J. G. and Proctor, B.A., *Nature, London*, Vol. 196, 1962, p. 1082.
5. Brenner, S. S., *Growth and Perfection of Crystals*, edited by R. H. Doremus et al., Doremus, Wiley, 1958, p. 157.
6. Evans, C. C., Gordon, J. E., Marsh, D. M. and Parratt, J. N., *Tube Invest. Res. Lab. Report* No. 133, 1961.
7. Bayer, P. D. and Cooper, R. E., *J. Mater. Sci.*, Vol. 2, 1967, p. 233.
8. Gordon, J . E., *Proc. R. Soc.*, Vol. A282, 1964, p. 16.

9. Mehan, R. L. and Herzog, J. A., *Whisker Technology*, edited by A. P. Levitt, Wiley, 1970, p. 157.
10. Shaffer, P. T. B., *Ceramic Age*, Vol. 82(5), 1966, p. 46; Webb, W. W., Batha, H. D. and Shaffer, P. T. B., *Strengthening Mechanisms, Metals and Ceramics*, edited by J. J. Buke et al., Syracuse University, Press, 1966, p. 239.
11. Gatti, A., Mancuso, C., Feingold, E. and Mechan, R., *Proc. Int. Conf. on Crystal Growth*, Boston, 1966, edited by H. S. Peiser, Pergamon Press, (Suppl. No 1 to *J. Phys. Chem. Solids*, Vol. 28, 1967, p. 317.
12. Sandulova, A. V., Bogoyavlenskii, P. S. and Dronyuk, M. I., *Soviet Phys. Solid State*, Vol. 5, 1964, p. 1883.
13. Newkirk, H. W. and Smith, D. K., *Amer. Miner.*, Vol. 50, 1965, p. 22.
14. Ryschkewitch, E., *Int. Sci. Tech.*, February 1962, p. 54.
15. Davies, T. J. and Evans, P. E., *Nature, Lond.*, Vol. 207, 1965, p. 254.
16. Tavadze, F. N., Surmava, G. G., Nikolaishvili, A. A. and Makovets, S. E., *Soviet Phys. Solid State*, Vol. 15, 1973, p. 901.
17. Zukowski, R. and Gaze, R., *Nature, Lond.*, Vol. 185, 1959, p. 35.
18. Orowan, E. Z., *Phys.*, Vol. 82, 1933, p. 235.
19. Proctor, B., *Phys. Chem. Glasses*, Vol. 3, 1962, p. 7.
20. Proctor, B. A. and Whitney, I., *Proc. R. Soc.*, Vol. R297, 1967, p. 534; Morley, J. G., Andrews, P. and Whitney, I., *Phys. Chem. Glasses*, Vol. 5, 1964, p. 1; Morley, J. G., *Proc. R. Soc.*, Vol. A282, 1964, p. 43.
21. Talley, C. P., *J. Appl. Phys.*, Vol. 30, 1959, p. 1114; Talley, C. P. and Clark, W. J., AD296575, 1959.
22. Yajima, S., Kayano, H., Okamura, K., Omori, M., Hayashi, J., Matsuzawa, T. and Akutsu, K., *Bull. Am. Ceram. Soc.*, Vol. 55, 1976, p. 1065.
23. Dow Corning Corp., Midland, MI, Form No. 19-073-83, 1983.
24. Dinghra, A. K., *Phil. Trans. R. Soc.*, Vol. A294, 1980, p. 411.
25. Birchall, J. D., *Trans. J. Br. Ceram. Soc.*, Vol. 82, 1983, p. 143.
26. Reynolds, W. N. and Moreton, R., *Phil. Trans. R. Soc.*, Vol. A294, 1980, p. 451.
27. Hughes, J. D. H., Morley, H. and Jackson, E. E., *J. Phys. D.: Appl. Phys.*, Vol. 13, 1981, p. 921.
28. Roberts, D. A., Defense Met. Info. Center Memo. 80, Battell Mem. Inst., Columbus, OH, 1961.
29. Nixdorf, *J.*, *Proc. R. Soc.*, Vol. A319, 1970, p. 17.
30. Ulitovskiy, A. V., *Pribory Tekh. Eksp.*, Vol. 3, 1957, p. 115.
31. Smith, G. D. W., Smity, D. A., Taylor, G. S., Goringe, M. J. and Esterling, K., *High Voltage Electron Microscopy*, edited by Swann, P. R., Humphreys, C. J. and Goringe, M. J., Acad. Press, NY, 1974, p. 240.
32. Rumbles, W. W., Watanabe, S. F., Hayes, E. J. and Petrick, E. N., *Proc. SAMPE Filament Winding Conf.*, 1961, p. 122.
33. Fulap, R. J., *Mat. Des. Eng.*, Vol. 55, 1962, p. 10.
34. Taylor, G. F., *Phys. Rev.*, Vol. 23, 1924, p. 655; Taylor, G. F., US Pat. No. 1, 739,529, 1931.
35. Kunzi, H. U., *Glassy Met. II, Topics in Appl. Phys.*, Vol. 53, edited by H. Beck and H. J. Guntherodt, Springer-Verlag, 1983, p. 169.
36. Matsubara, H., Nishida, Y., Yamada, M., Shirayanagi, I. and Imai, T., *J. Mater. Sci. Letts.*, Vol. 6, 1987, pp. 1313–1315.
37. Imai, T., Nishida, Y., Yamada, M., Shirayanagi, I. and Matsubara, H., *J. Mater. Sci. Letts.*, Vol. 6, 1987, pp. 1257–1258.
38. Nysten, B., Piraux, L. and Issi, J-P, *J. Phys D: Appl. Phys.*, Vol. 18, 1985, pp. 1307–1310.
39. Taylor, A. and Jones, R. M., *Proc. of the Conf. on Silicon Carbide*, edited by J. R. O'Connor and J. Smiltens, Pergamon Press, 1960, p. 147.

Appendix C Eshelby's tensor for elasticity S_{ijkl}

Eshelby's tensors S_{ijkl} for istropic matrix are a function of the geometry of ellipsoid with axes a_1, a_2, and a_3 and Poisson's ratio of the matrix ν. In the following are the Eshelby's tensors for the ellipsoids with simple geometry. The reader can find complete information of the Eshelby's tensors for general cases of ellipsoid in Mura's book, *Micromechanics of Defects in Solids.*

The domain of an ellipsoidal inclusion is bounded by

$$\frac{x_1^2}{a_1^2}+\frac{x_2^2}{a_2^2}+\frac{x_3^2}{a_3^2}=1$$

where a_1, a_2 and a_3 are the principal axes of the ellipsoid and coincide with the x_1, x_2, and x_3 axes, respectively. S_{ijkl} satisfies the symmetry; $S_{ijkl} = S_{jikl} = S_{ijlk}$

(1) Sphere: $a_1 = a_2 = a_3$ (Fig. C1)

$$S_{1111} = S_{2222} = S_{3333} = \frac{7-5\nu}{15(1-\nu)}$$

$$S_{1122} = S_{2233} = S_{3311} = S_{1133} = S_{2211} = S_{3322} = \frac{5\nu-1}{15(1-\nu)}$$

$$S_{1212} = S_{2323} = S_{3131} = \frac{4-5\nu}{15(1-\nu)}$$

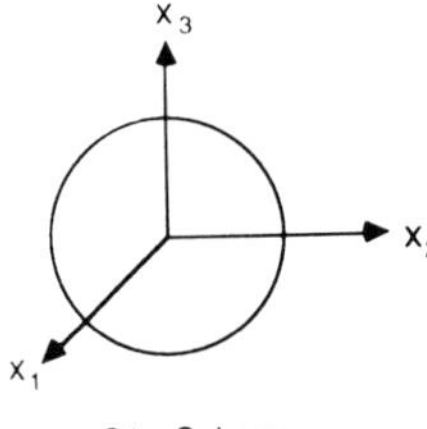

C1 : Sphere

(2) Elliptic cylinder: $a_1, a_2 \ll a_3 \to \infty$ (Fig. C2)

$$S_{1111} = \frac{1}{2(1-\nu)}\left\{\frac{a_2^2+2a_1a_2}{(a_1+a_2)^2}+(1-2\nu)\frac{a_2}{(a_1+a_2)}\right\}$$

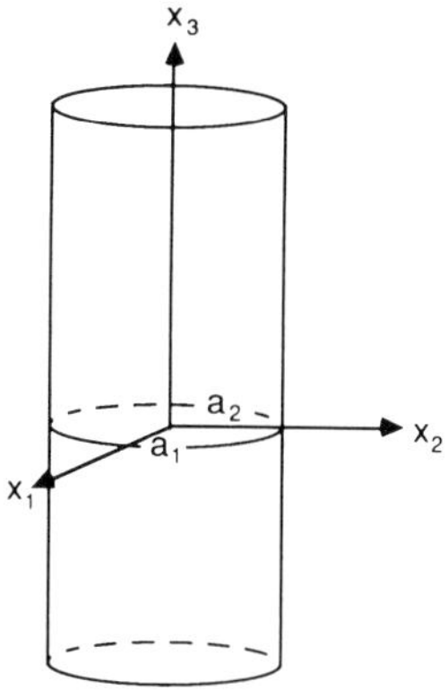

C2 : Elliptical Cylinder

$$S_{2222}=\frac{1}{2(1-v)}\left\{\frac{a_1^2+2a_1a_2}{(a_1+a_2)^2}+(1-2v)\frac{a_1}{(a_1+a_2)}\right\}$$

$$S_{3333}=0$$

$$S_{1122}=\frac{1}{2(1-v)}\left\{\frac{a_2^2}{(a_1+a_2)^2}-(1-2v)\frac{a_2}{(a_1+a_2)}\right\}$$

$$S_{2211}=\frac{1}{2(1-v)}\left\{\frac{a_1^2}{(a_1+a_2)^2}-(1-2v)\frac{a_1}{(a_1+a_2)}\right\}$$

$$S_{2323}=\frac{a_1}{2(a_1+a_2)}$$

$$S_{2233}=\frac{1}{2(1-v)}\cdot\frac{2va_1}{(a_1+a_2)}$$

$$S_{3311}=0$$

$$S_{1133}=\frac{1}{2(1-v)}\cdot\frac{2va_2}{(a_1+a_2)}$$

$$S_{3322}=0$$

$$S_{1212}=\frac{1}{2(1-v)}\left\{\frac{a_1^2a_2^2}{2(a_1+a_2)^2}+\frac{1}{2}(1-2v)\right\}$$

$$S_{3131}=\frac{a_2}{2(a_1+a_2)}$$

Please note that continuous fibers are usually considered to be circular cylinders, $a_1=a_2\ll a_3=\infty$, for which the above S_{ijkl} are further simplified, i.e., function only of Poisson's ratio, v.

(3) Penny shape: $a_1 = a_2 \gg a_3$ (Fig. C3)

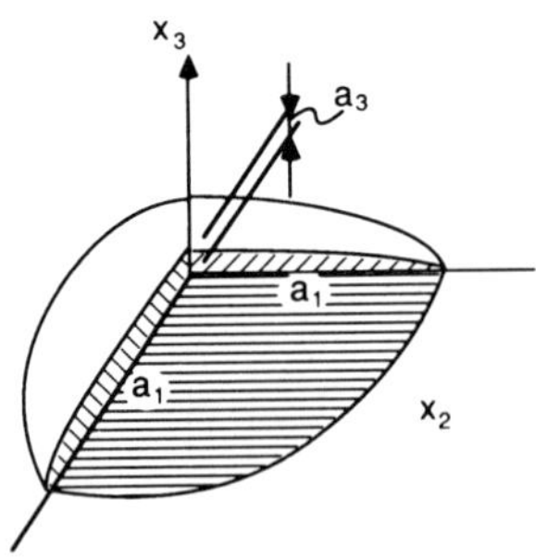

C3 : Penny Shape

$$S_{1111} = S_{2222} = \frac{\pi(13-8\nu)}{32(1-\nu)}\frac{a_3}{a_1}$$

$$S_{3333} = 1 - \frac{\pi(1-2\nu)}{4(1-\nu)}\frac{a_3}{a_1}$$

$$S_{1122} = S_{2211} = \frac{\pi(8\nu-1)}{32(1-\nu)}\frac{a_3}{a_1}$$

$$S_{1133} = S_{2233} = \frac{\pi(2\nu-1)}{8(1-\nu)}\frac{a_3}{a_1}$$

$$S_{3311} = S_{3322} = \frac{\nu}{(1-\nu)}\left\{1 - \frac{\pi(1+4\nu)}{8\nu}\frac{a_3}{a_1}\right\}$$

$$S_{1212} = \frac{\pi(7-8\nu)}{32(1-\nu)}\frac{a_3}{a_1}$$

$$S_{1313} = S_{2323} = \frac{1}{2}\left\{1 + \frac{\pi(\nu-2)}{4(1-\nu)}\frac{a_3}{a_1}\right\}$$

(4) Oblate spheroid: $a_1 = a_2 > a_3$ (Fig. C4)

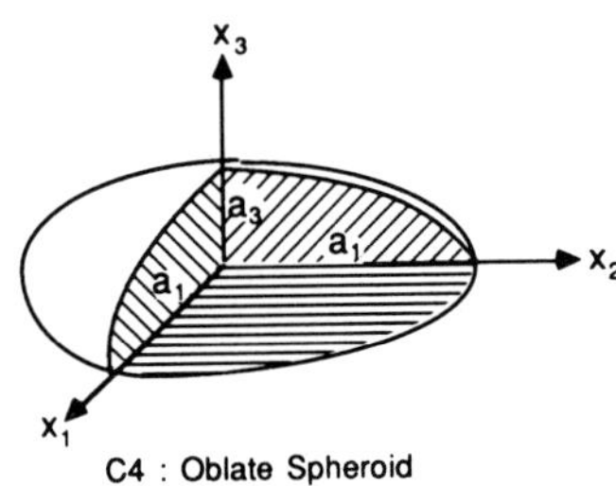

C4 : Oblate Spheroid

$$S_{1111} = S_{2222} = -\frac{3}{8(1-\nu)} \cdot \frac{\alpha^2}{(1-\alpha^2)} + \frac{1}{4(1-\nu)} \left\{ 1 - 2\nu + \frac{9}{4(1-\alpha^2)} \right\} g$$

$$S_{3333} = \frac{1}{2(1-\nu)} \left\{ 4 - 2\nu - \frac{2}{(1-\alpha^2)} \right\} + \frac{1}{2(1-\nu)} \left\{ -4 + 2\nu + \frac{3}{1-\alpha^2} \right\} g$$

$$S_{1122} = S_{2211} = \frac{1}{8(1-\nu)} \left\{ 1 - \frac{1}{(1-\alpha^2)} \right\} + \frac{1}{16(1-\nu)}$$

$$\times \left\{ -4(1-2\nu) + \frac{3}{(1-\alpha^2)} \right\} g$$

$$S_{1133} = S_{2233} = \frac{1}{2(1-\nu)} \cdot \frac{\alpha^2}{(1-\alpha_2)} - \frac{1}{4(1-\nu)} \left\{ 1 - 2\nu + \frac{3\alpha^2}{(1-\alpha^2)} \right\} g$$

$$S_{3311} - S_{3322} = \frac{1}{2(1\nu)} \left\{ -(1-2\nu) + \frac{1}{1-\alpha^2} \right\} + \frac{1}{4(1-\nu)}$$

$$\times \left\{ 2(1-2\nu) - \frac{3}{(1-\alpha^2)} \right\} g$$

$$S_{1212} = -\frac{1}{8(1-\nu)} \cdot \frac{\alpha^2}{(1-\alpha^2)} + \frac{1}{16(1-\nu)} \left\{ \frac{3}{(1-\alpha^2)} + 4(1-2\nu) \right\} g$$

$$S_{1313} = S_{2323} = \frac{1}{4(1-\nu)} \left\{ 1 - 2\nu + \frac{1+\alpha^2}{1-\alpha^2} \right\} - \frac{1}{8(1-\nu)}$$

$$\times \left\{ 1 - 2\nu + 3\left(\frac{1+\alpha^2}{1-\alpha^2} \right) \right\} g$$

where

$$\alpha = \frac{a_3}{a_1} < 1$$

$$g = \frac{\alpha}{(1-\alpha^2)^{3/2}} \{ \cos^{-1}\alpha - \alpha(1-\alpha^2)^{1/2} \}$$

(5) Prolate spheroid: $a_1 = a_2 < a_3$ (Fig. C5)

$$S_{1111} = S_{2222} = \frac{3}{8(1-\nu)} \frac{\alpha^2}{(\alpha^2-1)} + \frac{1}{4(1-\nu)} \left\{ 1 - 2\nu - \frac{9}{4(\alpha^2-1)} \right\} g$$

$$S_{3333} = \frac{1}{2(1-\nu)} \left[1 - 2\nu + \frac{3\alpha^2-1}{\alpha^2-1} - \left\{ 1 - 2\nu + \frac{3\alpha^2}{(\alpha^2-1)} \right\} g \right]$$

$$S_{1122} = S_{2211} = \frac{1}{4(1-\nu)} \left[\frac{\alpha^2}{2(\alpha^2-1)} - \left\{ 1 - 2\nu + \frac{3}{4(\alpha^2-1)} \right\} g \right]$$

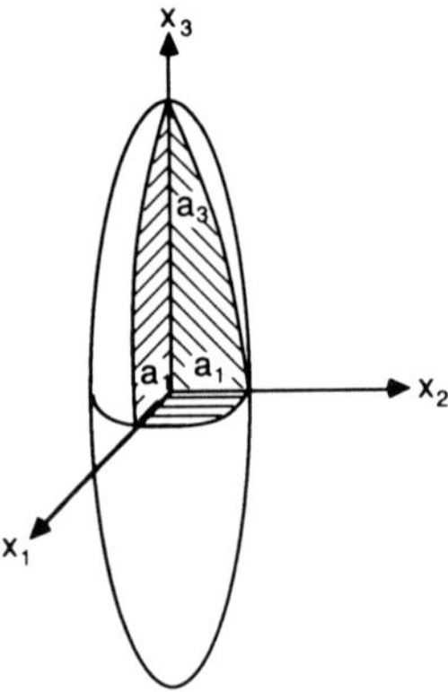

C5 : Prolate Spheroid

$$S_{1133} = S_{2233} = -\frac{1}{2(1-\nu)}\frac{\alpha^2}{(\alpha^2-1)} + \frac{1}{4(1-\nu)}\left\{\frac{3\alpha^2}{(\alpha^2-1)} - (1-2\nu)\right\}g$$

$$S_{3311} = S_{3322} = -\frac{1}{2(1-\nu)}\left\{1-2\nu+\frac{1}{\alpha^2-1}\right\} + \frac{1}{2(1-\nu)}$$

$$\times\left\{1-2\nu+\frac{3}{2(\alpha^2-1)}\right\}g$$

$$S_{1212} = \frac{1}{8(1-\nu)}\frac{\alpha^2}{(\alpha^2-1)} + \frac{1}{4(1-\nu)}\left\{1-2\nu-\frac{3}{4(\alpha^2-1)}\right\}g$$

$$S_{2323} = S_{1313} = \frac{1}{4(1-\nu)}\left\{1-2\nu-\frac{(\alpha^2+1)}{(\alpha^2-1)}\right\} - \frac{1}{8\,(1-\nu)}$$

$$\times\left\{1-2\nu-\frac{3(\alpha^2+1)}{(\alpha^2-1)}\right\}g$$

where

$$\alpha = \frac{a_3}{a_1} > 1$$

$$g = \frac{\alpha}{(\alpha^2-1)^{3/2}}\left\{\alpha(\alpha^2-1)^{1/2} - \cosh^{-1}\alpha\right\}$$

Appendix D Eshelby's tensors for heat conduction, S_{ij}

A semi-detailed description of S_{ij} is given in the Appendix of Ref. 27 of Chapter 5. Please note that $S_{ij} = 0$ for $i \neq j$.

(1) Sphere: $a_1 = a_2 = a_3$ (Fig. C1)

$$S_{11} = S_{22} = S_{33} = \tfrac{1}{3}$$

(2) Elliptic cylinder: $a_1, a_2 \ll a_3\ \infty$ (Fig. C2)

$$S_{11} = \frac{a_2}{a_1 + a_2},\ S_{22} = \frac{a_1}{a_1 + a_2},\ S_{33} = 0.$$

(3) Penny shape: $a_1 = a_2 \gg a_3$ (Fig. C3)

$$S_{11} = S_{22} = \frac{\pi a_3}{4a_1},\ S_{33} = 1 - \frac{\pi a_3}{2a_1}.$$

(4) Oblate spheroid: $a_1 = a_2 > a_3$ (Fig. C4)

$$S_{11} = S_{22} = \frac{a_1^2 a_3}{2(a_1^2 - a_3^2)^{3/2}} \left\{ \cos^{-1}\frac{a_3}{a_1} - \frac{a_3}{a_1}\left(1 - \frac{a_3^2}{a_1^2}\right)^{1/2} \right\},$$

$$S_{33} = 1 - 2S_{22}$$

(5) Prolate spheroid: $a_1 = a_2 < a_3$ (Fig. C5)

$$S_{11} = S_{22} = \frac{a_1^2 a_3}{2(a_3^2 - a_1^2)^{3/2}} \left\{ \frac{a_3}{a_1}\left(\frac{a_3^2}{a_1^2} - 1\right)^{1/2} - \cosh^{-1}\frac{a_3}{a_1} \right\},$$

$$S_{33} = 1 - 2S_{22}.$$

Author Index

Subject Index